Pests and parasites of horses

Pests and parasites of horses

Erika T. Machtinger, Emma N.I. Weeks,
Christopher J. Geden and Erica Lacher

Wageningen Academic
Publishers

EAN: 9789086863716
e-EAN: 9789086869237
ISBN: 978-90-8686-371-6
e-ISBN: 978-90-8686-923-7
DOI: 10.3920/978-90-8686-923-7

Photo cover: Erika Machtinger,
Pennsylvania State University

First published, 2022

Table of contents

Chapter 1. Introduction to 'Pests and parasites of horses'

1.1 Overview of the book

Transportation and agriculture were intimately dependent upon horses before the age of the automobile. Households that could afford a horse had one, and many owned stables to care for as well. Managing pests of horses was part of everyday husbandry practices. Although no longer used primarily for these purposes, horses and the equine industry maintain some original agricultural elements while also serving as companions and competition animals.

Even though horses are not integral to society anymore, horse owners, caretakers, riders, and enthusiasts want to know more about caring for their animals including pest management practices. However, because horses are no longer considered production animals in many parts of the world, they are typically neglected in pest management research and education. Often pest management guidelines for equine farms and horse owners come from research that was conducted on other animal facilities, such as dairy cattle or poultry farms. Because of the appreciable differences between production animal facilities and horse facilities, these recommendations are sometimes not appropriate or are difficult to interpret for horses or horse farms. This book was written to compile recommendations specifically for controlling pests of horses found in Australia, Canada, Europe, New Zealand, and the United States of America.

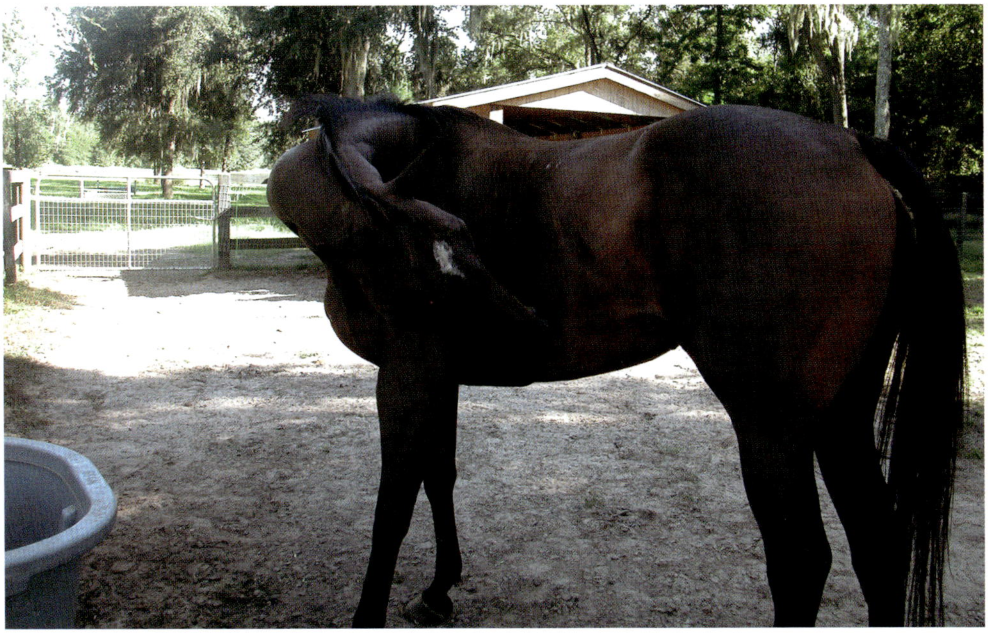

Figure 1.1. Arthropod pests can be a serious nuisance causing horses pain and distress.

Pests that affect horses, donkeys, mules, or even captive zebras (grouped and called 'horses' throughout this book) include flies that can be found in the stable or in the pasture, ectoparasites on horses, such as ticks, lice, and mites, and even a few internal pests, such as warbles and bot fly larvae. These pests can have a negative influence on horse welfare. Flies, especially biting flies, can cause horses to move around as they try to avoid fly bites, reducing the available time horses have to graze. Mares with foals may not be able to stand still long enough to nurse when aggravated by flies. Other biting arthropods (animals with exoskeletons, segmented bodies, and paired legs) that live on the horse body, like lice and mites, can cause skin irritation and hair loss. Some arthropods can transmit pathogens (such as bacteria, viruses, worms, and microscopic parasites) that can cause diseases. For example, mosquitoes can transmit the virus that causes West Nile encephalitis and *Ixodes* species ticks can transmit the bacterium that causes Lyme disease. For all these reasons, and because pests can be a general nuisance, it is important to manage pests on and around horses.

In the early 1900s, pests were managed with chemicals that were largely ineffective, expensive, and at times hazardous. With the development of synthetic insecticides in the mid-1900s, there was a trend away from understanding pest biology to manage them in favour of a strictly chemical approach to pest control. However, this approach caused environmental damage and harmed non-target animals like birds and beneficial insects like pollinators. In addition, pests demonstrated an astonishing ability to develop resistance to the applied insecticides. This resistance meant that more chemicals would need to be applied, or chemicals would need to be applied more often

Figure 1.2. Horses, donkeys, zebras, and other equids can all be affected by insects and other arthropod pests.

to get the same result, or chemicals would stop working completely (see Chapter 10 for more information on insecticide resistance). Later in the 1900s, this continuous application of chemicals was recognized as unsustainable, and the 1962 publication of Rachel Carlson's *Silent Spring* marked a turning point that led to public demand for reduced insecticide use. Environmental agencies and institutions were established to better understand and regulate pesticides, and methods were developed to better manage pests without relying on chemicals alone.

Integrated pest management, often abbreviated to IPM, was one of the methods developed to manage insecticide resistance and was first used in agriculture in the 1970s. The approach emphasised the integration of pest biology and cultural practices in controlling insect pests in crops but has been extended to any pest situation. The goal of IPM is to increase control effectiveness in the safest way possible for humans, other animals, and the environment.

IPM is a multi-step process to pest management that includes:
- *Identifying the pest* so that you understand what options are available for control; many control options are pest species-specific.
- *Understanding the pest's life cycle* so that you can target multiple life stages (e.g. egg, larvae or nymph, pupae, adult) with control methods resulting in better and longer-lasting control.
- *Choosing multiple control methods* that are suitable for your situation.
- *Monitoring* your pests and control methods to ensure that the control methods you have put in place are effective.

Monitoring pest population numbers is recommended as part of an IPM plan. By monitoring pests, you can determine if control methods are working and if you need to increase or decrease control efforts. You can also learn when pests numbers increase seasonally, which can help you plan for implementing control before numbers are too high the following year, or if they are associated with specific on-farm practices. Some pests can be easily monitored by examining the coat carefully when grooming (e.g. lice, mites, ticks, keds, bot fly eggs, warble flies), examining fresh faeces (e.g. worms, equine bot fly larvae), examining bedding and older faeces (e.g. filth fly larvae and pupae), and using simple traps (e.g. filth fly adults). However, for many pests monitoring requires pest identification training or specialized equipment so monitoring methods have not been provided for some pests in this book. Pest-specific monitoring resources have been included for the pests where monitoring can be conducted easily in Chapter 10.

When using IPM, incorporating several methods of pest control that are suitable for your pest of concern that can kill the pest at several life stages should result in effective and sustainable control. Control methods that are used in IPM include cultural control practices, physical and mechanical control practices, biological control, and chemical control.
- *Cultural control* practices are preventative and least invasive in the environment. For example, keeping stalls clean and dry to prevent pests from developing in wet manure, or dumping and cleaning water troughs to prevent mosquitoes from developing.
- *Physical and mechanical control* practices kill pests directly or make the environment unsuitable or inaccessible for the pests, such as using screens to prevent pests from entering the stable.

- *Biological control* uses natural organisms such as predators, parasites, and pathogens (e.g. bacteria, fungi, and viruses) to control pests.
- *Chemical control* options use natural or synthetic chemicals to harm or kill pests. These chemicals can be used as repellents or pesticides, or as veterinary treatments. The chemicals may be applied to the environment, barns, stables, or other premises, or used directly on horses.

The focus of any pest control should be cultural control to prevent pests from establishing. While the use of chemical control options is not restricted, the goal of IPM is to use chemical control options cautiously and only when necessary.

One of the greatest challenges with understanding how to control pests on horse facilities is the wide range of husbandry practices in the equine industry. Depending on facility size, region location, or sometimes even discipline, management practices can vary widely. Each reader will need to determine what methods of control should work best for them and their horse and take into consideration their current husbandry practices, time, facility, cost, location, and in many cases after consultation with their veterinarian. The options provided should be considered tools and not a formula for control. The more tools that can be used, the more effective a control program might be, but even a few changes or control options may be enough to relieve pest pressure.

Each chapter focuses on a group of major external or internal pests of horses, and there is one chapter on beneficial insects found on equine facilities. This book is structured to follow the steps of IPM including a *pest summary*, and sections on the *distribution and seasonality, description*, and *life history* of each pest. Many common names for equine pests differ by region and as many of these names have been provided as possible, but scientific names or groups have also been given for clarity. *Control* options are provided in the categories of cultural, physical and mechanical, biological, and chemical control methods, when relevant. For each pest, there is a section discussing associated *veterinary concerns* and at the conclusion of each chapter is a section with *continuing education* and *resources,* which include other reading materials to augment learning about specific pests or veterinary issues, and sources for specific materials referenced in the chapter. For this chapter, IPM and equine medicine reference books and resources have been provided in the 'Continuing education' section.

This book is structured so relevant information can easily be found and used to make this information available to a novice or an expert without prior knowledge of technical terminology. However, in some cases language will be used, especially regarding arthropod life cycles or veterinary concerns, which may require some additional explanation. Every effort has been made to enable the reader to gain the information they need without being overcome by technical nomenclature. A chapter on basic information on pests and parasites of horses including overviews of life cycles and anatomy is included. A glossary of important terms that are found throughout the book is provided in the appendices to help with any gaps in knowledge.

Figure 1.3. Using IPM can reduce pest numbers on and around horses to improve equine health and comfort.

Along with horse riders, trainers, owners, caretakers, and general enthusiasts, students in animal sciences, veterinary sciences, parasitology, horse care courses, or horse training programs can find this book a resource to their educational programs. This book is not meant to provide detailed veterinary information, such as diagnosis or treatment of arthropod-associated conditions, but an overview of veterinary concerns is provided as well as continuing education options. Relevant images have been included to support the provided information. We hope that this information can be used to empower the reader to manage pests on and around horses, ultimately improving equine health, comfort, and performance.

Continuing education

Integrated pest management: principles and practice

Abrol, D.P. and Shankar, U., 2016. CABI, Wallingford, UK. ISBN 978-1786390318

- This book provides a critical evaluation of the management strategies involved in ecologically-based pest management and a balanced overview of environmentally safe and ecologically sound approaches. Topics covered include biological control with fungi and viruses, conservation of natural predators, use of botanicals, and how effective pest management can help promote food security.

Integrated pest management: concepts, tactics, strategies, and case studies

Radcliffe, E.B., Hutchinson, W.D. and Cancelado, R.E., 2009. Cambridge University Press, Cambridge, UK. ISBN 978-1444187496

- This book covers a broad and comprehensive range of topics in integrated pest management, focused primarily on theory and concepts. Written with an international audience in mind, this text is a valuable resource for researchers, extension specialists, and IPM practitioners worldwide.

A history of pesticides

Matthews, G.A., 2018. CABI, Wallingford, UK. ISBN 9781786394873

- This book reviews the history of the development and use of chemicals for the control of pests, weeds, and vectors of disease. The book examines the development of legislation designed to control and restrict the use of pesticides, and the emergence of IPM, and the use of biological control agents.

Manual of equine dermatology

Marsella, R., 2019. CABI, Wallingford, UK. ISBN 9781786395085

- This book deals exclusively with the equine dermatological problems that a clinician would see in practice. Illustrated in colour throughout, coverage progresses from the approach to a case, which covers history, signalment, identification of primary or secondary lesions, distribution of lesions, with examples of differential diagnoses based on the detection of specific lesions, to therapy and to chapters of diseases organized by presenting problems and signs.

Handbook of equine parasite control

Nielsen, M.K. and Reinemeyer, C.R., 2018. Wiley Blackwell, Hoboken, NJ, USA. ISBN 9781119382782

- This second edition gives all the details needed to appropriately manage endoparasites in equine patients and support discussions between horse owners and their veterinarians. It offers comprehensive coverage of internal parasites and factors affecting their transmission, principles of equine parasite control, and diagnosis and assessment of parasitologic information.

Current therapy in equine medicine

Sprayberry, K.A. and Robinson, N.E., 2014. 7th ed. Elsevier, Amsterdam, the Netherlands. ISBN 9781455745555

- With coverage of current issues and the latest therapeutic advances, this book provides a concise, all-new reference for the management of equine disorders and conditions. Chapters emphasize the practical aspects of diagnosis and treatment and provide details for therapeutic regimens. Cutting-edge topics include emerging and exotic infectious diseases that may endanger horses in North America; biosecurity strategies; imaging updates; medical genetics; multimodal pain management; and regenerative, geriatric, and oncologic medicine.

Knottenbelt and Pascoe's colour atlas of diseases and disorders of the horse
McAuliffe, S.B., 2014. 2nd ed. Elsevier, Amsterdam, the Netherlands. ISBN 9780723436607
• A unique collection of photographic illustrations of the major equine disorders, providing both qualified practitioners and veterinary students with an invaluable guide to greater diagnostic accuracy, treatment options, and a wider understanding of the processes and signs of equine disorders. Conditions presented are gathered from around the world, making for a resource of universal application that is a major aid to the rapid visual recognition and interpretation of clinical signs that are vital elements of success in veterinary practice.

Photo credits

Figure 1.1. Erika Machtinger, Pennsylvania State University.
Figure 1.2. Marie Sacha, Shutterstock.
Figure 1.3. Brian Adams, Flickr, CC BY 2.0, https://www.flickr.com/photos/22746348@N02/3712372951.

Chapter 2. Basic information on pests and parasites of horses

Chapter overview

Pests discussed in this book are either arthropods (Chapters 3-8) or parasitic worms (Chapter 9). There are some characteristics of both pest groups, like physical descriptions and life cycles, that can help with understanding how to identify and control them. This chapter will briefly discuss some basic information on arthropods and parasitic worms that can be used to complement each of the subsequent chapters.

Insects are a type of arthropod, but there are other pests of horses like ticks and mites that are not insects but are classified as arthropods because they are arachnids. Arthropods have an exoskeleton (external skeleton), a body made up of segments, and jointed paired legs.

2.1 Basic arthropod anatomy

Insects have an external skeleton made of a tough substance called chitin, a body with three regions, three pairs of legs with joints (6 total), compound eyes, and one pair of antennae (Figure 2.1). Some insects have one pair of wings, some have two pairs of wings, and some do not have any wings. The three-body regions of insects are the head, thorax, and abdomen. The head has mouthparts, eyes, and antennae. The thorax is the section behind the head from where the legs and wings originate in adults. The abdomen is the last region, it is typically made up of 11 to 12 segments and it usually is softer than the other body parts.

Ticks and mites are arachnids. Like insects, this group has an exoskeleton made of chitin but unlike insects, this group only has two body regions, four pairs of legs as adults (8 total), and do not have antennae or wings (Figure 2.2). The two body regions of ticks and mites are called the cephalothorax (a combination of the head and thorax) and the abdomen. Adult mites and ticks have four pairs of legs, but larvae have only three (6 total).

Arthropods discussed in this book have several different types of mouthparts. Some insects have sponging mouthparts used to soak up liquid food on surfaces (Figure 2.3). Others penetrate the host skin and feed on blood and other tissues in different ways. For example, mosquitoes have piercing-sucking mouthparts used to pierce skin and blood capillaries and suck blood (Figure 2.4), biting midges and horse flies have cutting and sponging, or biting-sucking mouthparts used to enter the skin, form a pool of blood, and then suck the blood from the pool (Figure 2.5). Some pests do not feed on animals or animal secretions. Many of these pests, like ants, have chewing mouthparts for eating plants, seeds, or other materials (Figure 2.6). Chewing mouthparts have distinct mandibles, or jaws, that operate in a sideways motion and are used for gripping, biting, and cutting food and other materials. Some of the species of lice have chewing mouthparts that rasp away at skin and hair.

Figure 2.1. Insects have three body regions, three pairs of legs, and a pair of antennae. Shown here is *Muscidifurax raptor*, a parasitoid that can be used as a biological control agent of some filth flies.

Figure 2.2. Arachnids, like the brown dog tick (*Rhipicephalus sanguineus*) shown here, have two body regions, four pairs of legs as adults, and do not have antennae or wings.

Figure 2.3. Sponging mouthparts are common in house flies and some other filth flies. They cannot chew but instead soak up liquids. House flies are notorious for regurgitating on solid food to liquefy it before sponging.

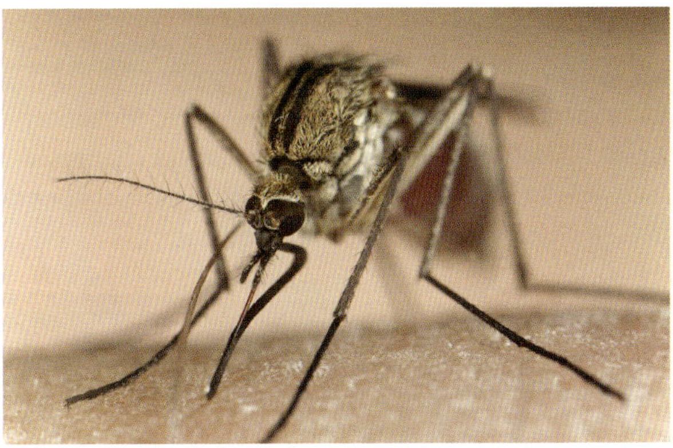

Figure 2.4. Mosquitoes have piercing-sucking mouthparts. These arthropods will pierce the skin with their tube-like mouthparts and suck out blood from within the host animal.

Figure 2.5. Horse and deer flies (A) have cutting-sponging and stable flies (B) have biting-sucking mouthparts. The long bayonet-like mouthparts have rasping teeth at the tip to cut the skin and create a pool of blood, which is then consumed.

Figure 2.6. Ants, beetles, bees, some lice, and other arthropods have chewing mouthparts. Distinct mandibles, or jaws, are generally present.

2.2 Arthropod life cycles

Insects and arachnids undergo metamorphosis as they grow. Metamorphosis is a biological process where the animal develops through relatively abrupt changes in the body structure through cell growth and rearrangement. Arthropods that are pests of horses can have one of two types of metamorphosis: holometabolous or hemimetabolous.

Holometabolous metamorphosis is also called complete metamorphosis because there is a profound change in appearance and function from the immature to adult stages (Figure 2.7). These life cycles typically have four stages: egg, larva, pupa, and adult. Examples of arthropods with holometabolous life cycles include the flies, such as mosquitoes, house flies and stable flies, moths like the Eastern tent caterpillar, and bees like the Africanized honey bee. Once a holometabolous arthropod reaches the adult stage, that individual will not grow or change in size. The terminology

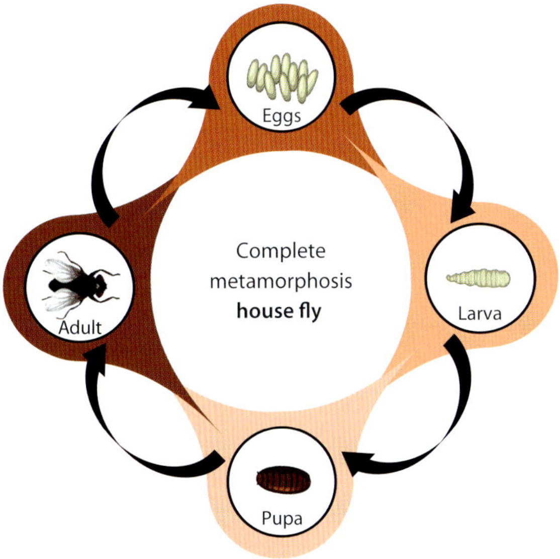

Figure 2.7. Holometabolous metamorphosis is also called complete metamorphosis. Arthropods that undergo complete metamorphosis, like house flies, have four life stages: egg, larva, pupa, and adult.

used in discussing insect larvae can be confusing. All holometabolous insects have a larval stage, but some specialised terms are used to describe particular groups. The term caterpillar is used to describe the larvae of all butterflies and moths (Lepidoptera). Larvae of some beetles, especially scarab beetles, are referred to as grubs. These are not to be confused with cattle grub larvae, which are a type of fly larvae. Larvae of some flies are sometimes called maggots (e.g. house flies, blow flies, and their relatives), whereas this term is never used to describe larvae of mosquitoes or other so-called primitive flies.

Hemimetabolous metamorphosis is also called incomplete metamorphosis, so there is less difference between the immature and adult stages as they change gradually each time they moult (Figure 2.8). These life cycles usually have three to four main stages including egg, larva, nymph, and adult. Arthropods with hemimetabolous life cycles include lice, ticks, and mites. There may be a single nymphal stage (ticks, mites) or many (lice). Like arthropods with holometabolous life cycles, adults do not grow or change in size once they have reached this final stage.

Arthropods are ectotherms, meaning they do not regulate their body temperatures within a narrow range like mammals. Instead, they rely on environmental temperature for heat. This is important in their development from egg to adult. Each arthropod has a temperature below and above which they will not continue development. That does not necessarily mean that the arthropod will die, but that development stops. Generally, development occurs faster as the temperatures increase until the top threshold is reached. This faster development is why many pest species, like flies, increase during warm periods in the spring and summer months. Adult activity also depends on ambient temperatures, so that most species cannot fly, bite, or lay eggs unless it is warm enough for their muscles to work properly. Throughout the pest life cycle descriptions in

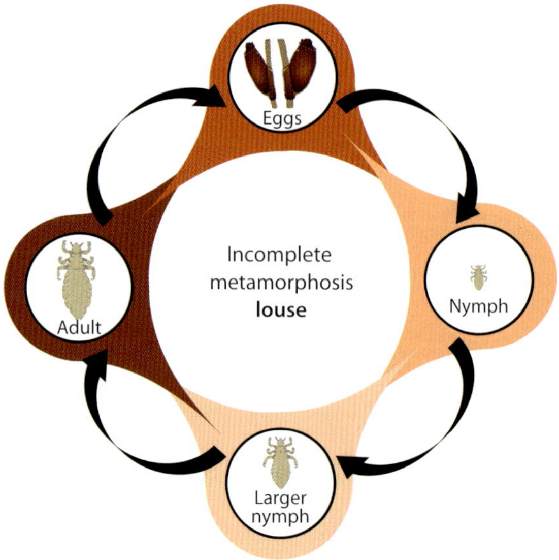

Figure 2.8. Hemimetabolous metamorphosis is also called incomplete metamorphosis. Arthropods that undergo incomplete metamorphosis, like lice, have three life stages, eggs, nymph, and adult. The nymphs get progressively bigger until they moult into the adult form. A fourth stage, the larval stage, may also be present in some organisms before the nymph stage, for example, mites and ticks.

this book, the general development time for each pest is given. It is important to recognize that this development time may decrease or increase with environmental conditions.

Arthropods are also dependent on nutrition throughout their life cycle. In many cases, pests develop together in suitable habitats. If there is overcrowding or a lack of suitable nutrition, this can kill the pest as it develops or, as often happens, reduce its final body size. For example, some house flies that are crowded with inadequate nutrition will still complete their development but adults will be significantly smaller than flies that developed with adequate nutrition. Because of the influence of the environment on arthropod development, it is not recommended to use size as the only characteristic for species identification.

2.3 Parasitic worm identification and life cycles

Parasitic worms that affect horses include flatworms (Phylum Platyhelminthes) and roundworms (Phylum Nematoda). The basic life cycle of a parasitic worm of horses involves ingestion of eggs or larvae (directly or indirectly), migration to a preferred location (usually in the digestive system), and development to the adult stage before production of eggs that are shed in faeces (Figure 2.9). The larvae develop through different growth stages internally before maturing into adults. Details of this process are species-specific.

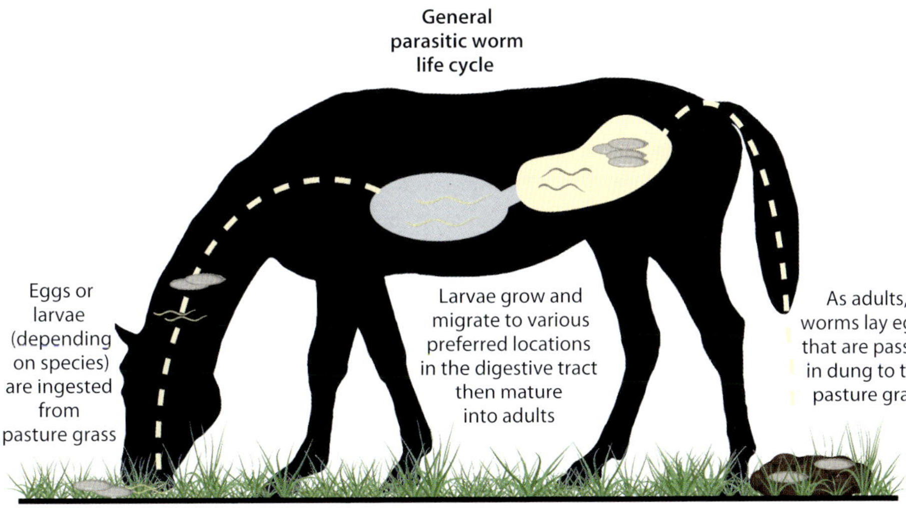

Figure 2.9. The basic parasitic worm life cycle begins when horses ingest eggs or larvae from pasture grasses. The parasitic worms mature in the gut and adults lay eggs that are passed with dung back onto pasture grasses restarting the cycle.

The tapeworms in the class Cestoda are flatworms, they are large worms of 8 to 25 cm (3 to 10 in) and have a head with four suckers. The eggs are released in packets and are D-shaped. The three tapeworm species that infest horses have an indirect life cycle that involves a mite that is ingested by horses during grazing. After development inside the horse to the adult stage, the eggs containing larvae (oncosphere) are released in packets in the faeces and then ingested by the mites.

Three main groups of nematodes affect horses, the large and small *Strongyles*, the ascarids, and the pinworms. These nematodes are small and thin round worms (hence their other name roundworms). Small strongyles are red and 1.0 to 1.5 cm (0.4 to 0.6 in), large strongyles are also red and 2.5 to 5.0 cm (1 to 2 in) long. Eggs are thin-shelled and oval in shape with the developing embryo visible. Ascarids are off-white to yellow and about 8 cm (3 in). The eggs of the ascarid *Parascaris equorum* are circular and brownish. Pinworm adults are similar in size to ascarids but they are white with one-third of the body tapering to a point and variable in translucency. Pinworm eggs are translucent and sticky; they are stuck to the area around the anus of horses by the female.

The three parasitic worm groups follow the basic life cycle described above with the following exceptions. Change to roundworms develop from eggs to early larvae in the environment before being ingested by horses. Large strongyles larvae migrate through organs causing damage, whereas small strongyles limit their damage to the intestine. The ascarids develop into an infectious stage within the eggs and are ingested as eggs by the horse. They migrate to the liver and lungs before being coughed up and swallowed. Pinworm larvae develop within the egg into the infectious stage, which are ingested by horses. The worms mature in the colon and rectum, the females emerge from the anus and lay eggs around the anus.

2.4 Quick identification key to pests of horses

Sometimes it can be a challenge to determine what type of pest is the problem. Below is a quick reference guide that will help to determine which group of pests to focus on and which chapter is most relevant. Review the characteristics of the pests and associated photos and determine which pest best represents the pest of interest. It is important to remember that multiple pests may be present at the same time, so make sure to consider all the potential pests.

2.4.1 Fly pests that live in dung, manure, and waste (Chapter 3)

Filth flies develop in organic waste like manure, decomposing hay or feed, or other organic waste. Some are seen in the environment, like in stalls or run-in sheds. This group of pests is almost always seen on the horse at some point. Filth flies cause extreme annoyance and have some characteristic behaviours that can be used to help identify them.

Characteristics of filth flies

- flies are clearly seen on the horse but can be on the face, body, or legs (Figure 2.10);
- flies look black or dark grey, often with dark bands on the thorax (Figure 2.11);
- flies are usually seen during the day;
- flies are highly mobile;
- flies are frequently found on horses;
- flies common around wounds and mucosal areas like the eyes, nose, and mouth (house flies, face flies);
- horses twitching shoulders or shaking head, swishing tail (house flies, face flies, horn flies);
- horse rubbing legs, stomping feet, or biting sides (Figure 2.12) (stable flies);
- flies found around or in the stables or waste accumulation areas (Figure 2.13) (house flies, stable flies).

Risk factors for filth flies

- cattle, poultry, or other animal facilities are nearby (within several km/mi);
- high stocking rate (horses per acre);
- accumulation of manure (e.g. manure piles or pits);
- use of straw or shavings as bedding.

2.4.2 Fly pests that develop in water (Chapter 4)

These insects develop in aquatic and semi-aquatic areas but are not usually seen consistently on the horse with some exceptions. Often these pests are determined by conditions seen on the horse, and not the pest themselves. Humans may be bitten as well because many of these pests do not exclusively bite horses.

Characteristics of aquatic and semi-aquatic arthropod pests

- skin conditions like hives and losses of hair around the head, neck, or midline of the belly (Figure 2.14);

Figure 2.10. Filth flies are often seen on the horse during the day and can be on the head, body, or legs.

Figure 2.11. Filth flies look black or dark grey and are strong fliers.

Figure 2.12. Filth flies are often found on horses causing stomping, head shaking, nipping at sides (shown), tail swishing, muscle twitching, or leg rubbing.

Figure 2.13. Filth flies can be found on horses, in the stable and on bedding, and around waste accumulation areas like manure piles, depending on the fly species.

- horses appear restless when turned out, especially in the early morning, late evening, or evening hours;
- small areas of blood on the horse as a result of insect bites (Figure 2.15);
- head shaking, twitching shoulders, swishing tail when outside;
- large flies (with sometimes vibrantly coloured eyes) seen repeatedly attacking animals (horse flies and deer flies);
- white scabs or plaques in the ears (Figure 2.16);
- humans being bitten by small or delicate-looking insects;
- the appearance of small 'wriggling' creatures in water troughs.

Risk factors for aquatic and semi-aquatic arthropod pests

- recent heavy rainfall;
- water (standing or moving) nearby;
- nearby marshlands, either fresh or saltwater.

2.4.3 Pests that develop and live on horses (ectoparasites) (Chapter 5)

Ectoparasites live on the horse. Some of these pests are small and may go unnoticed. However, similar to aquatic and semi-aquatic arthropods, if horse(s) start exhibiting skin conditions, mites and lice should be considered. Other ectoparasites can be seen biting the horse.

Characteristics of ectoparasites

- areas of hair loss, matted hair, and/or skin scabs (mites and lice);
- small insects seen in the hair (lice);
- horses are repeatedly scratching or rubbing their body on fences or structures (Figure 2.17);
- biting arthropod found embedded in the skin (ticks);
- fast-moving insects are seen on or in the hair (keds or louse flies).

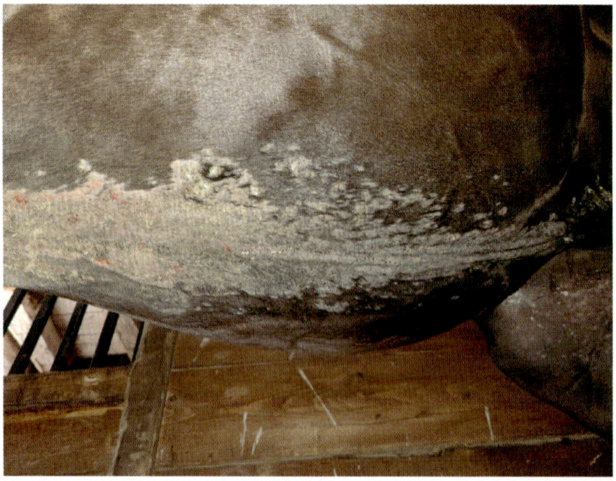

Figure 2.14. Sweet itch is a condition that can be caused by biting midges. This can appear as hair loss and sores in several areas of the body including the midline of the belly (shown).

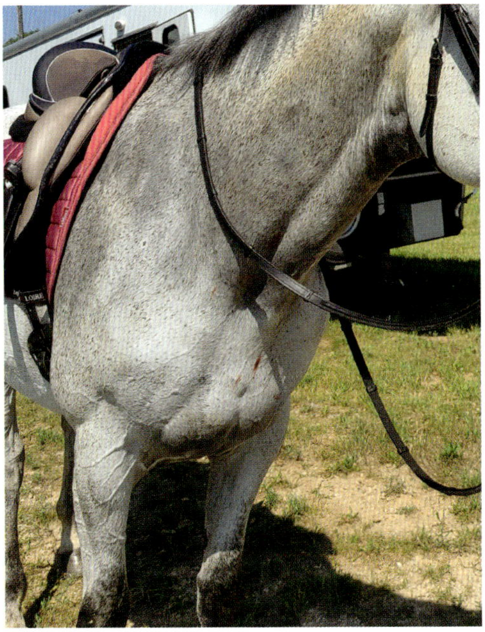

Figure 2.15. Horse and deer flies are pool feeders, meaning they cut through the skin to lap up the pool of blood that results. This can leave small blood marks on horses.

Figure 2.16. White scabs or plaques can appear in the ear when horses are being bitten by black flies.

Risk factors for ectoparasites

- winter conditions (mites and lice);
- groups of horses together with heavy winter coats (mites and lice);
- new horse(s) to the property;
- shared equipment with infested horses (lice, mites);
- heavily feathered horse breeds (mites);
- travel with horses to new areas where different ectoparasite pests may be active (i.e. ticks);
- tick habitat near pastures or riding in tick habitat (tall grass or brush, forested areas) (Figure 2.18);
- stress and other health conditions (mites and lice).

2.4.4 Pasture and forage pests (Chapter 6)

Pasture and forage pests are a bit more of a challenge to identify because they can be found in the pasture or around a horse facility, but not normally on the horse. Some can bite or sting, and some are pests because they can cause conditions in horses if they are encountered or ingested. The risks associated with this group of arthropods are highly dependent on geographic location.

Figure 2.17. Frequent scratching or rubbing may be an indication of ectoparasite infestation because many ectoparasites can cause extreme itching.

Figure 2.18. Tall grass, brush, and forested habitats are all areas where horses may be at risk for tick bites.

Characteristics of pasture and forage pests

- horses 'island grazing,' i.e. leaving areas of ungrazed grass, and/or mounds of raised soil (fire ants) (Figure 2.19);
- horses aggressively stomping with inflamed lower legs (fire ants);
- very small flies around the corners of the eye, lips, or other areas with secretions or fluids like scratches or wounds (eye gnats) (Figure 2.20);
- high populations of caterpillars with long hairs around pastures, stables, and other horse areas (tent or processionary caterpillars) (Figure 2.21);
- aggressive bees (Africanized bees);
- bees in small cavities or exposed on tree branches (Africanized bees);
- decline in performance in a region with tent caterpillars;
- spontaneous abortion (tent or processionary caterpillars);
- symptoms of kidney failure (blister beetles);
- signs of colic after feeding second or third season cuttings of alfalfa (blister beetles).

Risk factors for pasture and forage pests

- sand soils (fire ants);
- feeding alfalfa (blister beetles);
- host plants of tent or processionary caterpillars nearby.

2.4.5 Other insect pests of horses (Chapter 7)

Like pasture and forage pests, it can be difficult to determine if these pests are present before the conditions they cause are observed. In addition, the risks associated with this group of arthropods are also highly dependent on geographic location.

Figure 2.19. Large fire ant mounds appear as raised mounds in pastures, but horses also will avoid small fire ant colonies. Because of this, pastures can have clumps of ungrazed grasses.

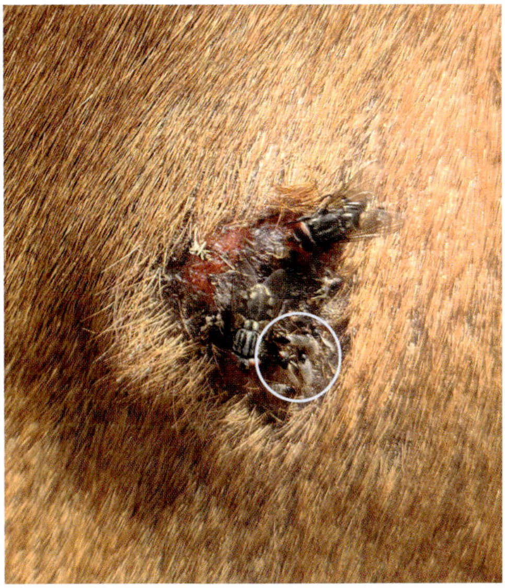

Figure 2.20. Eye gnats are commonly seen around the eyes, lips, nose, and scratches and open wounds, as seen here (circled) alongside larger filth flies.

Figure 2.21. Tent caterpillars are often found in silken nests in trees. However, when they leave these nests and are ingested by horses, they can cause spontaneous abortions and pericarditis.

Characteristics of other arthropods that impact horses

- hairy, bee-like flies around animals but not feeding (warble flies);
- raised bumps along the neck, withers, or back of the horse that are painful to touch (caused by warble flies);
- fly larvae (maggots) developing in open sores (screwworm);
- small white dots (eggs) on the hair, primarily on the legs, shoulder, or head (bot flies) (Figure 2.22);
- mass numbers of delicate insects with long thread-like tails around horse areas (mayflies) (Figure 2.23);
- crusts or nodules, which open to weeping sores in the ears, scrotum or udder, neck, and legs (sandflies).

Risk factors for other arthropods of concern

- horses pastured with or near cattle (warble flies);
- horses near a body of water (mayflies);
- rodent and rabbit burrows on or near horse facilities (sand flies).

2.4.6 Parasitic worms of horses (Chapter 9)

Internal parasites by their very nature cannot be identified externally. The eggs and larval stages are too small to be seen by the naked eye in pastures or manure. Horses on pasture are susceptible to internal parasites, while horses who are only kept in stables or dry lots have minimal parasite exposure. These parasites are best identified by close, microscopic examination of faeces following specific processing procedures. Each of the different species has different processing requirements to best locate them via microscope. Ultrasound examination is also helpful to identify ascarids in the small intestine of young horses.

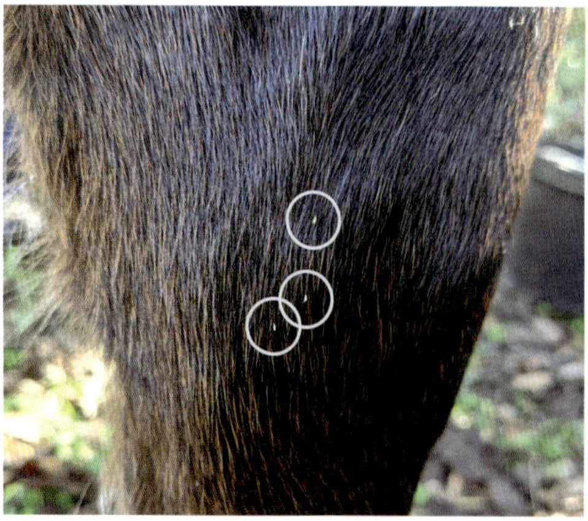

Figure 2.22. Bot fly eggs appear as little white dots and are most often found on the legs, shoulders, or head.

Figure 2.23. Mayflies emerge from the water in large groups. They only live a short time but can transmit the agents that cause equine neorickettsiosis (Potomac horse fever).

Characteristics of parasitic worm infestation

- weight loss;
- dull, rough coat (Figure 2.24);
- potbelly;
- decreased stamina or lethargy;
- coughing;
- diarrhoea;
- colic;
- tail rubbing (Figure 2.25).

Risks for parasitic worm infestation

- very young or old horses (Figure 2.26);
- stress;
- malnutrition;
- high egg shedders;
- lack of an internal parasite management program;
- good quality pasture with grass clippings or other vegetative mats (Figure 2.27).

2.5 Useful pest identification tools

Some pests of horses can be recognised by behaviour (like pest flies) and some can be recognised by the signs of the conditions they cause (like lice and mites). Some are more challenging to identify. Even closely related species of pests can have very different biologies. Therefore, it is important to identify the pest of interest before attempting control. Arthropod pests of horses can be small, even microscopic, so identification can be a challenge and some internal parasites are recognized by the presence of eggs. However, some tools can be useful to aid in identification.

Figure 2.24. Characteristics of parasitic worm infestations include dull, rough coats and weight loss.

Figure 2.25. Frequent tail scratching or rubbing can be a sign of parasitic worm infestations, particularly pinworms.

Figure 2.26. Very young and old horses are more susceptible to parasitic worm infestations.

Figure 2.27. Good quality pasture, especially with a layer of grass clippings that can protect the worms, can be a risk for parasitic worm infestations.

Pest identification supplies

- *Gloves* (Figure 2.28). Disposable gloves are helpful to have when checking for bot fly eggs, treating louse, mite, or other arthropod infestations, applying pour-ons or other repellents, or grooming off lice or mites, or for any other reason that it would be beneficial to protect the skin.
- *Magnifying glass* (Figure 2.28). Arthropods, especially ectoparasites, can be very small and difficult to see much less identify. A magnifying glass can help with species identification of flies, ticks, or other larger arthropods. A magnifying glass will not be powerful enough to identify lice or mites, but may help in determining if they are present in combed from hair.

- *Forceps* (Figure 2.28). Forceps or tweezers can be useful for removing ticks, mites, and lice from horses. Forceps or tweezers should have a fine tip that can be straight or bent. Fine tipped forceps make it easier to properly grasp a tick for removal. Capturing lice and mites is easier with fine tip forceps as well. Specialised tick removal tools (see Chapter 5) may be useful if hair makes it difficult to remove ticks that have already attached to the horse.
- *Tape* (Figure 2.29). Tape can be used to remove small arthropods from flea or louse combs or directly from the horse to help identify the pest of interest. The tape can be placed under a microscope or a high magnification magnifying glass for identification.
- *Sticky lint remover* (Figure 2.29). Like tape, sticky lint removers can be used to roll over some ectoparasites, like keds or louse flies. These lint removers are less sticky than tape and can also be used after riding through areas where ticks may be active to remove ticks walking on the legs before they bite.
- *Clippers* (Figure 2.30). Clippers can be helpful to remove body hair to make tick checks easier, especially when winter guard hairs are long in the autumn and spring. They can also be used to remove hair and feathers if mange or a lice infestation is suspected to improve the likelihood of treatment success.
- Lice/flea combs (Figure 2.31). Lice and flea combs can be used to remove several pests from horses that can move quickly through the hair or that are small and challenging to see including keds or louse flies, ticks, and lice.
- Headlamp (Figure 2.31). While a headlamp is certainly not a requirement, it can be helpful when looking for nits (lice eggs), inspecting wounds for pests, checking for ticks, or other grooming activities that may be conducted indoors or in low light. Some magnifying glasses also come with lights that can serve the same purpose.

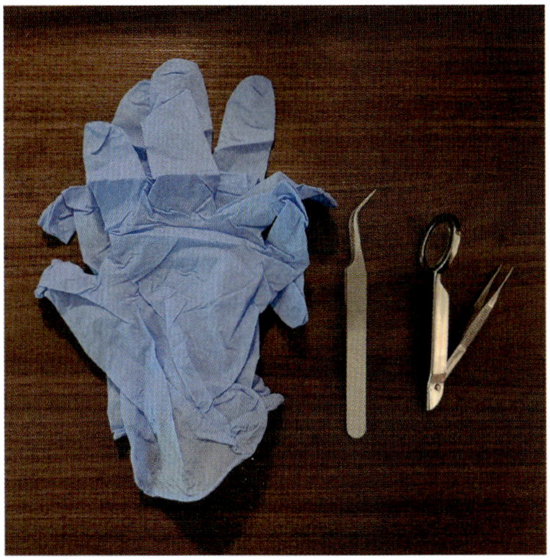

Figure 2.28. Gloves, fine-tipped forceps, and a magnifying glass can be helpful for tick identification and for removing ticks that have bitten a horse.

Figure 2.29. Tape and lint rollers can be used to remove arthropods from combs and forceps or to roll over legs after riding through tick or ked habitat.

- *Microscope* (Figure 2.32). A microscope may seem expensive or be intimidating for some people, but newer microscopes are relatively easy to use and even inexpensive models can be used to help identify internal parasites and small pests like lice and mites. It can also be used to see defining characteristics of any other pest species, like wing veins, markings, and mouthparts, that may be useful when developing an integrated management plan.

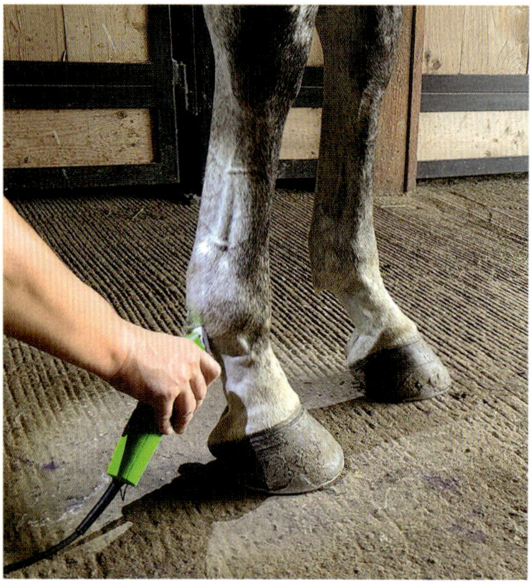

Figure 2.30. Clippers can be used to remove body hair to make tick checks easier or to remove hair infested with lice or mites to improve the likelihood of successful treatment.

Figure 2.31. A headlamp can be helpful when looking for small ectoparasites like lice and mites or inspecting wounds for pests. A flea/louse comb can be used to help part the hair and remove small arthropods for further inspection.

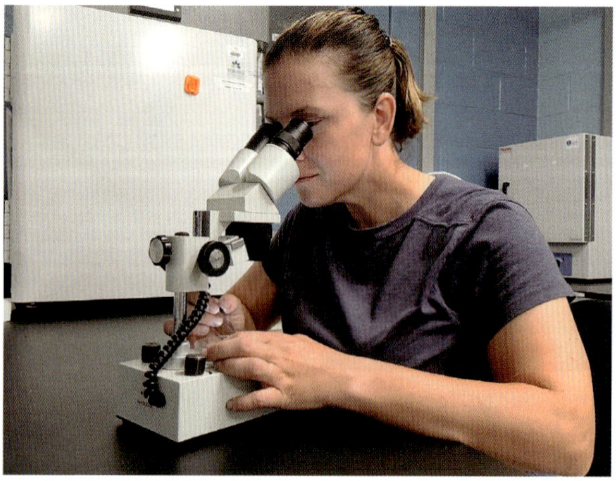

Figure 2.32. A simple, inexpensive microscope can be useful for identifying arthropod pests removed from horses.

Photo credits

Title page	Vinicius R Souza, Shutterstock.
Figure 2.1.	Lyle Buss, University of Florida.
Figure 2.2.	Erika Machtinger, Pennsylvania State University.
Figure 2.3.	Garo Manjikian, Shutterstock.
Figure 2.4.	Henrik Larsson, Shutterstock.
Figure 2.5.	(A) Kondor83, Shutterstock, (B) Cosmin Manci, Shutterstock.
Figure 2.6.	Jiade, Shutterstock.
Figure 2.7.	Erika Machtinger, Pennsylvania State University.
Figure 2.8.	Erika Machtinger, Pennsylvania State University.
Figure 2.9.	Erika Machtinger, Pennsylvania State University.
Figure 2.10.	Erika Machtinger, Pennsylvania State University.
Figure 2.11.	Erika Machtinger, Pennsylvania State University.
Figure 2.12.	Erika Machtinger, Pennsylvania State University.
Figure 2.13.	Erika Machtinger, Pennsylvania State University.
Figure 2.14.	Erika Machtinger, Pennsylvania State University.
Figure 2.15.	Erika Machtinger, Pennsylvania State University.
Figure 2.16.	Heather Maples.
Figure 2.17.	Tamara L Sanchez, Shutterstock.
Figure 2.18.	vio0orel, Shutterstock.
Figure 2.19.	Liz Weber, Shutterstock.
Figure 2.20.	Erika Machtinger, Pennsylvania State University.
Figure 2.21.	Daniel Potter, University of Kentucky.
Figure 2.22.	Erica Lacher, Springhill Equine Veterinary Clinic.
Figure 2.23.	Hugh Landsdown, Shutterstock.
Figure 2.24.	Anton Kasatkin.
Figure 2.25.	Doug McLean, Shutterstock.
Figure 2.26.	Erika Machtinger, University of Florida.
Figure 2.27.	Barry Fowler, Shutterstock.
Figure 2.28.	Erika Machtinger, Pennsylvania State University.
Figure 2.29.	Erika Machtinger, Pennsylvania State University.
Figure 2.30.	Erika Machtinger, Pennsylvania State University.
Figure 2.31.	Erika Machtinger, Pennsylvania State University.
Figure 2.32.	Erika Machtinger, Pennsylvania State University.

Chapter 3. Fly pests that develop in manure and waste

Chapter overview

This chapter covers the groups of true flies (order Diptera) that are considered filth flies. Filth flies develop in decaying organic matter like waste hay and feed, manure, soiled bedding, and even grass clippings. There are five primary species of filth flies associated with horses: house flies, stable flies, *Fannia* flies (Section 3.1), horn flies, and face flies (Section 3.2). Two of these species, horn flies and face flies, do not reproduce on equine farms because of their inability to develop in any material other than cattle manure. However, because horses are often located in rural areas where there may also be cattle, these flies can also become a pest problem on equine farms. Unfortunately, many of these filth flies look very similar and can be a challenge to tell apart by appearance alone. Some fly management options are species-specific so identification is critical. There are some distinct behavioural differences that, coupled with slight differences in appearance, can be used to identify species of fly pests. These differences are described in each respective section of this chapter. Filth flies are some of the most common flies associated with animal production and as such, there has been extensive research on this group of insect pests. Most research has been conducted on poultry and cattle production facilities. However, in many cases, this information can be used to develop management practices for fly control around horses as well. Filth flies often cause annoyance, and in the case of stable flies and horn flies, can inflict painful bites. In addition, because of their extensive and repeated visitation among animals and their development and feeding in waste and decomposing matter, filth flies can carry and transmit pathogens that cause diseases in horses including the pathogens and parasites that cause pigeon fever, summer sores, eyes worms, sarcoids, and respiratory viruses.

3.1 Filth flies that can develop in equine manure

House flies, stable flies, and *Fannia* flies are considered generalist filth flies because they develop in a wide range of decomposing organic matter, including horse manure. They are some of the most common pest flies associated with animal production and are frequently encountered on equine facilities throughout the world.

3.1.1 House fly (*Musca domestica*)

3.1.1.1 Pest summary (Table 3.1)

The house fly (Figure 3.1) is a single species, *Musca domestica*, that is present throughout the world. It is a medium-sized grey fly and the most well-known and important member of the insect family Muscidae. The house fly has been one of the primary pests of humans and their domesticated animals for as long as humans have kept records. The house fly appears to have evolved along with our non-human primate ancestors, moving along with them and developing in faeces and food waste. This association strengthened as humans settled into permanent communities and domesticated animals. The house fly moved along with human populations as they migrated across the planet. These flies play an important role in the decomposition of organic material. However, their attraction to this material leads to a close association between these flies and humans and their domesticated animals, including horses. While they can be a nuisance due to sheer abundance, they do not bite, and their pest status comes from annoyance and their ability to move pathogens that may cause disease (e.g. bacteria and viruses) in the environment, for example, from manure to feed or between wounds on different horses. They can be a particular irritation to horses around sores and wounds (Figure 3.2).

3.1.1.2 Distribution and seasonality

The house fly has one of the widest distributions of any insect in the world. It is present from the Arctic to the tropics and is very common wherever humans and their animals are present. In cooler areas, house flies do not hibernate during the winter, but rather continue developing at very slow rates in areas that do not freeze and remain frozen. Adults become active as temperatures warm and are common in the summer months in temperate regions and year-round in subtropical and tropical regions.

3.1.1.3 Description

Adult house flies have sponging mouthparts for feeding on liquids. They are 3 to 8 mm (0.12 to 0.30 in) long, have four dark stripes on top of the thorax where the wings attach, and a distinct bend in the fourth wing vein. Female house flies have eyes that are much more widely separated than those of males (Figure 3.3). Adult house flies rest with their bodies held horizontally, unlike stable flies (Section 3.1.2). Larvae are cream coloured, elongated, and develop into pupae that resemble rodent droppings and vary in colour from light red when young to dark brown when older (Figure 3.4).

Table 3.1. Pest summary of house fly, *Musca domestica*.

Distribution[1]	Australia: Yes Canada: Yes Europe: Yes New Zealand: Yes United States: Yes
Size	Adult flies are 6-7 mm (0.24-0.28 in)
Key identification characteristics	Grey, four black stripes, bend in the wing vein, found off the horse indoors and outdoors especially around waste; can be on the body of the horse and not limited to the face
Mouthparts	Sponging (non-biting)
Feed on	Decaying organic matter like manure, soiled bedding, wet feed, etc.
Development time[2]	8-10 days in summer
Adult life span	1-2 weeks
Total eggs laid	Up to 900, but usually about 100
May be mistaken for	Stable flies (Section 3.1.2), *Fannia* flies (Section 3.1.3), face flies (Section 3.2.1), horn flies and buffalo flies (Section 3.2.2), blow flies (Chapter 8), flesh flies
Impact on horses	Annoyance, potential vectors of pathogens causing pigeon fever, summer sores, and sarcoidosis

[1] The pest distribution in the regions targeted by this publication are listed as of the date published. Areas of primary pest concern (if any) are in bold. Global distributions are listed in Section 3.1.1.2.
[2] Development times are approximate and can change with environmental conditions including temperature, humidity, and availability of food resources.

Figure 3.1. House fly, *Musca domestica*.

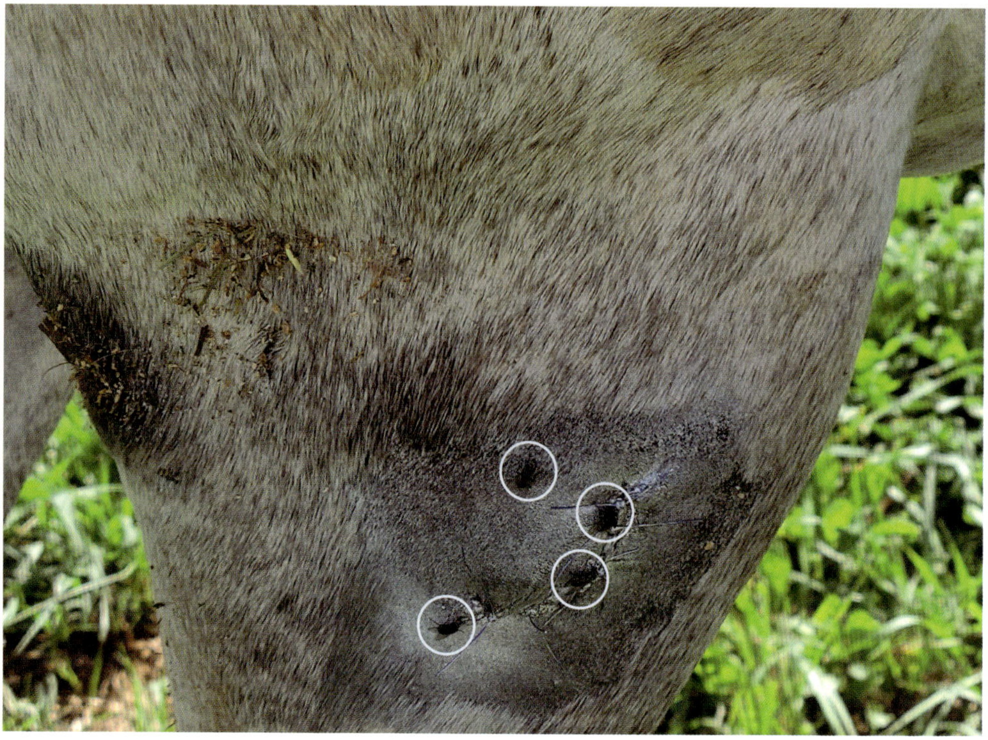

Figure 3.2. House flies (circled) can be irritating to horses and wounds with their persistent feeding behaviour.

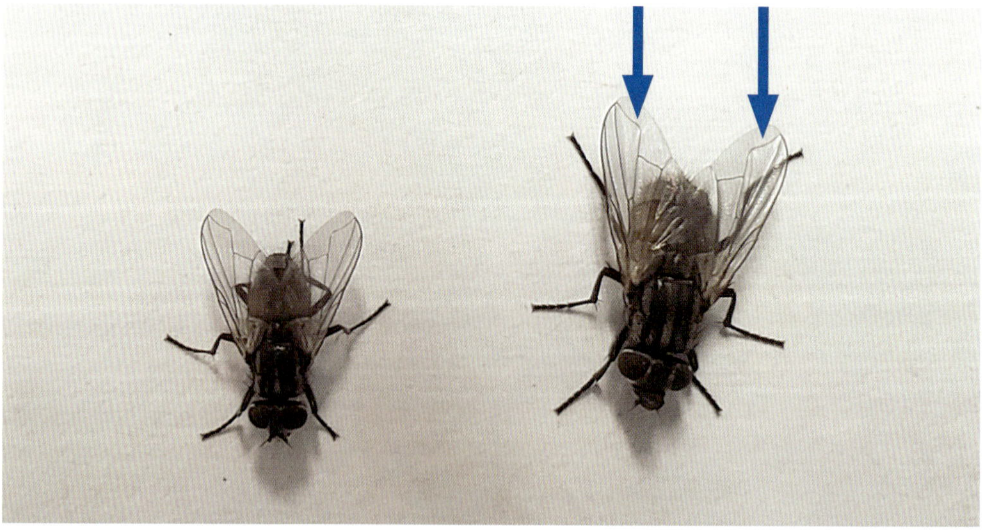

Figure 3.3. House flies have four dark stripes on the top of the thorax and a distinct bend in the fourth wing vein (arrows). There is less distance between the eyes of male house flies (left) than the eyes of female house flies (right). Although the male appears smaller here, size is not a good indication of fly sex because adult size is dependent on nutrition and environment during development.

Figure 3.4. House fly larva (A) and pupae of different ages with the youngest on the left and oldest on the right (B).

Because of their small, nondescript greyish colouring, house flies may be confused for other filth flies including stable flies, *Fannia* flies, face flies, and horn flies described in this chapter. However, behavioural characteristics like where they are found, how they rest, and how they eat can provide a practical way to distinguish among these species. Blow flies may also be confused for house flies, but these are usually brightly metallic coloured flies that are commonly only seen around an attractant like garbage, carrion, a fly trap, or a manure pile (Figure 3.5), and these flies generally do not become pests (Chapter 8). Flesh flies (family Sarcophagidae) (Figure 3.5) can also be found around waste and other areas of an equine farm, but these flies often have bright red eyes, a checkerboard pattern on their abdomen, are usually larger than house flies, and are not present in high numbers.

Figure 3.5. House flies can be confused for some other species of fly like blow flies (A) or flesh flies (B) that are attracted to waste and manure, but these flies are generally not pests.

3.1.1.4 Life history

House flies have complete metamorphosis with four life stages: egg, larva, pupa, and adult. Development from egg to adult can be completed in 8 to 10 days under warm weather conditions but can take over 30 days if the weather is cooler or if where the larvae are developing is less suitable for development. Larvae move into drier substrates like dry soil or bedding to pupate. House flies mate soon after emergence as an adult, with females only needing to mate once in their lifetimes. Female flies will begin laying eggs 4 to 7 days after emergence. Wild flies generally live less than 10 days and lay under 100 eggs per female. Because of their quick development time, very high fly populations are possible when seasonal temperatures increase.

Adult house flies are found in indoor and outdoor situations. They are opportunistic feeders that feed on a wide range of materials including horse manure and body secretions, horse feed, food waste, plant nectars, and a variety of decomposing and fermenting organic materials. Adults require both carbohydrates for energy and protein for egg development. Their sponging mouthparts lack teeth and limit them to consuming liquid foods, but house flies can feed on solids by first regurgitating liquid from their gut to soften it. This feeding behaviour along with faeces deposition leads to the deposition of faecal and regurgitant spots on surfaces wherever house flies are abundant (Figure 3.6). Moreover, their habit of repeatedly regurgitating and ingesting food is the primary reason the house fly is notorious as a vector for many pathogens that can cause disease to people and horses. Fly spots are frequently found on the resting areas of house flies, and because face flies and horn flies will not leave spots indoors, this can help with fly species identification.

House fly larvae can develop on an astonishingly wide variety of decomposing organic matter, feeding on the microbes in the material. On equine farms, house flies lay eggs on and develop in fresh manure, manure mixed with bedding, hay, or feed, or hay and feed that has become wet. Relatively little development occurs in piles of manure that are a week or more old (Machtinger *et al.*, 2014).

Figure 3.6. House flies defecate and vomit when they rest, leaving small spots on these surfaces, as seen here (circled) on this wooden frame.

3.1.2 Stable fly (*Stomoxys calcitrans*)

3.1.2.1 Pest summary (Table 3.2)

The stable fly, *Stomoxys calcitrans* (Figure 3.7), is another species in the insect family Muscidae. Other common names for the fly include barn fly, dog fly, biting house fly, and power mower fly. Like the house fly, the stable fly is a medium-sized grey fly that is present throughout the world. But unlike the house fly, the stable fly has piercing-sucking (i.e. biting) mouthparts and feeds on blood. Stable flies will readily bite and feed on horses, dogs, and humans as well as other farm animals, such as pigs, cattle, and goats. Nuisance biting by stable flies on horses results in avoidance behaviour by horses like stomping and running, and reduced performance. This species may also transmit pathogens during feeding between wounds on different horses.

Table 3.2. Pest summary of stable fly, *Stomoxys calcitrans*.

Distribution[1]	Australia: Yes Canada: Yes Europe: Yes New Zealand: Yes United States: Yes
Size	4-7 mm (0.16-0.28 in)
Key identification characteristics	Forward-facing mouthparts, checkerboard pattern on their abdomen, rests at an angle; prefers to feed on the legs; horse stomping generally suggests stable flies
Mouthparts	Biting-sucking
Feed on	Blood and plant nectars
Development time[2]	About 20 days under summer conditions
Adult life span	Average of about 8 days
Total eggs laid	Up to 800 but usually about 100
May be mistaken for	House flies (Section 3.1.1), *Fannia* flies (Section 3.1.3), face flies (Section 3.2.1), horn flies and buffalo flies (Section 3.2.2), blow flies (Chapter 8), flesh flies
Impact on horses	Annoyance, agitation, pathogen transmission

[1] The pest distribution in the regions targeted by this publication are listed as of the date published. Areas of primary pest concern (if any) are in bold. Global distributions are listed in Section 3.1.2.2.

[2] Development times are approximate and can change with environmental conditions including temperature, humidity, and availability of food resources.

Figure 3.7. Stable fly, *Stomoxys calcitrans*.

3.1.2.2 Distribution and seasonality

The stable fly is found worldwide. They are typically pests around stables, on pastured animals, in a variety of animal agriculture facilities, on beaches, and near cropland where plant waste is allowed to accumulate. Stable flies are sensitive to high temperatures and populations rise and fall twice during the year in temperate areas, peaking in early summer and again in late summer. Cooler locations may only have a single peak in midsummer, whereas warmer locations may see a single peak in late winter through early spring. Because stable fly development sites are mostly in outdoor locations, rainfall is needed to keep such sites wet enough for immature flies.

3.1.2.3 Description

Adult stable flies are 4 to 7 mm (0.16 to 0.28 in) long, grey, and have a dark biting-sucking proboscis, or pointed mouthpart, extending from the head, two pairs of broad dark thoracic stripes, and large spots on the top of the abdominal segments that appear as a checkerboard pattern (Figure 3.7). Unlike house flies that rest in a horizontal position, stable flies have a characteristic resting posture with their heads higher than the abdomens and with the wings partially spread (Figure 3.8). Like house flies, adult stable flies can be confused for other filth flies (e.g. house flies, face flies, horn flies), or other flies like blow flies (Chapter 8) or flesh flies (Section 3.1.1.3). Their larvae and pupae are very similar in appearance to house fly larvae and pupae, and it can be difficult to tell them apart (see the key of Scott and Littig, 1962 for identification tips).

3.1.2.4 Life history

Like house flies, stable flies have complete metamorphosis with four life stages: egg, larva, pupa, and adult. Immature stable fly development is slower than the house fly. Development from the larval to the pupal stage requires about 13 days under warm conditions, with an additional 4 to 5

Figure 3.8. Stable flies rest with their heads higher than their abdomens, which is different from the horizontal resting positions of other filth flies, like house flies seen in Figure 3.6.

days before adult emergence. Pupae are generally found near the larval development site. As with house flies, these development times can be extended to many weeks in cooler weather or if the areas where the flies develop have poor nutrition or are overpopulated. Stable flies mate and begin feeding soon after emergence and are ready to lay their first eggs after several days of successful blood feeding. Each female has the potential to lay several successive batches of about 100 eggs for up to 800 eggs over her lifetime, but few flies live long enough for maximum production.

Although adult stable flies will occasionally feed on nectar, the vast majority of their feeding is on blood, and they inflict a painful bite. Stable flies may feed several times per day because of the stomping, tail swishing, and other efforts of horses to dislodge them with each attempted bite. Biting is often most prominent on the legs, bellies, and sides of horses. In response, horses frequently stomp, swish their tail, run their heads on their legs, and swing their head around to their belly. Horses may also bunch together in the field to try to avoid stable fly bites. Adult stable flies are highly mobile and can travel 1.5 km (0.9 miles) in less than a day, and many flies may immigrate to equine farms from surrounding cattle farms to feed on the animals (Pitzer *et al.*, 2011). Stable flies are more common outdoors and are not frequently found in large stables or other indoor spaces. However, they will rest and bite horses in stable entrance ways, and shallow covered buildings like run-in sheds or outward-facing stalls like shedrow and other row barns (Figure 3.9).

Stable fly larvae develop in fermenting plant materials that may or may not be mixed with horse manure. Stable flies can also develop in horticultural crop residues, such as waste vegetables. Unlike house flies, stable flies generally do not lay eggs on or develop in fresh horse manure, preferring older and stockpiled manure or manure that is mixed with hay, bedding, or urine (Machtinger *et al.*, 2014). Often, many of the stable flies that visit horses are produced off the equine farm, especially from nearby cattle facilities. The waste from hay bales or repeated feeding of hay in the same area over time is a common stable fly source, where larvae develop in the resulting mixture of scattered hay, urine, manure, and rainwater (Figure 3.10).

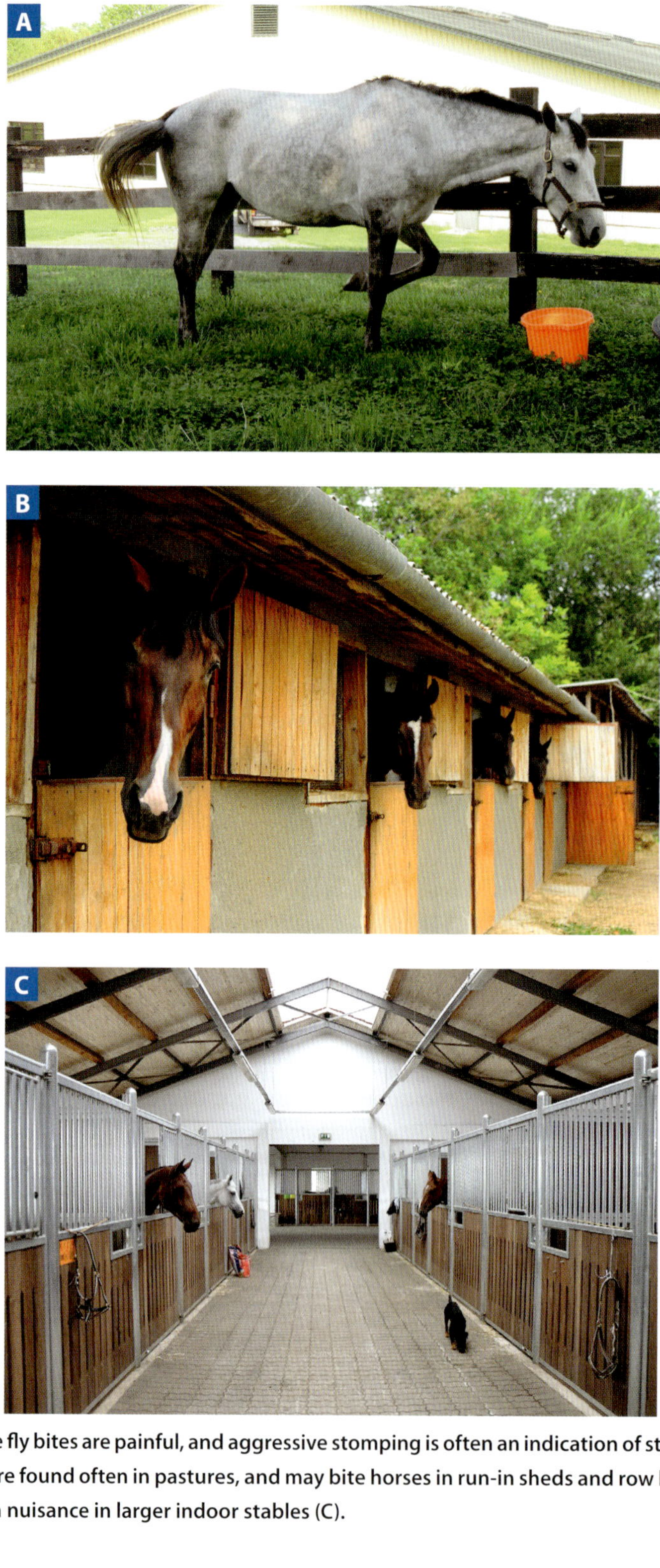

Figure 3.9. Stable fly bites are painful, and aggressive stomping is often an indication of stable fly presence (A). Stable flies are found often in pastures, and may bite horses in run-in sheds and row barns (B), but are less likely to be a nuisance in larger indoor stables (C).

Figure 3.10. Stable flies develop readily in manure mixed with waste hay, especially from large round or square bales in the pasture (stable fly pupae shown).

Stable flies can develop on equine farms or on other animal facilities, so control efforts should incorporate both prevention and methods to reduce bites.

3.1.3 *Fannia* flies (*Fannia* spp.)

3.1.3.1 Pest summary (Table 3.3)

The fly family Fanniidae includes four genera of small flies. Within the genus *Fannia* there are approximately 288 species worldwide. However, only a few of these species have the potential to be pests on equine farms, including the little or lesser house fly *Fannia canicularis* (Figure 3.11), latrine fly *Fannia scalaris,* and the canyon flies of the *Fannia benjamini* group of subspecies. Like house flies, these flies are attracted to secretions from the eyes, nose, ears, and open wounds. Several species of *Fannia* flies are associated with myiasis (immature fly development in living or necrotic tissue of animals) although this has not been documented in horses.

3.1.3.2 Distribution and seasonality

As a genus, *Fannia* is present worldwide. *Fannia* flies are present worldwide and are common in poultry houses. The species that are typically considered equine pests are the lesser house fly *Fannia canicularis,* latrine fly *Fannia scalaris,* and the canyon flies of the *Fannia benjamini* complex. The latrine fly is present in holarctic regions, and the canyon flies are a group of species that occur in western North America. *Fannia* flies were rarely encountered in the 20th century until the expansion of high-density poultry housing systems.

Table 3.3. Pest summary of *Fannia* flies, *Fannia* spp.

Distribution[1]	Australia: Yes Canada: Yes Europe: Yes New Zealand: Yes United States: Yes
Size	Adults are 3-6 mm (0.12-0.24 in)
Key identification characteristics	Smaller than the house fly, three instead of four black stripes on the thorax, wings overlap when resting, a distinct aimless hovering flight pattern of males
Mouthparts	Sponging (non-biting)
Feed on	Body secretions, manure, other decaying organic matter
Development time[2]	20-35 days
Adult life span	Variable, depending on the species
Total eggs laid	Variable, depending on the species
May be mistaken for	House flies (Section 3.1.1), stable flies (Section 3.1.2), face flies (Section 3.2.1), horn flies (Section 3.2.2), flesh fly, blow fly (Chapter 8)
Impact on horses	Annoyance, transmission of the eyeworm *Thelazia*

[1] The pest distribution in the regions targeted by this publication are listed as of the date published. Areas of primary pest concern (if any) are in bold. Global distributions are listed in Section 3.1.3.2.
[2] Development times are approximate and can change with environmental conditions including temperature, humidity, and availability of food resources.

3.1.3.3 Description

Fannia flies bear a superficial resemblance to the common house fly (Section 3.1.1) but can be distinguished by several characteristics including a smaller body (about 2/3 the size of house fly), the presence of three (rather than four) longitudinal stripes along the thorax, and the habit of folding the wings in an overlapping position when at rest (Figure 3.12). Moreover, the wings of *Fannia* flies do not have a wing vein with a strong bend near the tip, as do those of the house fly. Larvae of *Fannia* flies are also distinct from house flies in being brown, somewhat flattened, and possessing an array of hairs or spikes down the length of the larva and longer stalks at the tail end (Figure 3.13). Like other filth flies, *Fannia* flies can be confused for other small, greyish flies like house flies, stable flies, face flies, or horn flies, or other flies like blow flies (Chapter 8) or flesh flies.

3.1.3.4 Life history

Like the other flies in this group, *Fannia* flies have complete metamorphosis with four life stages: egg, larva, pupa, and adult. Although the larvae can develop in a wide range of decaying organic substrates from carrion to horse manure, this group often thrives particularly well in poultry manure. Some *Fannia* species develop in the soil, with some species being particularly abundant in the leaf litter

Figure 3.11. Lesser house fly, *Fannia canicularis*.

Figure 3.12. *Fannia* flies look similar to house flies, but are small, have three stripes on the thorax instead of four, and fold their wings at rest.

Figure 3.13. The lighter brown *Fannia* fly larvae (A) and the darker brown pupae (B) do not look like those of house fly and stable fly (Figure 3.7). These *Fannia* fly immatures are somewhat flattened, brown, and have an array of hairs or spikes that run down the length of the body.

and soil complex associated with the non-native plant called the baby sun rose *Mesembryanthemum cordifolium* found in the western United States of America (Mullens and Gerry, 2014).

Fannia flies develop into adult flies in 20 to 35 days under warm conditions. Development times are longer in cooler weather, but this species is also sensitive to hot conditions, with temperatures over 30 °C (86 °F) inhibiting egg hatching. Because *Fannia* flies are sensitive to heat, populations are most common in spring and early summer. Adult *Fannia* flies feed on many of the same foods as house flies and are similar pests with one key exception. Males of *Fannia* flies form mating swarms and hover in circles about 2 m (6 ft) above the ground under some sort of marker object such as tree branches outdoors or overhead light fixtures indoors. Adult flies can be severe pests because of their habit of swarming around the heads of horses and people.

3.1.4 Control of fly pests that can develop in equine manure

It is important to be able to identify the species of filth fly before establishing control methods, as control of each of these fly species differs because of their distinct biology and habits. Control of generalist filth flies, especially house and stable flies, should be focused on cultural control methods (i.e. prevention and sanitation). Waste management is the most important preventative step, as removing the development sites for flies often reduces pest numbers. However, because these flies are strong fliers, if equine farms are located in rural areas or areas with neighbouring animal farms, adult flies may still be pests after immatures are managed with sanitation due to immigration, and the focus may need to be on adult fly control. In addition, some farms may not be able to adjust management practices due to time or financial constraints. In those cases, integrating different methods as part of an IPM plan (Chapter 1, Appendix B) is important to improve overall control. While house flies and stable flies are common on equine farms, *Fannia* flies are not a common pest and tend to be spillovers from neighbouring livestock or poultry production facilities. Please see Geden *et al.* (2021), Rochon *et al.* (2021), and Murillo *et al.* (2021) in 'Continuing education' for more information on house flies, stable flies, and *Fannia* flies, respectively.

Because of the rapid development times and ability of generalist filth flies to increase population numbers quickly, it is important to start implementing control methods before flies become a problem. Keeping records of when filth flies start appearing each year and population numbers throughout the year can help with predicting when to start control the following year. See Gerry (2020) and Machtinger (2019) in 'Continuing education' and Chapter 10 for more information on monitoring filth flies.

Cultural control

Moisture management

House flies and stable flies develop best in materials that are 70 to 80% moisture, so moisture management can be a simple way to reduce fly populations. Moisture may come naturally from manure or rain, or artificial sources like leaky waterers and hoses, or overflowing water troughs. Water management, like proper farm drainage or vegetation buffers, can reduce many natural sources of water. Leaky waterers, hoses, or other spigots should be repaired quickly. Sufficient ventilation in the stable can reduce moisture in stall areas as well. The addition of outdoor-rated fans to stalls can aid in the quick drying of bedding and stored manure. *Fannia* flies can develop in material with low moisture content, even as low as 30%, so moisture management will likely not be effective for these species.

Manure and waste management

Both house flies and stable flies are attracted to and can readily develop in horse manure. Effective control starts by eliminating or reducing available manure for these flies. Larvae can hatch from house fly and stable fly eggs in under a day, so timely removal of manure waste and soiled bedding is necessary to limit developing flies. Manure should be removed from horse stalls and congregation areas daily. In addition, manure should not be allowed to sit uncovered, such as in waste buckets or wheelbarrows (Figure 3.14). It is important to keep small areas between rubber mats in stalls and trailers clean as these are protected areas where immature flies can successfully develop (Figure 3.15). Wet, decomposing spilled feed can be a suitable habitat for fly development so feed rooms should be cleaned frequently and spilled feed removed quickly. *Fannia* fly development has not been evaluated in horse manure, but these manure management practices will likely keep these flies from developing.

Bedding changes

Bedding choices can influence the development of some filth flies in both stalls and manure piles. Sawdust or sand typically support fewer developing flies than other bedding choices like shavings and straw (Schmidtmann, 1989; Schmidtmann *et al.*, 1991) (Figure 3.16). Studies on other materials like paper shavings, pelleted bedding, and newer bedding options have not been investigated, but bedding that dries fast and allows for a thorough cleaning will be better for fly reduction than those that hold moisture or are more difficult to clean.

Figure 3.14. Flies can develop quickly in manure stored in areas like grooming or wash stalls (A). Keep manure covered (B) to prevent flies from developing in the manure or being attracted to the area. Emptying manure collections daily in areas away from the stable can help reduce fly numbers inside the stable.

Figure 3.15. Clean all waste from between and under rubber mats in stalls and trailers. These areas hold moisture and are good locations for fly development.

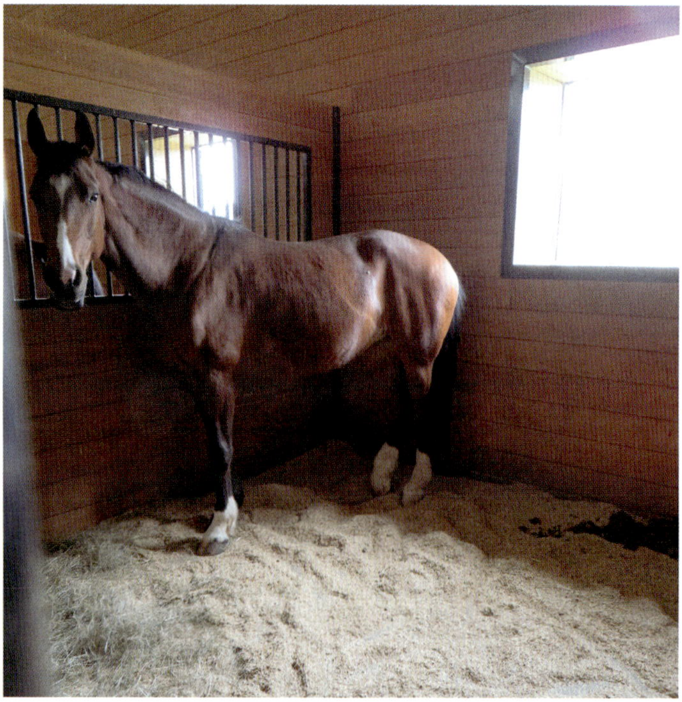

Figure 3.16. Bedding choice can help reduce fly development. Fewer flies develop in sawdust bedding (shown) than straw or flake shavings.

Forage management

Large round or square hay bales in pastures and other areas with loose hay piles are a major development site for pest flies, particularly stable flies (Figure 3.17). Basic management strategies, such as protecting round bales with metal or plastic rings, moving round bale sites when adding a new bale, or feeding hay off the ground will not only prevent hay from being soiled and trampled leading to suitable fly habitat but can also save money by reducing hay waste.

Figure 3.17. Large piles of loose hay from round bales or piles left in the field (A) are ideal development areas for pest flies, especially stable flies. Protecting bales or hay (B), feeding off the ground, or using slow feeders (C) can reduce the risk of fly development where hay is fed.

Food source reduction

House flies require sugar as part of their diet and are attracted to sweet odours. Keeping feed storage areas clean, removing spilled feed quickly, and covering feed in buckets or bins (Figure 3.18) will eliminate many sugar resources for flies and may reduce the number of flies that are attracted to the area. Using plastic feed wagons and frequent cleaning can reduce feed residues and odours that are attractive to flies.

Vegetation management

Stable flies can develop in many kinds of decomposing vegetation, not just hay or materials mixed with manure. They can be a particular nuisance in areas where grasses or other taller vegetation has been recently cut and left to decompose. Cut vegetation should be removed and disposed of or composted to prevent potential stable fly development.

Feed additives

There is no evidence that additives to feed, including garlic, essential oils, diatomaceous earth, or apple cider vinegar, help to repel flies. In fact, some may be unsafe for horses. Garlic can potentially cause anaemia in horses (Pearson *et al.*, 2005), and so these additives should be avoided unless recommended by a veterinarian. This does not include feed additives that contain insect growth regulators discussed in the chemical control section.

Physical and mechanical control

Exclusion

House flies are frequently found indoors and *Fannia* flies may be as well. Fly screens on windows in stable areas can reduce house fly presence in the stable. Fans in the stable area can circulate air that helps to dry bedding and also interrupt flight, preventing flies from landing (Figure 3.19).

Figure 3.18. Covering food in buckets or bins can reduce the food sources available for flies as well as attraction to the area.

Figure 3.19. Fans in the stable or stalls can be used to help prevent flies from landing on horses. Ensure that fans are for outdoor use before installing to reduce the risk of fire..

Fly sheets, boots, and masks

On-horse fly problems can be reduced by using on-horse protection methods. Fly masks and sheets are good options for house flies as they tend to land on the mucosal areas of the face, like the eyes, and on the neck, shoulders, and withers of the horse (Figure 3.20). Fly boots are of particular use for stable flies that prefer to bite the legs of horses. See Chapter 10 for more information on using fly sheets, boots, and masks.

Figure 3.20. Fly sheets (shown), fly boots, and fly masks can be used to keep flies off horses to reduce annoyance and bites.

Manure management

Once manure is removed from an area, mechanical methods can be used to process the manure to reduce the potential for fly development. If the manure is stored in a manure storage or holding area, sometimes called a manure pit, it can be actively composted. This requires frequent turning and covering to maintain a high internal temperature that will kill developing flies. If active composting is not feasible due to time or quantity of waste, a tarp or plastic sheeting can be used to cover the area and prevent fly access (Figure 3.21). Covering will help heat the manure to encourage quicker decomposition, produce temperatures too hot for fly development, and will also exclude females from laying eggs in these areas. Particular attention should be paid to the edge areas of the manure pit or stockpile, as those fresher areas (up to 7 days old) are the most likely areas for fly development. Storing manure under a roof structure keeps the manure moist and cooler, extending the decomposition process and allowing fly development (Figure 3.22). If manure cannot be covered with a tarp or similar material, allow the pile to be exposed to the sun or remove the manure at least weekly. The heat from the sun will aid in the decomposition and drying of the pile.

Figure 3.21. Covering a manure storage area with a tarp or plastic sheeting can help exclude flies from laying eggs and developing in these areas and increase the internal temperature to make the manure less suitable for fly development. Tarps should cover the edges of the manure pile (A). Fly development can still occur if tarps do not reach the edges (B).

Figure 3.22. Manure should not be stored under a covered structure unless it is removed at least weekly or more frequently, as this type of storage protects manure from the sun and air, which prevents suitable drying and decomposition.

Ideally, manure pits or piles should be as far away from the stables as possible. However, because filth flies are highly mobile and can travel among farms quickly and easily, manure management is also important for neighbouring facilities. Without management, these piles of manure can be an excellent development site for filth flies; house flies and stable flies in particular. Cases of filth fly production and subsequent neighbourhood annoyance from improper waste management on several types of animal facilities have been litigated in the USA, so it is important to think about the management of fly development sites from a community standpoint as well.

Other than being stocked in a designated area, manure can be spread on a field after it is removed (Figure 3.23). Spreading allows for quicker drying of manure so it cannot support fly development. For pastured horses, dragging can similarly break up manure piles in congregation areas, which permits nutrients to return to the soil as well. Spreading and dragging should be done at least weekly to kill developing flies before they can emerge as adults, although in warm temperatures biweekly may be needed to keep up with the fast development of these flies. It is important to understand and follow local or regional regulations concerning manure and nutrient management.

Manure additives

Even if cleaned regularly, flies will often find small protected places, such as between rubber stall mats or at the edges of stall walls, to develop. Adding a drying agent or another additive can keep stall floors unsuitable for fly development. Sodium bisulfate is safe for use around stalled horses and has been found to lower the pH of soiled bedding enough to reduce fly development when applied daily at 1.1 kg (2.4 lbs) per stall (Sweeney *et al.*, 1997, 2000). An added benefit of sodium bisulfate is it can help with the management of ammonia levels. Regularly applying calcium oxide (quicklime) or calcium hydroxide (hydrated lime) to kill developing flies has had mixed results and generally is not considered effective. However, it can be applied to dry out wet or moist areas where manure is stored. Other additives, such as zeolite or calcium cyanamide can also reduce moisture and change pH levels, which could influence fly development, but haven't been extensively studied.

Figure 3.23. Collecting manure and spreading it on fields allows for quick drying of the manure which prevents fly development. Collected manure should be spread at least weekly, although more frequent manipulation may be needed in warmer temperatures.

Traps

Jug (Figure 3.24) or bag traps are usually species-specific in the fly pests that they target. Many of these traps are pheromone-based and target house flies, although they may also incidentally trap other fly species like blow flies and flesh flies. Stable flies will not be captured in jug or bag traps. The pheromone or odour that comes with the trap is mixed with water and the trap is hung where house flies are seen congregating. These traps can have odours that are unpleasant to people. Traps need to be replaced at least every 7 days, if not sooner, in the summer months, as dead flies will enrich the mixture allowing it to serve as a suitable place for fly development. There is no evidence that bags of water hung around pastures, barns, stables, sheds, or stalls can be used to repel or control flies.

Sticky traps like sticky ribbons or sticky cards (Figure 3.25) can be purchased commercially and hung in areas where flies congregate. Sticky traps are useful in areas that have few flies that need to be removed, but not for controlling large populations. Feed rooms, tack rooms, or lounge spaces are good areas to hang sticky traps. Sticky traps can be unsightly depending on where they are hung and can collect dust and other particulate debris that make them less sticky, so they should be checked and replaced as needed. Sticky traps target house flies, but other species, such as *Fannia* flies can be trapped as well.

Stable fly traps are sight-based instead of odour-based like the house fly traps. Several commercial traps are available to capture stable flies, including the Olson Biting Fly Trap, BugJammer Knight Stick™, and the Starbar BiteFree™ trap (Figure 3.26) in the USA. These can be hung around

Figure 3.24. Jug traps typically target house flies, but other fly species can be attracted as well that are not usually considered pests, like blow flies and flesh flies. Traps are most effective when not saturated with flies (A). Traps should be replaced at least weekly or if traps are full (B).

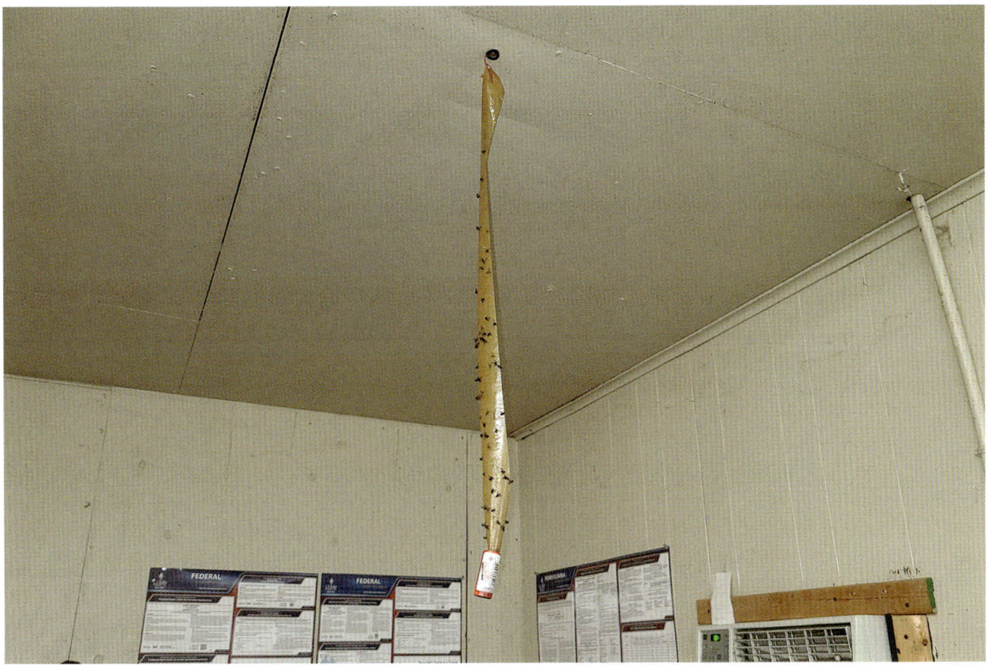

Figure 3.25. Sticky ribbons (pictured) and sticky traps can be useful in smaller areas where fly control is desired, such as feed rooms, tack rooms, or lounge areas. Ribbons dry quickly and are less effective when covered with dust, so they should be replaced often, even if they are not covered with flies.

the pastures where stable flies are a problem. Because these traps operate like sticky traps, it is important to keep them out of the reach of pastured horses. Curious animals may end up with traps stuck on their face or body. If traps are placed within the pasture, protect them with electric fencing and keep vegetation around these traps low so that the flies can see them. Although these traps target stable flies, other flies like house flies may be trapped incidentally.

Like sticky ribbons and traps, ultraviolet zappers can be used in smaller, targeted spaces like tack rooms, or lounge spaces (Figure 3.27). These traps may kill flies, but can also attract flies and other unwanted pests to an area. They may also kill many beneficial insect species, so these traps should not be used outside. In general, traps are not suitable for controlling high populations of flies, but may be effective for removing small numbers of flies from small areas. It is important to recognise that the electrocution of flies in confined spaces can disperse fly body parts (Ananth *et al.*, 2008) and aerosolized bacteria (Broce, 2008) from the exploding bodies of flies, so these should not be used in break rooms or places where people eat, and should be used with caution in animal feed rooms.

Biological control

Parasitoid wasps

Parasitoid wasps (Figure 3.28) are the most commonly used biological control for filth flies. These tiny, stingless wasps can be purchased commercially and will not harm humans, horses, or other vertebrate animals. The wasps attack and kill the pupal stage of developing filth flies. Parasitoid wasps are almost unnoticeable because they are extremely small, 1.5 to 3.2 mm (0.06 to 0.13 in).

Figure 3.26. BugJammer Knight Stick™ traps (A) and Starbar BiteFree® trap (B) are commercially available traps that target stable flies and can also trap house flies and other fly pests. These traps should be placed outside of pasture areas or protected with electric fences if placed in pastures.

They can be found naturally on most farms with developing flies, just at very low numbers. Female parasitoid wasps lay eggs in the fly pupae. The eggs hatch and the tiny wasp larvae consume the developing flies, effectively killing them. The immature wasps develop within the pupae and then emerge as adult wasps to repeat the cycle.

Figure 3.27. Fly zappers are generally not suitable for controlling high populations of flies, but may help control flies in smaller spaces like feed rooms, lounge rooms, and tack rooms.

Figure 3.28. Parasitoid wasps, like *Spalangia cameroni* (pictured), can be used to help control some fly species on equine farms. This parasitoid is shown on a house fly pupa.

To be effective, the species of pest fly should be identified. Parasitoids may not work to control *Fannia* flies, and will not be effective against other filth flies like horn flies and face flies, or horse or deer flies (Chapter 4). Parasitoids should start to be applied before fly populations increase to where they are being noticed as a problem, usually in the spring. Many companies sell these wasps and will provide them via mail-order monthly. However, it is recommended that releases are made at least bi-weekly to keep up with rapidly developing fly populations throughout warmer seasons. While parasitoid releases can be very effective at keeping populations of flies down, the parasitoids are only effective at killing immature flies so flies must be developing on-site and not immigrating from other farms in the area for them to work. They are also most effective when used with other fly control methods, in particular cultural, mechanical, and physical methods.

Not all parasitoid species are effective for fly control. Parasitoids in the genera *Spalangia* and *Muscidifurax* have been used successfully to control pest filth flies in other livestock and poultry facilities and are the appropriate species to release on equine farms. There is no evidence that parasitoids in the genus *Nasonia* will be effective at controlling pest flies. The parasitoid supplier should provide information about the species of parasitoids that will be sent before parasitoids are purchased.

Parasitoids are shipped as immature wasps developing within their fly host pupa, and often come packed in wood shavings to avoid damage in transit (Figure 3.29). They are spread in the environment as soon as the first parasitoids can be seen moving around in the packaging. Because these tiny insects do not fly very far, it is important to place them in areas where they can find fly pupae, such as manure stockpiles, edges of stalls, around waste hay, and sheds in the pasture. Keep in mind that horses walking over the pupae may kill the parasitoids, and birds like chickens and guinea fowl will eat fly pupae, so it is best to release them in protected areas. Release stations can also be made. To make a release station, shake a portion of the contents of the bag with the parasitoids into a small disposable cup (Figure 3.29). These can be placed in protected areas in the barn, in run-in sheds or around manure storage areas. Packages of parasitoids can also be placed in a mesh pouch and hung in stalls or around other areas to further protect the emerging parasitoids. Please see the 'Continuing education' section in this chapter for a detailed guide on the use of parasitoids on equine farms (Machtinger *et al.*, 2015).

Other biological control agents

In addition to parasitoids, several other biological control agents have been suggested for fly management. Beetle and mite predators that live in manure and bedding eat fly eggs and young larvae but have not been tested on equine facilities and most have not been developed into commercial products for practical use (see Chapter 8). The fungal pathogen *Beauveria bassiana* can be effective against adult house flies and is available commercially but has not yet been evaluated on horse facilities. Black dump flies (*Hydrotaea,* previously called *Ophyra,* Chapter 8) have long been suggested as a biocontrol agent because of the predaceous nature of their larvae on pest fly species. They are available from commercial insectaries but mostly have been used in concentrated animal systems that are infamous for high fly production, such as swine and poultry houses. It is unclear if they would be efficient on equine facilities or if they would become pests after release. Parasitic nematodes also are marketed and sold for biological fly control but there is no information demonstrating their effectiveness for fly management.

Figure 3.29. Parasitoids used as biological control of flies are sold commercially. They are shipped when they are still within their host pupae (A) and can be released once several wasps are seen within the packaging. The contents of the bag can either be scattered where flies are developing or placed in a release station like a cup (B) or a mesh pouch.

Chemical control

Many chemical products have been developed for house and stable fly control. However, fly resistance to active ingredients has been increasing, and it is becoming more difficult to find products that work. In addition, while pesticides or repellants may reduce adult flies that come into contact with the compound, they do not address the source of fly development. In some cases, chemical control methods may be required, such as when immediate control of large populations is necessary. Examples are after unexpected wet weather, interruption to standard manure management, flies coming to the property from other animal facilities or locations, or to reduce fly pressure on individual horses in the short-term such as for a single ride, farrier or vet visit, or a competition. See Chapter 10 for more information on chemical resistance management, active ingredients, and chemical rotation.

Residual pesticides

Residual pesticide premise sprays (Figure 3.30) are applied to walls, ceilings, and other places where flies rest. These are typically sprayed and allowed to dry, and affect the flies when they land on the treated surface. Residual pesticides are most effective on clean surfaces and can last a few weeks. Reapplication should be at intervals indicated by the chemical label directions or when house flies and stable flies are seen resting on treated areas. If targeting house flies, an indication of when to reapply is when fly spots can be seen.

Space sprays

Space sprays, including automatic spray systems and mist foggers, provide a quick knockdown of adult insects but have little residual activity. Caution must be taken to protect horses and other animals and remove animal water sources during application of pesticides. Treatment failure can be minimised by rotating active ingredients to prevent localised resistance of pests to these sprays. However, with indiscriminate use, such as with automatic sprayers, localised resistance will likely occur quickly resulting in product failure while repeatedly exposing humans, horses, and other animals to pesticide vapour and droplets. For this reason, automatic spray systems are strongly discouraged.

Baits

Fly baits combine an attractant and a food source (usually sugar) with a pesticide. Baits are most effective when competing food sources are limited, but many have other odour attractants that increase fly visitation. It is important to note that baits will attract and kill house flies and potentially some *Fannia* flies, but stable flies will not be attracted to, or feed on, baits. Four active ingredients from three pesticide classes are currently represented in granular house fly baits: carbamate, neonicotinoid, and ryanoid classes. Active ingredients in these classes that are used for baits are methomyl, imidacloprid, dinotefuran, and cyantraniliprole. In recent field trials, products containing cyantraniliprole and dinotefuran performed better against house flies than the other active ingredients (Murillo *et al.*, 2014). However, resistance to the chemicals used in these baits can develop quickly in house flies. To reduce the risk of resistance development, baits should not be used as the primary method of control, and bait chemical classes should be rotated after a few fly generations or approximately every 3 to 4 weeks as described in Chapter 10.

Figure 3.30. Residual pesticides are sprayed on surfaces where flies rest and can help control flies in stables and pasture sheds.

Since the pesticides in house fly baits are toxic to a variety of other animals, it is necessary to prevent other animals and children from being exposed. To reduce the risk of poisoning, baits should be placed in areas where flies congregate and in a bait station made of a plastic container. If there is no risk of bait being encountered by children or animals, baits can be placed on a plate or a food storage container. Baits should not be scattered haphazardly in the environment (Figure 3.31). Bait stations for house flies can be purchased commercially or made at home. An easy way to build a homemade trap is to cut four 5 cm (2 in) diameter holes in the upper part of the sides of a plastic jug. The bait is placed in the bottom of the jug and the jug should be hung in the stable in the eaves where flies generally rest or on fence lines (Figure 3.31). Indoor jugs should be placed in areas where fly activity can be seen. Because baits tend to liquify under hot conditions, outdoor jugs will work for longer if they are hung in the shade.

Larvicides

Larvicides are pesticides that kill flies in the larval stage. They are applied to fly development sites where large numbers of flies are predicted to emerge. These can be used when the area cannot be cleaned as frequently as needed to reduce fly development (see manure management, Section 3.1.4). Larvicides can be insect growth regulators (IGRs) or contact pesticides. Common active ingredients of the IGR larvicides are diflubenzuron, novaluron, cyromazine, and pyriproxyfen. Some IGRs allow the larvae to reach the pupal stage before killing them. Contact larvicide active ingredients are typically tetrachlorvinphos, either alone or in combination with dichlorvos, and spinosad.

While larvicides are commercially available, many of these products are more frequently used in other animal production, such as cattle and poultry, than with horses, so it is important to check the label to make sure that they are safe to use around horses.

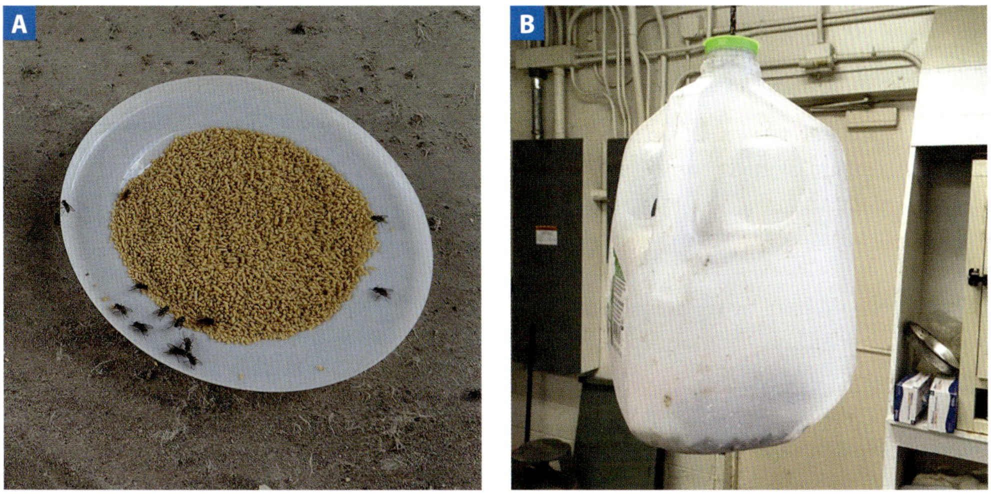

Figure 3.31. Fly baits should not be scattered on the ground, but instead placed in bait stations like a container or plate (A) or a milk jug trap (B) in designated areas where house flies congregate.

Larvicidal sprays are applied directly on manure or other fly development sites. The intent is to kill fly larvae that are actively developing before they emerge as adults. For this to be successful as a fly control method, flies must be developing onsite and not immigrating from other areas. Also, the larvicide must penetrate the manure and contact the larvae. There are some challenges with this type of application. Most larvicides break down quickly in the environment, and spraying larvicides over stalls or manure storage areas could kill beneficial insects. Because of these challenges, this type of chemical fly control is best used in specific circumstances and not as part of a general management plan. For example, larvicides can be used in specific areas like the cracks in between stall mats or on manure that has received greater than normal amounts of water.

Treated fly sheets, boots, and masks

Fly sheets, boots, and masks with permethrin bound to the fabric fibres can be purchased to repel flies. The repellent action of these products generally lasts 20 to 40 washes, depending on the product. The duration of effectiveness may change with exposure to sunlight and weather patterns, so observe horses periodically to make sure it is still providing the protection desired. Fly sheets purchased without pesticide treatment can also be sent to a company that specializes in binding permethrin to fabric for treatment. It is important to recognize that do-it-yourself treatment of fabrics with pesticides that do not list fabric treatment for horses on the label may be a violation of the law. In addition, horses are sensitive to chemical pesticides, and off-label treatment may cause skin reactions.

Topical products

On-animal fly sprays, roll-ons, and wipes are available with a variety of active ingredients, percentages (or concentrations), and trade names. These may or may not be effective depending on the toxicity of the active ingredient to flies, local pest resistance, and method of application. House flies, in particular, are becoming increasingly tolerant to the most active ingredients found in pesticides and repellents that are currently sold. However, recently newer formulations using fatty acids (e.g. octanoic, nonanoic, and decanoic acid) or combinations of plant extracts like neem, geraniol, and peppermint oil may be more effective and longer-lasting than synthetic ingredients in the laboratory (Tuorinsky and Machtinger, 2020).

Topical repellents can be sold as ready-to-use formulations (often in a spray) or as a concentrate that is mixed and diluted before use. It is extremely important to not use concentrated products directly on the animal as this may result in extreme irritation, hair loss, and skin sloughing (see Chapter 10 for more information on using repellents).

Many formulations of fly repellents degrade quickly in ultraviolet light during turnout, wash off with sweat or rain, or can be hidden under layers of dirt after rolling. Some formulations may only retain an acceptable level of repellency for a few hours and others for up to a day (Tuorinsky and Machtinger, 2020). It is important to follow label application rates to ensure sufficient application amounts and full coverage over the body, head, and legs. The most effective applications are to clean animals and using a towel, applicator mitt, or brush to ensure coverage across the entire animal (Figure 3.32). If applying repellents in this way, it is important to consider using disposable gloves to protect the applicator's skin from the products. Roll-on or cream pesticides can also be used. Shampoos containing the active ingredient permethrin are drying to the skin and the residual effect of the active ingredient is short-lived like other topical products.

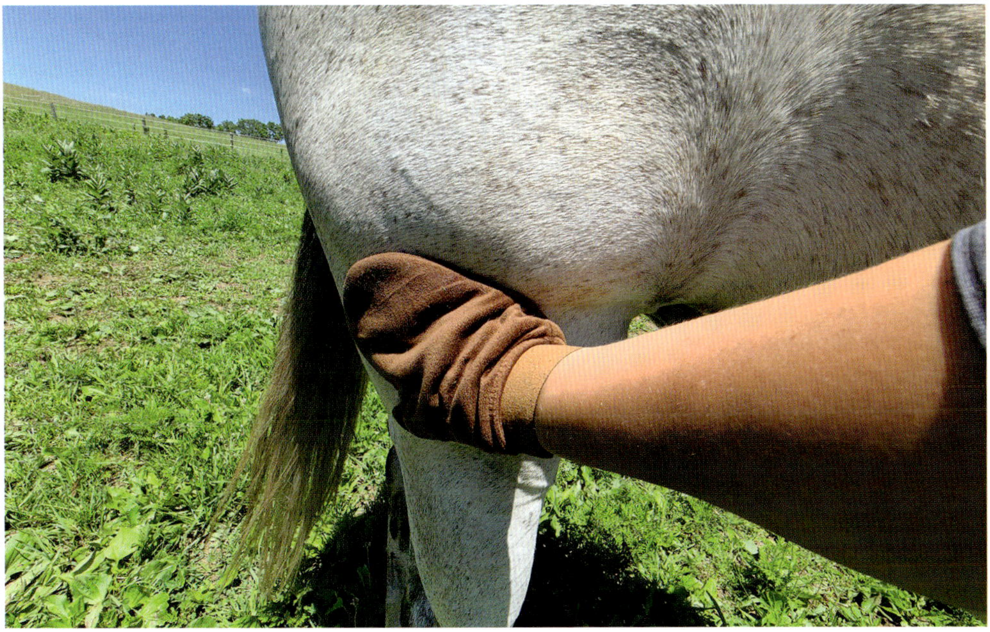

Figure 3.32. Applying topical repellents with a towel or applicator mitt (shown) ensures complete coverage of the horse with the repellent and reduces waste and potential inhalation irritation from airborne droplets.

Feed-through insect growth regulators and feed-through larvicides

Feed-through insect growth regulators are administered as supplements that are fed to the horse, pass through the digestive system, and are excreted in faecal matter. There are three compounds found in most commercial products: diflubenzuron, cyromazine, and novaluron. In some cases, these are sold as ration balancers or feed or mineral blocks that include the compound. These compounds interact with the physiology of the developing flies and prohibit them from emerging as adults. They can be 50 to 95% effective depending on local conditions. The action of some of these compounds is also exclusive to flies so will not harm beneficial insects of other groups, like dung beetles. However, for these products to be effective, they must be fed to every horse at a property, or flies will be able to develop in the manure of untreated animals. Flies can become resistant to these active ingredients, so products with different active ingredients should be rotated every month. Feed-through growth regulators do not need to be fed to horses when flies are not a problem, such as the cooler months in temperate climates.

3.2 Filth flies that cannot develop in equine manure

Face flies, horn flies, and buffalo flies can be found on equine facilities in some parts of the world. However, these flies are distinct from filth flies in Section 3.1 in that they do not develop in horse manure or horse-associated substrates. All three species are limited to developing in very fresh cattle manure including cattle, buffalo, and yaks (herein grouped as cattle). The presence of these flies on a horse facility indicates cattle in the area, although not necessarily on or neighbouring the farm, as these flies can travel many kilometres (miles) across the landscape after emerging as adults.

3.2.1 Face fly (*Musca autumnalis*)

3.2.1.1 Pest summary (Table 3.4)

The face fly is a single species, *Musca autumnalis* (Figure 3.33), in the genus *Musca,* like the house fly, and belongs to the family Muscidae, like the house and stable fly. It is a medium-sized (7 to 8 mm, 0.28 to 0.30 in) grey fly that prefers to feed around the face of cattle and horses. Face flies feed on animal secretions and manure in the pasture. When feeding on secretions the flies cluster around the eyes, nose, and mouth, causing great annoyance to the animals. Additionally, these flies can transfer pathogens with their sponging mouthparts.

3.2.1.2 Distribution and seasonality

The face fly is present in Europe, from the British Isles in the north to North Africa in the south, and east to Pakistan, Kazakhstan, northern India and Nepal. Face flies were introduced to North America in the 1940s and have spread into all areas south of 53°N except for Arizona, New Mexico, Texas, Louisiana, and Florida. They are present in pastures during the warm months and overwinter in barns and other outbuildings in groups of adults. Face fly does not occur in Australia, but the Australian bush fly, *Musca vetustissima*, has similar habits but tends to congregate on humans.

Table 3.4. Pest summary of face fly (*Musca autumnalis*).

Distribution[1]	Australia: Yes (Bush fly) Canada: Yes Europe: Yes New Zealand: No United States: Yes
Size	7-8 mm (0.28-0.30 in)
Key identification characteristics	Very similar to house flies, but cluster around the face of the horse; not found indoors or off the horse
Mouthparts	Sponging (non-biting)
Feed on	Tears and secretions from the eyes, nose, and mouth
Development time[2]	15-25 days
Adult life span	4-8 weeks
Total eggs laid	up to 130
May be mistaken for	House flies (Section 3.1.1)
Impact on horses	Annoyance, transmission of pinkeye and *Thelazia* eyeworms

[1] The pest distribution in the regions targeted by this publication are listed as of the date published. Areas of primary pest concern (if any) are in bold. Global distributions are listed in Section 3.2.1.2

[2] Development times are approximate and can change with environmental conditions including temperature, humidity, and availability of food resources.

Figure 3.33. Face fly, *Musca autumnalis*, has a silver colouring around the eyes.

3.2.1.3 Description

The face fly is a medium-sized fly that bears a strong resemblance to its relative, the house fly (Figure 3.1), and has similar mouthparts that are designed for sponging up liquid food. Face flies are somewhat larger and darker than house flies and have a larger silverish area that borders the eyes (Figure 3.33). Identification of individual specimens requires careful examination by someone familiar with the subtle characteristic differences between the species (Figure 3.34), but face flies on animals are easily recognized by their behaviour. On equine farms, female face flies spend most of their time around the eyes, nose, and mouth of pastured horses, where they feed on tears and other secretions (Figure 3.35). If an animal has groups of large flies feeding around the eyes that cause the animal to shake its head and switch the tail over the face, then they are almost certainly face flies.

Unlike house flies, the eggs have a respiratory tube or horn that protrudes just above the manure surface (Figure 3.36). Larvae are lemon-yellowish in colour and develop rapidly. An unusual feature of face flies is that the larvae store calcium that is incorporated into their puparium, resulting in pupae that are distinctly white (Figure 3.37) rather than the reddish-brown colour like house and stable flies. However, immature face flies will not be found on equine farms unless cattle are also present.

3.2.1.4 Life history

Face flies have complete metamorphosis with four life stages: egg, larva, pupa, and adult. Unlike the house fly and stable fly, the face fly has very specific requirements for larval development. Their eggs resemble bananas and are laid exclusively in fresh cattle droppings on pastures (Figure 3.36). Because of this, any flies found on horses are immigrants from farms with cattle. Larvae leave the manure as early as 4 days after eggs are laid and pupate in the soil, emerging as adults about 1 week later under hot summer conditions.

Figure 3.34. The eyes of male face flies (right) are very close together, whereas female house flies (left) have a wider separation.

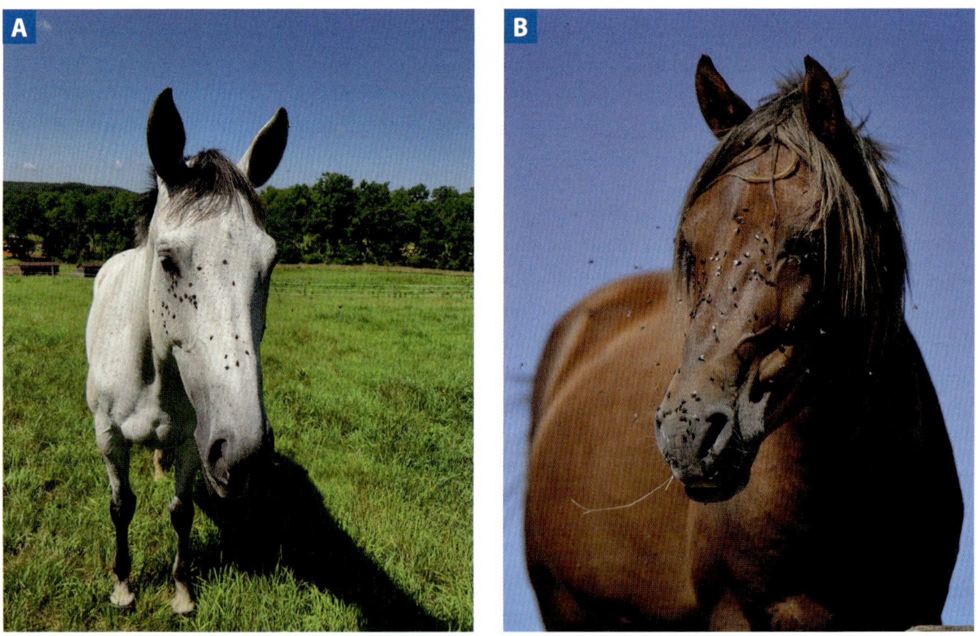

Figure 3.35. Although face flies develop only in cattle manure, adults can be found congregating on the faces of horses as well.

Both male and female flies will feed on plant nectar, but on equine facilities, female flies spend their daylight hours almost exclusively feeding on secretions on the faces of horses to obtain protein for egg development. While face flies do not bite, they are also opportunistic blood feeders that feed avidly on blood from wounds or injury from biting flies like horse flies. Face flies are strictly an outdoor pasture pest, and most leave the horses when they enter stables or other shelters.

Figure 3.36. The respiratory horn of face fly eggs gives them a banana-like appearance. Eggs are deposited in fresh cattle manure with the respiratory horn at the manure surface.

Figure 3.37. Face fly pupae are unique in that the fly incorporates calcium into the pupa, which turns it white instead of the reddish-brown colour of other pest filth fly species.

Depending on local temperature conditions, face flies can complete 4 to 11 generations per year, with populations usually highest in mid- to late-summer. As the days shorten and temperatures drop, the newly emerged adults enter a hibernation state. These flies feed to build up their fat reserves but do not develop their ovaries for egg production. Hibernating flies then enter barn or stable lofts, attics, tree cavities, church steeples, and any structure that provides refuge for the winter months. Overwintering flies can survive freezing temperatures and are usually inactive, but will become active and feed on warm days and sometimes become pests in houses when they leave their hibernation areas and enter the living quarters of the home.

3.2.2 Horn fly (*Haematobia irritans irritans*) and the buffalo fly (*Haematobia irritans exigua*)

3.2.2.1 Pest summary (Table 3.5)

There are two subspecies of *Haematobia irritans* that can affect horses. The horn fly, *Haematobia irritans irritans* (Figure 3.38), and the buffalo fly, *Haematobia irritans exigua*, of the insect family Muscidae. The horn fly is present on cattle worldwide where it also feeds on horses and other animals. The buffalo fly is found in Australia where it is primarily a pest of cattle but can also feed on horses. Both are small, blackish-grey flies that are about 3.5 to 5.0 mm (0.14 to 0.20 in) with piercing-sucking mouthparts to feed on blood. Like the face fly, they are typically present in pastures and not in covered structures, like stables or run-in sheds in pastures. The larvae develop in fresh manure. Biting by these flies, which is typically on the back or belly, agitates horses.

3.2.2.2 Distribution and seasonality

Horn flies are present throughout the Americas, from Canada to temperate Uruguay and Argentina, Europe, Asia, and non-tropical regions of Africa. Buffalo flies are invasive to Australia from Asia

Table 3.5. Pest summary of horn fly (*Haematobia irritans irritans*) and the buffalo fly (*Haematobia irritans exigua*).

Distribution[1]	Australia: Yes (Buffalo fly) Canada: Yes (Horn fly) Europe: Yes (Horn fly) New Zealand: No United States: Yes (Horn fly)
Size	Adult size is 3.5-5.0 mm (0.14-0.20 in)
Key identification characteristics	Forward facing mouthparts; found on the body of the horse (back and belly) and usually not the legs; not found indoors
Mouthparts	Biting-sucking
Feed on	Blood
Development time[2]	10-20 days
Adult life span	up to 7 days
Total eggs laid	up to 20 eggs
May be mistaken for	House flies (Section 3.1.1), stable flies (Section 3.1.2)
Impact on horses	Annoyance, horn fly implicated in the transmission of *Corynebacterium pseudotuberculosis* the causative agent for pigeon fever

[1] The pest distribution in the regions targeted by this publication are listed as of the date published. Areas of primary pest concern (if any) are in bold. Global distributions are listed in Section 3.2.2.2

[2] Development times are approximate and can change with environmental conditions including temperature, humidity, and availability of food resources.

Figure 3.38. Horn fly, *Haematobia irritans irritans*.

and are found throughout northern Western Australia, the Northern Territory, Queensland, and north-eastern New South Wales. These flies are primarily a pest of cattle but will feed on horses and spend most of their time on their animal host, feeding frequently in pastures. In temperate climates, horn fly populations are typically highest in early summer and can decline if hot and dry conditions later in the season cause manure pats to dry out before the larvae can complete development. A second peak sometimes occurs in late summer to early autumn as conditions once again become favourable for larval survival. Fly populations can occur year-round in warm locations such as the south-eastern USA. In cooler climates, the flies overwinter in the pupal stage in or under manure pats. In colder climates the horn flies become dormant in the winter in the pupal state, ready to emerge when the weather warms in the spring. Buffalo flies are less of a pest in southern inland areas of Queensland and inland New South Wales because of the shorter periods of warm weather. Peaks of buffalo flies are in late summer and autumn in these areas.

3.2.2.3 Description

Adult horn flies and buffalo flies are similar in appearance. They have a rigid, forward-pointing proboscis (mouthparts) similar to the stable fly and feed on the blood of large mammals, especially cattle. They can be recognised as brown-grey flies that are about half the size of a house fly or stable fly and have biting-sucking mouthparts (Figure 3.38). As with face flies, they are most easily recognised by their behaviour and location on horses. Horn flies and buffalo flies are usually present in groups along the back and sides of horses, sometimes moving to the belly during the hottest time of day. They are remarkably persistent biters that return immediately to the animal if they are dislodged by a muscle twitch or the swish of a tail. Populations on cattle can exceed 1000 flies per side (Figure 3.39), but because horses are accidental hosts for horn flies, they are generally present at much lower numbers. Horn flies and buffalo flies are most likely to be confused with stable flies or house flies but are usually much smaller than either fly and are not found on the legs like stable flies.

Figure 3.39. Horn fly numbers can be in the thousands on cattle seen here along the back and legs, but typically there are far fewer on horses.

3.2.2.4 Life history

Like other flies, horn flies and buffalo flies have complete metamorphosis with four life stages: egg, larva, pupa, and adult. Both males and females are exclusively blood feeders and do not require any other source of nutrition. On pastured horses, they remain on the animal nearly all the time, with females only leaving to deposit eggs if cattle are nearby. As with face flies, horn flies and buffalo flies require fresh manure pats of cattle for development. Larvae develop in the manure pat, feeding on microorganisms in the material, and begin to leave the pat to pupate in the soil after 4 to 8 days, depending on temperature. Horn fly and buffalo fly larvae and pupae resemble house fly and stable fly larvae and pupae, but will not be found in horse manure. Adults emerge 6 to 8 days later and begin feeding, mate 3 to 5 days after emergence, and begin laying eggs. Females lay an average of about 80 eggs during their 6- to 7-day lifespans, with some flies laying as many as 200 eggs.

3.2.3 Control of filth flies that cannot develop in equine manure

Face flies, horn flies, and buffalo flies are associated with cattle and are only incidental pests on horses. They tend to remain close to where they emerged but can disperse several kilometres (miles) away from cattle to local equine farms, perhaps aided by weather patterns. Horn flies and buffalo flies can cause serious annoyance to horses even at low population levels, but their status as immigrants from nearby cattle farms and behaviour to remain on the body of the horse limits management to control of adult flies on horses.

Manure management and other cultural control practices, as well as other control methods like baits, larvicides, and feed-throughs used for house flies and stable flies, will not be effective for these species. This underscores the need to identify the species of flies when considering pest management options. If the facility is mixed-use and has cattle onsite, see Boxler *et al.* (2021) and Kaufman and Weeks (2018) in 'Continuing education' for more information on managing horn flies in pastures, Spence (2007) for monitoring and managing buffalo flies, and Trout-Fryxell *et al.* (2021) in 'Continuing education' for more information on monitoring and managing face flies.

Physical and mechanical control

Fly sheets, boots, and masks

Face fly control is focused on protecting the mucous tissues of the eyes where the flies prefer to feed. Physical exclusion of these flies with the use of fly masks can prevent access to the eyes. Masks with longer nose guards can also protect nasal passages from fly landing to further prevent annoyance (Figure 3.40). Unlike chemical control methods, these masks are a sustainable way to reduce exposure to face flies for a long duration that is not affected by environmental factors. See Chapter 10 for more information on using fly sheets, boots, and masks.

Horn flies are found on the body of the horse, as opposed to the face like face flies. Generally, horn flies will be found on the back or shoulders of horses, or if it is hot or raining, they will move to the belly area (Figure 3.41). As blood-feeders, these flies spend most of their life on their animal host and feed in regular intervals, so physically excluding these flies from feeding with fly sheets will prompt migration away from the target horse. However, as previously mentioned, horn flies can feed on the belly region, so protective sheets should also cover that area, or topical repellents should be used to protect the belly along with fly sheets.

Figure 3.40. Fly masks (A) can be used to deter face flies and prevent eye irritation. Often, face flies will take advantage of exposed areas, so masks with longer nose guards (B) can protect the nostrils as well.

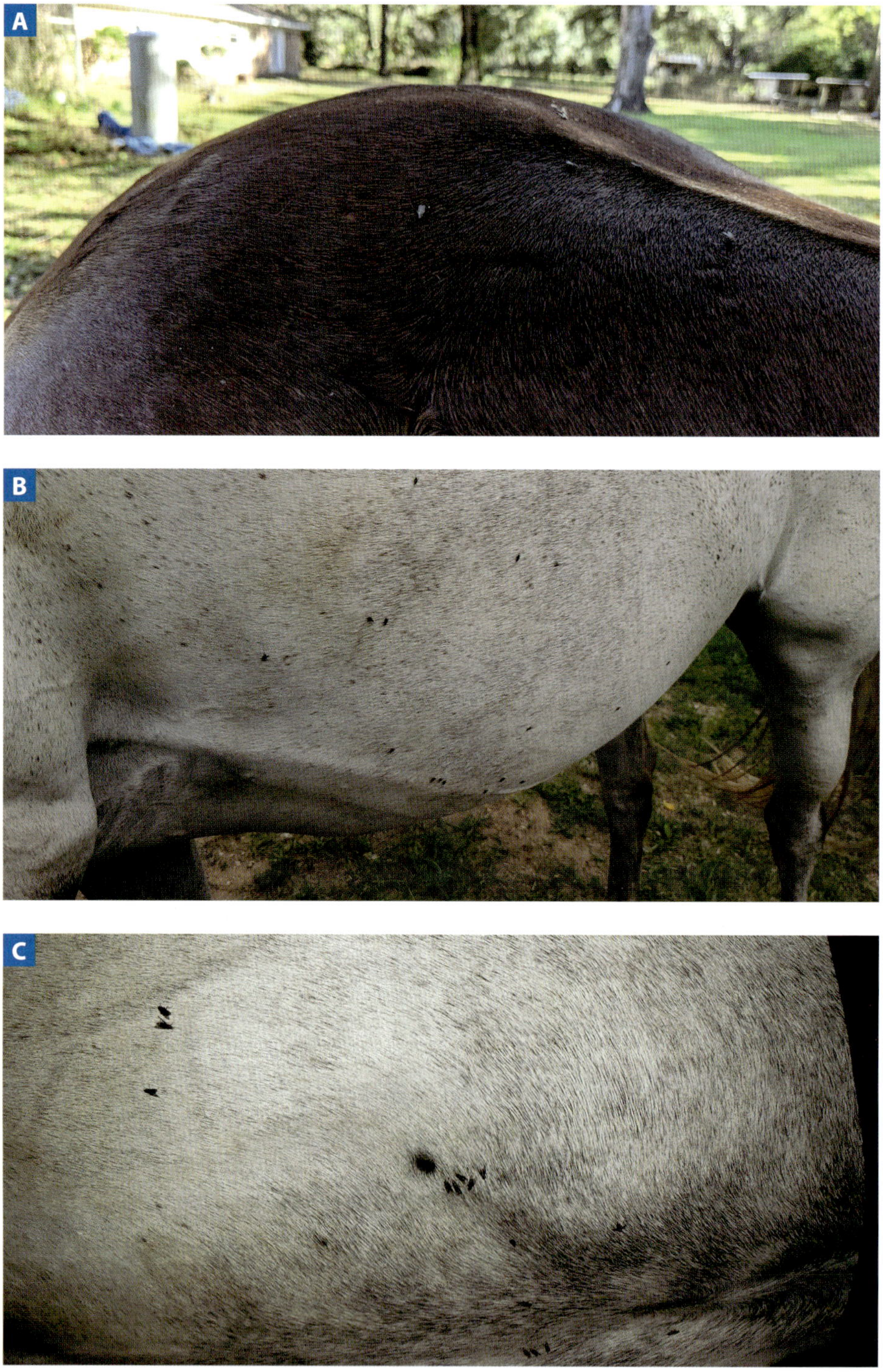

Figure 3.41. Horn flies are usually found on the back and sides (A and B), but will move to the belly (C) if the weather is hot or if it is raining.

Chemical control

Like the filth flies that develop in equine manure, horn flies, buffalo flies, and face flies have developed degrees of resistance to some active ingredients of chemicals used to control these flies. This has made some products ineffective or reduced how well others work. However, for flies that develop in cattle manure, it may be difficult to address the source of the flies and chemical control may be a key tool to relieve fly pressure. See Chapter 10 for more information on chemical resistance management, active ingredients, and chemical rotation to extend the use of chemical control options.

Topical products

Topical products are a common method for reducing face flies and horn flies on horses. Buffalo flies have demonstrated resistance to synthetic pyrethroids and organophosphates, which limits the effectiveness of many repellents.

Repellents can be sold as ready-to-use formulations (often in a spray) or as a concentrate that needs to be mixed and diluted before use. Concentrated products should not be used directly on the animal as this may result in extreme irritation, hair loss, and skin sloughing (see Chapter 10 for more information on using repellents).

While insect repellents are marketed for horses, many break down quickly in heat and rain, or their action is reduced due to perspiration, so it is likely the repellent will need to be reapplied daily. Repellents for horn fly control should be applied to the body of the animal. Applications should be made when the animal is clean and with a towel, applicator mitt, or brush to ensure even and thorough coverage on the top, sides, and belly of the horse (Figure 3.42). If applying repellents in this way, it is important to use disposable gloves to protect the applicator's skin from the products. Also, it is important to be extremely careful applying repellents to the face of any horse. Roll-on repellents or creams can be used for application to the face (Figure 3.43). Insecticide impregnated halter bands may also be used to control face flies, but there is no published data to support their effectiveness. Please see Chapter 10 for more information on how to apply repellents effectively.

Treated fly sheets, boots, and masks

As with generalist flies, fly sheets, boots, and masks treated with permethrin can help deter face flies, horn flies, and buffalo flies. These are described under chemical control in Section 3.1.4.

3.3 Veterinary concerns

Flies in the insect family Muscidae cause many problems for horse owners. Persistent feeding by flies leads to pain and irritation and can trigger natural pest avoidance responses from horses, including tail swishing, head and neck movements, twitching, stamping, and shelter seeking (McDonnell, 2003). Horses pressured by high fly populations may alter grazing behaviour, reduce grazing time, and lower forage intake (Keiper and Berger, 1982). Horses biting at their flanks, abdomen, shoulder, and chest, can be caused by intense and unrelieved physical discomfort (McDonnell, 2008), which can be seen with biting flies like horn flies and stable flies. Additionally,

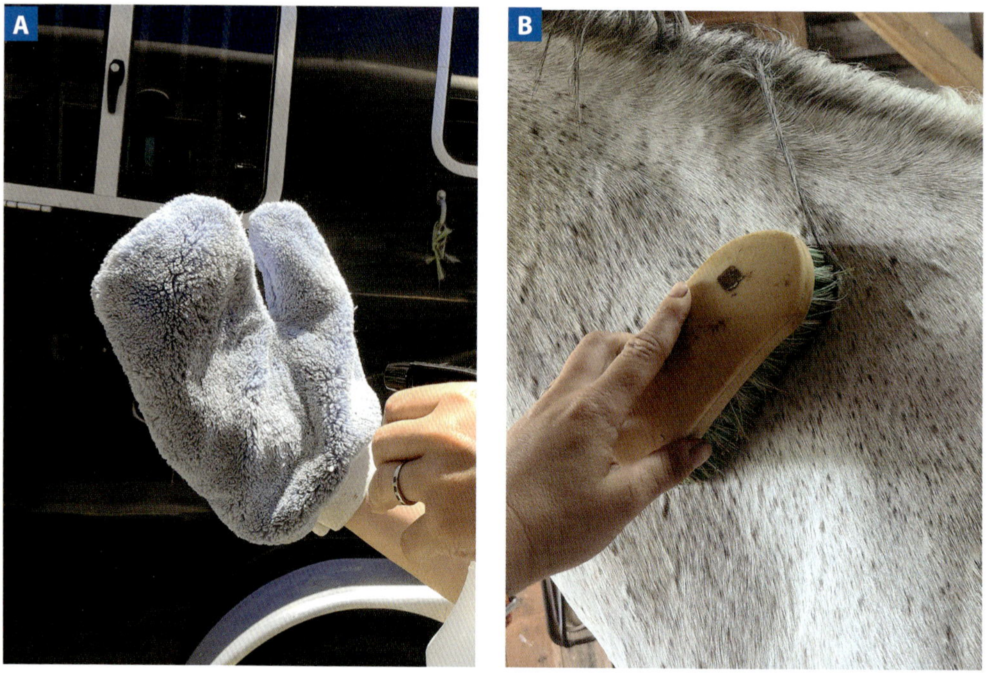

Figure 3.42. Topical repellents should be applied to clean and dry horses. The best coverage is achieved by spraying the repellent on a towel, applicator mitt (A), or brush (B), and wiping it over the horse to reach all areas flies may reach including the head and legs.

Figure 3.43. Roll on repellents can be used on the head, around wounds or sores, or other areas where a targeted application would be useful.

severe fly persistence may interfere with the growth and health of foals by not permitting mares to maintain appropriate distances or remain stationary long enough for foals to nurse (Keiper and Berger, 1982). Repeated stomping, movement, and reduced grazing caused by flies can have serious effects on the performance of pastured and show horses, including long or short-term injury and loss of condition.

In addition to these avoidance behaviour concerns, this family of pest flies can transmit pathogens causing disease or conditions in horses. Flies pick up pathogens from infected horses and transmit them to unaffected horses mostly through passive or mechanical transfer. The organisms simply hitch a ride from a horse or manure to another horse.

3.3.1 Pigeon fever

Corynebacterium pseudotuberculosis is the causative agent for pigeon fever. This syndrome causes large abscesses, often in the chest, which produce copious amounts of purulent material (thick discharge from a wound) (Figure 3.44). House, stable, and horn flies are attracted to this material, and their body parts become contaminated. When the flies land on the next horse, the bacteria have the opportunity to infect that horse. If possible, abscesses should be monitored until they have coalesced. At that time abscesses should be drained. Extreme care should be taken when opening *C. pseudotuberculosis* abscesses to remove and contain all of the purulent material. Any flies which land in this material can transmit *C. pseudotuberculosis* to nearby horses.

Figure 3.44. Pigeon fever is caused by *Corynebacterium pseudotuberculosis*, which can be transmitted by flies.

3.3.2 Summer sores

The causative agents for summer sores, *Draschia* and *Habronema* species, are both spread by house, face, and stable flies. Both *Draschia* and *Habronema* are stomach worms. However, their life cycle includes deposition in open sores on the skin by these fly species. When these nematodes penetrate the skin they cause a local hypersensitivity reaction resulting in papules, non-healing wounds, or exuberant granulation tissue (proud flesh) that is most common on the legs, ventral abdomen, prepuce, and medial canthus of the eye. Open sores are slow to heal and difficult to treat (Figure 3.45). Treatment is aimed at killing any migrating *Draschia* or *Habronema* through judicious use of deworming products, reducing inflammation in the area of the sore, and decreasing flies landing on the horse. Please see Pugh *et al.* (2014) in 'Continuing education' for a review on Habronemiasis.

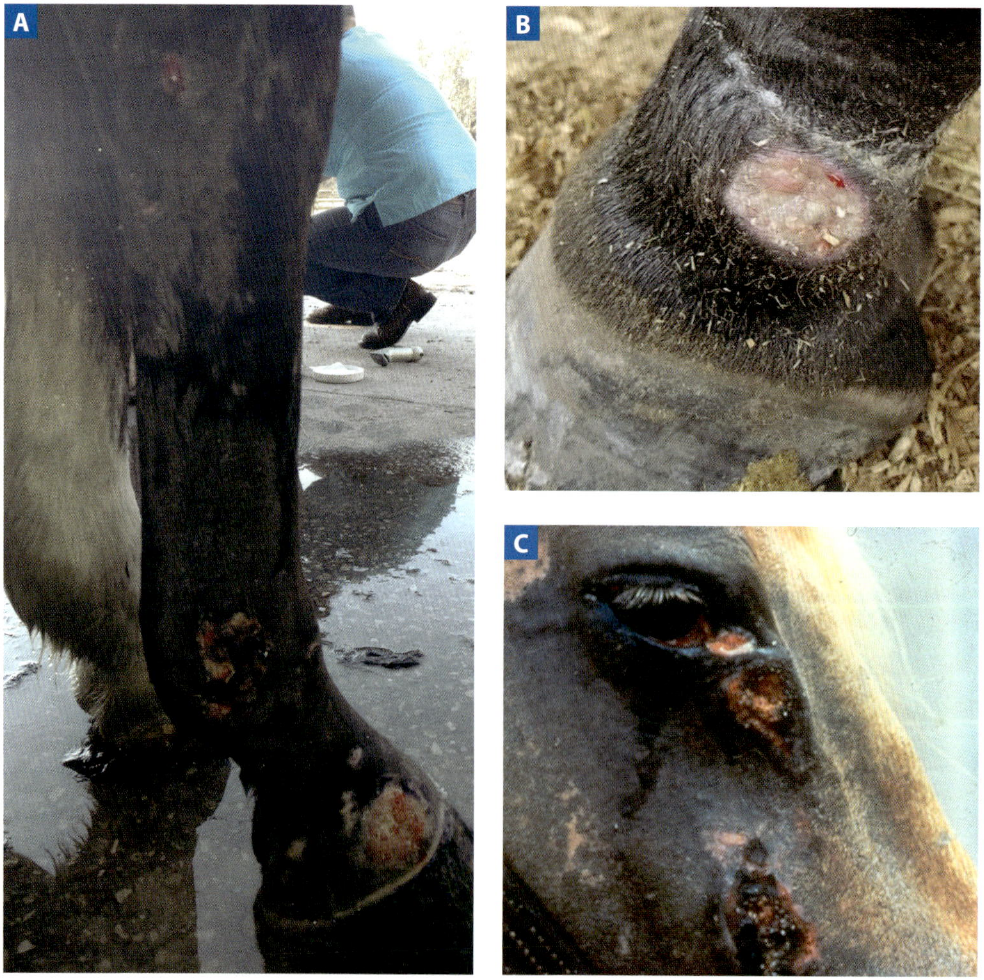

Figure 3.45. Large, open sores called summer sores are often difficult to treat and slow to heal. These are caused by worms that can be spread by filth flies.

3.3.3 Eyeworms

The equine eyeworm (*Thelazia lacrymalis*) is one of several different *Thelazia* species that parasitise large mammals. All can be transmitted by the face fly. Female *Thelazia* deposit newly hatched larvae near the tear ducts, conjunctival sac, and under the third eyelid (Figure 3.46), and can cause profound inflammation of all structures of the eye. *Thelazia* can be readily eliminated by common deworming therapies. Deworming therapeutics should be used only as needed due to increasing resistance by nematodes. Please see Otranto and Traversa (2021) in 'Continuing education' for a review on *Thelazia* eyeworms.

3.3.4 Bacteria and viruses

Filth flies are particularly good at transporting bacteria and viruses. House, stable, and face flies have been strongly implicated as vectors for respiratory viruses. These species are also effective vectors of bacteria, such as *Salmonella*, *Streptococcus*, *Staphylococcus*, and *Escherichia coli*. This allows them to spread a wide variety of diseases throughout a horse population. Stable flies have been definitively linked to transmission of anthrax and are suspected in the transmission of some equine infectious anaemia cases. Treatment for each of these diseases will require specifically targeted therapeutics. Veterinary care should be sought out if any are suspected.

3.3.5 Sarcoids

Recent evidence shows that there is a strong possibility that flies are implicated in the spread of sarcoids among horses. The *Fannia* flies, house flies, and stable flies have been shown to carry the bovine papillomavirus, which is believed to trigger sarcoid formation in equids (Figure 3.47). Features of sarcoids themselves point to the transmission of an infective agent carried by flies.

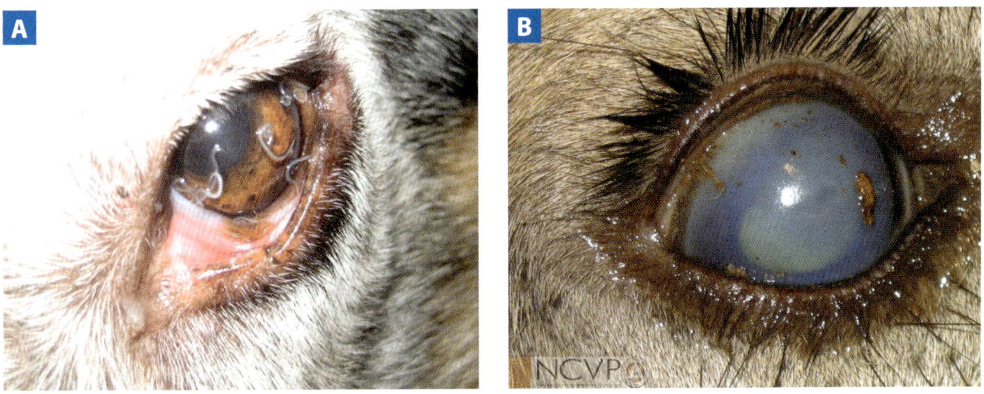

Figure 3.46. ***Thelazia*** **eyeworms in (A) the eye of a dog, and (B) mule deer. Thelazia species inhabit the conjunctival sac or lacrimal duct of mammals and birds.**

Sarcoids are often located in anatomical areas favoured by flies, and anecdotal evidence suggests more horses are affected on properties with large fly populations. More research is needed to confirm the link, but current work offers a strong rationale for aggressive fly control. Treatment involves debulking the sarcoid, then using therapeutics to stimulate an appropriate immune response. Therapeutics are often topical, but can also be injected into the affected area. Please see Ogłuszka *et al.* (2021) and Equine Medical Solutions in 'Continuing education' for a review on equine sarcoids.

Continuing education

House fly (*Musca domestica* [Diptera: Muscidae]) – biology, pest status, current management prospects, and research needs

Geden, C.J., Nayduch, D., Scott, J.G., Burgess IV, E.R., Gerry, A.C., Kaufman, P.E., Thomson, J., Pickens, V. and Machtinger, E.T., 2021. Journal of Integrated Pest Management 12: 39. https://doi.org/10.1093/jipm/pmaa021

- This review evaluates current literature on house flies relating to their biology and status as vectors of pathogens causing disease. Discussions of current control methods and research and education needs are provided for all major animal production systems.

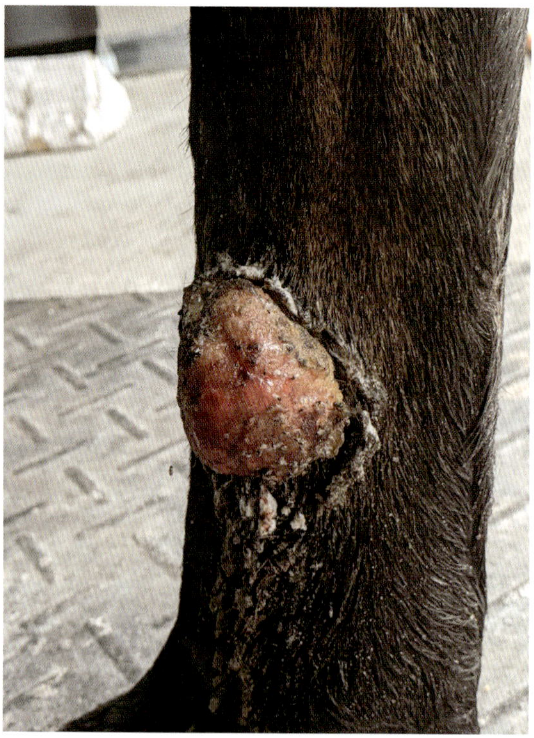

Figure 3.47. Sarcoids may be caused by bovine papillomavirus. This virus has been isolated in flies and may be transmitted from horse to horse by flies. Eye gnats (*Liohippelates* spp.) flies can also be seen in this image (Chapter 6).

Stable fly (*Stomoxys calcitrans* [Diptera: Muscidae]) – biology, management, and research needs

Rochon, K., Hogsette, J.A., Kaufman, P.E., Olafson, P.U., Swiger, S.L. and Taylor, D.B., 2021. Journal of Integrated Pest Management 12: 38. https://doi.org/10.1093/jipm/pmab029

- This review evaluates current literature on stable flies relating to their biology and status as vectors of pathogens causing disease. Discussions of current control methods and research and education needs are provided for all major animal production systems.

Big problems with the little house fly, *Fannia canicularis* (Diptera: Fanniidae)

Murillo, A.C., Hubbard, C.B., Gerry, A.C. and Hinkle, N.C., 2021.
Journal of Integrated Pest Management 12: 40. https://doi.org/10.1093/jipm/pmaa023

- This review evaluates current literature on *Fannia* flies relating to their biology and status as vectors of pathogens causing disease. Discussions of current control methods and research and education needs are provided for all major animal production systems.

Monitoring house fly (Diptera: Muscidae) activity on animal facilities

Gerry, A.C., 2020. Journal of Insect Science 20: 15.
https://doi.org/10.1093/jisesa/ieaa109

- House flies are one of the most common pests of equine facilities. Monitoring flies can tell you if your control methods are working. In this review, ways of monitoring house flies are discussed, which can improve control.

Filth fly control on horse farms

Machtinger, E.T., 2019.
https://extension.psu.edu/filth-fly-control-on-horse-farms

- This guide is an additional resource that can be used to assist in filth fly management on equine facilities.

Use of pupal parasitoids as biological control agents of filth flies on equine facilities

Machtinger, E.T., Geden, C.J., Kaufman, P.E. and House A.M., 2015. Journal of Integrated Pest Management 6: 16. https://doi.org/10.1093/jipm/pmv015

- The use of pupal parasitoids on equine farms is becoming more popular on equine farms for fly control. However, there is limited information on how to use parasitoids effectively. In this review, the biology of available parasitoids and recommendations for use are provided.

Face fly (*Musca autumnalis* [Diptera: Muscidae]) – biology, pest status, current management prospects, and research needs

Trout Fryxell, R.T., Moon, R., Boxler, D. and Watson, W., 2021.
Journal of Integrated Pest Management 12: 5. https://doi.org/10.1093/jipm/pmaa020

- This review evaluates current literature on face flies relating to their biology and status as vectors of pathogens causing disease. Discussions of current control methods and research and education needs are provided for all major animal production systems.

Horn fly, *Haematobia irritans irritans* (Diptera: Muscidae) – biology, management, and future research directions

Brewer, G.J., Boxler, D.J., Loftin, K.M., Domingues, L.D., Talley, J.L., Trout Fryxell, R.T., Holderman, C., Machtinger, E.T., Smythe, B. and Watson, W., 2021. Journal of Integrated Pest Management 12: 42. https://doi.org/10.1093/jipm/pmab019

• This review evaluates current literature on horn flies relating to their biology and status as vectors of pathogens causing disease. Discussions of current control methods and research and education needs are provided for all major animal production systems.

Horn fly management

Kaufman, P.E. and Weeks, E.N.I., 2018.
https://edis.ifas.ufl.edu/publication/IN952

• Horn flies may need to be managed if an equine farm also has pastured cattle. Horn fly biology, monitoring control is reviewed in this resource.

Buffalo flies and their control

Spence, S., 2007.
https://www.dpi.nsw.gov.au/__data/assets/pdf_file/0017/110915/buffalo-flies-and-their-control.pdf

• Buffalo flies are a major pest of cattle in Australia like horn flies in the United States and other areas of the world. This guide, although written for cattle, provides information on the basic biology and management of buffalo flies.

Habronemiasis: biology, signs, and diagnosis, and treatment and prevention of the nematodes and vector flies

Pugh, D.G., Hu, X.P. and Blagburn, B., 2014.
Journal of Equine Veterinary Science 34: 241-248. https://doi.org/10.1016/j.jevs.2013.06.004

• Habronemiasis is a parasitic disease of horses and other equids, which is caused by the invasion of nematodes and is transmitted by house flies, face flies, and stable flies. This article details the biology of nematode parasites, diagnosis and treatment of this condition, and methods of prevention and control of the transmitting flies.

Thelazia eyeworm: an original endo-and ecto-parasitic nematode

Otranto, D. and Traversa, D., 2005.
Trends in Parasitology 21: 1-4. https://doi.org/10.1016/j.pt.2004.10.008

• The *Thelazia* genus includes several parasites commonly called eyeworms. Transmission of eyeworms occurs from non-biting flies that feed on the eye secretions, tears, and conjunctiva of animals. This review discusses the relationship between eyeworms and their hosts, recent molecular insights, and the biology of *Thelazia* and their hosts.

Equine sarcoids – causes, molecular changes, and clinicopathologic features: a review

Ogłuszka, M., Starzyński, R.R., Pierzchała, M., Otrocka-Domagała, I. and Raś, A., 2021.
Veterinary pathology 58: 472-482. https://doi.org/10.1177/0300985820985114

• Equine sarcoids are the most common skin tumours of horses. Typically, sarcoids are not fatal but lesions tend to persist, resist therapy, and tend to recur after treatment, interfering with welfare. There is some evidence that flies may be involved in the transmission of the virus that causes sarcoids. This review discusses these flies and other causes, as well as clinical presentations and other factors associated with equine sarcoids.

Sarcoids

Equine Medical Solutions. https://equinesarcoid.co.uk/sarcoids

- An article with helpful photos on how sarcoids affect horses and what areas are typically affected.

Resources

Resources are listed in alphabetical order and may not be inclusive of all companies, materials, or products available.

Bite Free™ Stable Fly Trap. https://www.starbarproducts.com/all-products/traps/bite-free-stable-fly-trap
BugJammer Knight Stick®. https://bugjammer.com/
Insect Shield® Insect Repellent Fabric Treatment. https://www.insectshield.com/
Olson Biting Fly Trap. https://www.olsonproducts.com
ParlorPal® sodium bisulfate additive. https://www.jones-hamilton.com/jhproducts/parlorpal/

References

Ananth, G.P., Bronson, D.C. and Brown, J.K., 2008. Generation of airborne fly body particles by four electrocution fly traps and an electronic fly trap. International Journal of Environmental Health Research 2: 106-113.

Broce, A.B., 2008. Electrocuting and electronic insect traps: Trapping efficiency and production of airborne particles. International Journal of Environmental Health Research 3: 47-58.

Keiper, R.R. and Berger, J., 1982. Refuge-seeking and pest avoidance by feral horses in desert and island environments. Applied Animal Ethology 9: 111-120.

Machtinger, E.T., Geden, C.J., Leppla, N.C. and Hogsette, J.A., 2014. Oviposition preference and development of house flies and stable flies in six common substrates found in equine facilities. Journal of Medical Entomology 51: 1144-1150.

McDonnell, S.M., 2003. The equid ethogram: a practical field guide to horse behavior. Eclipse Publications, Lexington, KT, USA, pp. 105-168.

McDonnell, S.M., 2008. Practical review of self-mutilation in horses. Animal Reproduction Science 107: 219-228.

Mullens, B.A. and Gerry, A.C., 2014. Life history and seasonal abundance of *Fannia benjamini* complex (Diptera: Muscidae) in southern California. Journal of Medical Entomology 43: 192-199.

Murillo, A.C., Gerry, A.C., Gallagher, N.T., Peterson, N.G. and Mullens, B.A., 2014. Laboratory and field assessment of cyantraniliprole relative to existing fly baits. Pest Management Science 71: 752-758. https://doi.org/10.1002/ps.3847

Pearson, W., Boermans, H.J., Bettger, W.J., McBride, B.W. and Lindinger, M.I., 2005. Association of maximum voluntary dietary intake of freeze-dried garlic with Heinz body anemia in horses. American Journal of Veterinary Research 66: 457-465.

Pitzer, J.B., Kaufman, P.E., TenBroeck, S.H. and Maruniak, J.E., 2011. Host blood meal identification by multiplex polymerase chain reaction for dispersal evidence of stable flies (Diptera: Muscidae) between livestock facilities. Journal of Medical Entomology 48: 53-60.

Schmidtmann, E.T., 1991. Suppressing immature house and stable flies in outdoor calf hutches with sand, gravel, and sawdust bedding. Journal of Dairy Science 74: 3956-3960.

Schmidtmann, E.T., Miller, R.W. and Muller, R., 1989. Effect of experimental bedding treatments on the density of immature *Musca domestica* and *Stomoxys calcitrans* (Diptera: Muscidae) in outdoor calf hutches. Journal of Economic Entomology 82: 1134-1139.

Scott, H.G. and Littig, K.S., 1962. Flies of public health importance and their control. US Department of Health, Education, and Welfare, Public Health Service, Center for Disease Control, Publication no. 772. https://stacks.cdc.gov/view/cdc/7611.

Sweeney, C.R., McDonnell, S.M., Russell, G.E. and Terzich, M., 1997. Effect of sodium bisulfate on ammonia concentration, fly population, and manure pH in a horse barn. American Journal of Veterinary Research 57: 1795-1798.

Sweeney, C.R., Scanion, T., Russell, G.E., Smith, G. and Boston, R.C., 2000. Effect of daily floor treatment with sodium bisulfate on the fly population of horse stalls. American Journal of Veterinary Research 61: 910-913.

Tuorinsky, E.V. and Machtinger, E.T., 2019. Behavioral inhibition of the house fly, *Musca domestica* (Diptera: Muscidae), when exposed to commercial equine fly repellents. Journal of Economic Entomology 113: 518-526. https://doi.org/10.1093/jee/toz243

Photo credits

Title page Erika Machtinger, Pennsylvania State University.
Figure 3.1. Mike Houtz, Pennsylvania State University.
Figure 3.2. Erika Machtinger, Pennsylvania State University.
Figure 3.3. Erika Machtinger, Pennsylvania State University.
Figure 3.4. Mike Houtz, Pennsylvania State University.
Figure 3.5. (A) Erika Machtinger, Pennsylvania State University; (B) Frank Vassen, Flickr CC BY 2.0 https://www.flickr.com/photos/42244964@N03/35499330022.
Figure 3.6. Erika Machtinger, Pennsylvania State University.
Figure 3.7. Lyle Buss, University of Florida.
Figure 3.8. Erika Machtinger, Pennsylvania State University.
Figure 3.9. Erika Machtinger, Pennsylvania State University.
Figure 3.10. Erika Machtinger, Pennsylvania State University.
Figure 3.11. Janet Graham, Flickr CC BY 2.0 https://www.flickr.com/photos/130093583@N04/16737966188/in/photolist-rv5scW.
Figure 3.12. Pests and Diseases Image Library, bugwood.org.
Figure 3.13. Stephanie Leon, University of California Riverside.
Figure 3.14. Erika Machtinger, Pennsylvania State University.
Figure 3.15. Erika Machtinger, Pennsylvania State University.
Figure 3.16. Erika Machtinger, Pennsylvania State University.
Figure 3.17. (A) Erika Machtinger, Pennsylvania State University; (B) Daria-Borovleva, Shutterstock; (C) Daniel Requena Lambert, Shutterstock.

Figure 3.18. Erika Machtinger, Pennsylvania State University.
Figure 3.19. Erika Machtinger, Pennsylvania State University.
Figure 3.20. Erika Machtinger, Pennsylvania State University.
Figure 3.21. (A) Erika Machtinger, Pennsylvania State University; (B) Erica Lacher, Springhill Equine Veterinary Clinic.
Figure 3.22. Erika Machtinger, Pennsylvania State University.
Figure 3.23. Erika Machtinger, Pennsylvania State University.
Figure 3.24. Erika Machtinger, Pennsylvania State University.
Figure 3.25. Erika Machtinger, Pennsylvania State University.
Figure 3.26. (A) Jerome Hogsette, USDA, ARS; (B) Erika Machtinger, Pennsylvania State University.
Figure 3.27. Erika Machtinger, Pennsylvania State University.
Figure 3.28. Lyle Buss, University of Florida.
Figure 3.29. Erika Machtinger, Pennsylvania State University.
Figure 3.30. Erika Machtinger, Pennsylvania State University.
Figure 3.31. (A) Erika Machtinger, Pennsylvania State University; (B) Chris Geden, USDA, ARS.
Figure 3.32. Erika Machtinger, Pennsylvania State University.
Figure 3.33. Matt Bertone.
Figure 3.34. Chris Geden, USDA, ARS.
Figure 3.35. (A) Erika Machtinger, Pennsylvania State University; (B) Predrag Djunovic, Shutterstock.
Figure 3.36. Matt Bertone, North Carolina State University.
Figure 3.37. Mike Houtz, Pennsylvania State University.
Figure 3.38. Mike Houtz, Pennsylvania State University.
Figure 3.39. Erika Machtinger, Pennsylvania State University.
Figure 3.40. Erika Machtinger, Pennsylvania State University.
Figure 3.41. (A) Erica Lacher, Springhill Equine Veterinary Clinic; (B, C) Erika Machtinger, Pennsylvania State University.
Figure 3.42. Erika Machtinger, Pennsylvania State University.
Figure 3.43. Erika Machtinger, Pennsylvania State University.
Figure 3.44. Erica Lacher, Springhill Equine Veterinary Clinic.
Figure 3.45. (A) Erica Lacher, Springhill Equine Veterinary Clinic; (B) Erika Machtinger, Pennsylvania State University; (C) image provided courtesy of the National Center for Veterinary Parasitology, ncvetp.org.
Figure 3.46. (A) G. Miro; (B) image provided courtesy of the National Center for Veterinary Parasitology, ncvetp.org.
Figure 3.47. Erica Lacher, Springhill Equine Veterinary Clinic.

Chapter 4. Fly pests that develop in water

Chapter overview

This chapter covers true flies (order Diptera) that have some immature development in aquatic or semi-aquatic environments. These environments include rivers, lakes, ponds and wet or marshy soil, and range from clean freshwater to water heavily polluted by organic substances. The groups of flies covered include mosquitoes (Section 4.1), biting midges (Section 4.2), horse and deer flies (Section 4.3), and black flies (Section 4.4).

Annoyance caused by these flies can reduce performance and increase the risk of injury or accidents when horses try to avoid bites. All of these fly groups bite and feed on blood so can transmit pathogens that cause equine diseases due to their blood-feeding life strategies, and they all aggravate horses with their bites. Mosquitoes transmit many viral pathogens causing several forms of encephalitis that can be fatal. Biting midges are most well-known for the intense allergic responses to their saliva, known as sweet itch. Horse and deer fly bites are painful to horses and their owners, and they can transmit the virus that causes equine infectious anaemia. Black flies include insect vectors for two diseases of concern for horse owners: vesicular stomatitis and auricular papillomavirus. Management of the immature stages of these flies is complicated due to their presence in natural water sources that may be environmentally sensitive. In general, the most effective management strategy for these water-developing flies is to protect the horse from bites.

4.1 Mosquitoes (family Culicidae)

4.1.1 Pest summary (Table 4.1)

Mosquitoes (Figure 4.1) are small biting flies in the family Culicidae. There are 3,500 species and 110 genera in this insect group. The most important genera of Culicidae involved in pathogen transmission are *Anopheles*, *Culex*, and *Aedes*. Although small, they are larger than black flies (Section 4.4) and *Culicoides* biting midges (Section 4.2). Unlike the other two groups, they have slender bodies and long legs. Male and female mosquitoes feed on nectar and other plant secretions, but only female mosquitoes feed on blood to produce eggs. This blood-feeding is a nuisance to horses in areas with high mosquito abundance, but the biggest threat of mosquitoes is due to their status as vectors of viruses causing several forms of encephalitis.

4.1.2 Distribution and seasonality

Mosquitoes are found throughout the world except for Antarctica and Iceland. They are found at high and low altitudes, even below ground in caves and mines. Mosquitoes typically travel only a few kilometres/miles from their development sites after emerging as adults, but when assisted by wind can travel much further.

Table 4.1. Pest summary of mosquitoes (family Culicidae).

Distribution[1]	Australia: Yes Canada: Yes Europe: Yes New Zealand: Yes United States: Yes
Size	3-6 mm (0.12-0.23 in) body and long legs
Key identification characteristics	Long legs, slender body, long antennae, long proboscis (mouthpart)
Mouthparts of adults	Piercing-sucking proboscis
Feed on	Blood
Development time[2]	5-21 days
Adult life span	5-7 days for males, 1-2 weeks for females
Total eggs laid	50-500 about every 3 days
May be mistaken for	Crane flies (Chapter 8)
Impact on horses	Annoyance, vector of pathogens including flaviviruses (Japanese encephalitis, West Nile virus) and alphaviruses (including eastern, western, and Venezuelan encephalitis)

[1] The pest distribution in the regions targeted by this publication are listed as of the date published. Areas of primary pest concern (if any) are in bold. Global distributions are listed in Section 4.1.2.
[2] Development times are approximate and can change with environmental conditions including temperature, humidity, and availability of food resources.

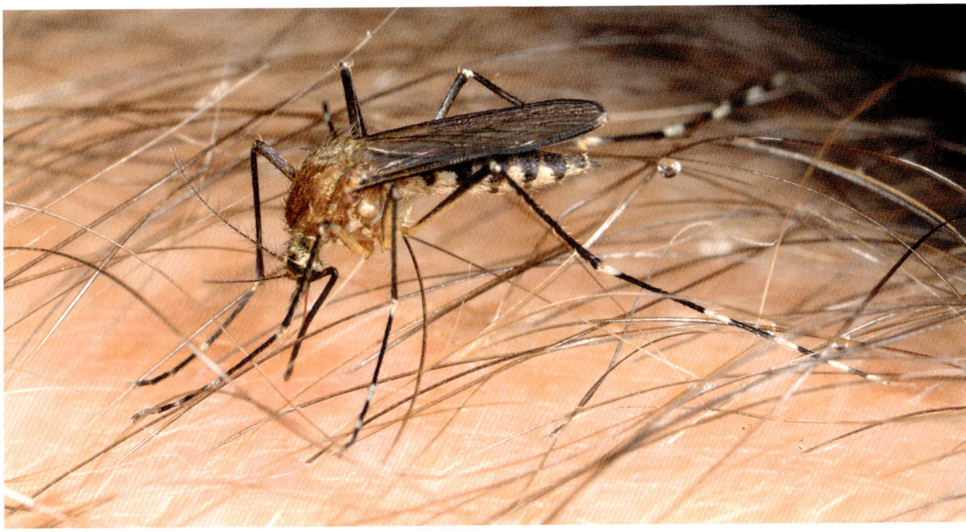

Figure 4.1. *Aedes (Ochlerotatus) canadensis*, a woodland pool mosquito, has been implicated in the transmission of eastern equine encephalitis to humans and horses (Armstrong and Andreadis, 2010).

4.1.3 Description

Mosquitoes that feed on horses primarily belong to three genera, *Aedes* (Figure 4.1), *Anopheles* (Figure 4.2) and *Culex*. *Psorophora, Mansonia,* and *Coquillettidia* species also bite horses, sometimes in large numbers.

Figure 4.2. *Anopheles* sp., most likely *A. quadrimaculatus*, resting in the characteristic *Anopheles* pose with the abdomen in the air.

Mosquitoes are true two-winged flies in the family Culicidae. These flies are slender with elongated bodies. The adult fly is covered with scales on the body and the veins of the wings. Mosquitoes also have long, fragile-looking legs and elongated mouthparts. Eggs are brown to black and elongated oval in shape. Eggs may be deposited individually (*Aedes* and *Anopheles*) or as part of a raft (*Culex*). Eggs may be in the water (*Anopheles* and *Culex*) or just above the waterline (*Aedes*). Larvae develop in water and have a large head and thorax and a narrow, wormlike abdomen (Figure 4.3). Larvae typically are found just below the surface of the water and wriggle downward when disturbed. Pupae look similar to larvae except for the first two body segments (head and thorax) are fused so they are comma-shaped. They do not feed but are mobile in the water and are often referred to as tumblers due to their way of moving (Figure 4.3).

Crane flies (Figure 4.4) are often confused with mosquitoes, but while these large flies develop in wet vegetation like several mosquito species, they do not bite and are not pests. They are often called mosquito killers, but they do not feed at all as adults and do not kill mosquitoes.

Figure 4.3. Two mosquito larvae (outside) and pupae (middle) develop in the water, with some species spending more time at the surface and others deeper in the water column.

Figure 4.4. Crane flies are often confused for mosquitoes or are called mosquito killers, but these flies do not bite and are neither pests nor biological control agents of mosquitoes.

4.1.4 Life history

Mosquitoes have complete metamorphosis with four life stages: egg, larva, pupa, and adult. While all mosquito larvae are aquatic, the types of aquatic systems in which they develop vary by species. *Aedes* species typically develop in temporary accumulations of relatively clean water in places such as tree holes, birdbaths, car tyres, or water in wheelbarrows, buckets, or other stable supplies (Figure 4.5). Eggs are laid individually just above the water line. *Aedes* eggs are resistant to drying out, and eggs can remain viable for months or years until suitable water conditions occur again. Some salt marsh species (e.g. *Aedes sollicitans*, *Aedes taeniorhynchus*, and *Aedes squamiger*) often occur in enormous numbers and are serious pests of horses, other animals, and people who live in or near coastal communities (Figure 4.6).

Anopheles are most commonly found along the water's edge in more permanent standing water, like marshy areas with emerging aquatic vegetation. Some species also will develop in small temporary sites such as puddles and flooded tyre tracks (Figure 4.7). In contrast to *Aedes*, *Anopheles* lay their eggs individually and directly on the water surface and have specialized structures to ensure buoyancy. Very large *Anopheles* populations sometimes occur in association with agricultural practices. This is especially true of areas with rice production, where flooded fields can provide vast acreages of suitable larval habitats (Figure 4.8).

Figure 4.5. Different mosquito species can develop in different types of aquatic habitats including streams (A), standing water in ponds or flooded pastures (B), or even standing water in wheelbarrows, buckets, troughs, and other containers (C).

Figure 4.6. Salt marshes can produce large numbers of mosquitoes such as *Aedes sollicitans* and *Aedes taeniorhynchus* in coastal areas.

Figure 4.7. Mosquito larvae (circled) can be seen in many types of aquatic habitats, even puddles like shown here.

Culex mosquitoes (Figure 4.9) often feed primarily on birds but can also feed on mammals. *Culex* larvae generally develop in stationary bodies of water with substantial nutrient loads and can tolerate high levels of organic pollution. For example, manure settling lagoons associated with pig and laying-hen production are often prolific breeders of the Southern house mosquito, *Culex quinquefasciatus.* Female *Culex* mosquitoes lay their eggs in batches of 100 to 300 eggs glued together in rafts that are deposited and float directly on the water surface.

Figure 4.8. Flooded rice fields can be important breeding sites for *Anopheles* mosquitoes.

Figure 4.9. *Culex pipiens*, the northern house mosquito (shown above), is one of the vectors for West Nile virus as well as other equine encephalitides in the Northern USA. A related species, *Culex quinquefasciatus* (the southern house mosquito), is the dominant *Culex* species in southern regions of the USA.

The habit of feeding on both birds and horses accounts for many infections with West Nile virus and the major equine encephalitis viruses. West Nile virus is typically vectored by *Culex*, particularly *Culex pipiens* and *Culex quinquefasciatus*, although species in other mosquito genera can also play a role in transmission. Eastern equine encephalitis is maintained in a cycle between birds and the mosquito *Culiseta melanura*. Bridge vectors (mosquitoes that feed on birds and mammals) move the virus from birds to humans and horses, species of *Aedes*, *Coquillettidia* and *Culex* have been implicated. Various species of *Culex* have been implicated in the transmission of Venezuelan equine encephalitis, St. Louis encephalitis, and western equine encephalitis, as well as *Culiseta* in the latter.

Mosquito larvae generally feed just below the water surface (Figure 4.3) and have a variety of morphological adaptations to ensure an adequate air supply. Some species spend more time in deeper water and only visit the surface occasionally to breathe. They are all filter feeders whose mouthparts are equipped with brushes and sweepers to collect and direct food particles. Larval development can take only 5 to 6 days for fast-developing species or up to 2 to 3 weeks for slower ones. Larvae moult into the short-lived, non-feeding pupal stage, which characteristically bobs at the water surface. Larvae and pupae can be seen together floating in water sources. Adults emerge from pupae on the surface of the water and fly once the wings have dried and hardened (Figure 4.10).

After emerging, both male and female mosquitoes feed on flower nectars for energy, and females of most pest species must also feed on blood to obtain nutrients necessary to develop their eggs. Female mosquitoes use cues such as carbon dioxide, body heat, and various skin and gland odours to guide them towards a host animal, which may include horses and people. Females will lay the first batch of eggs approximately 7 days after emerging as an adult. An additional blood meal is required for each subsequent batch of eggs.

4.1.5 Control

Mosquitoes are probably one of the most studied insect pests of horses. While there are several species of mosquitoes that are pests, control methods generally do not depend on which particular species are present. Horse farms usually have many wet areas that can support mosquito development and the best method of mosquito management is to reduce these areas as much as possible and protect horses from bites.

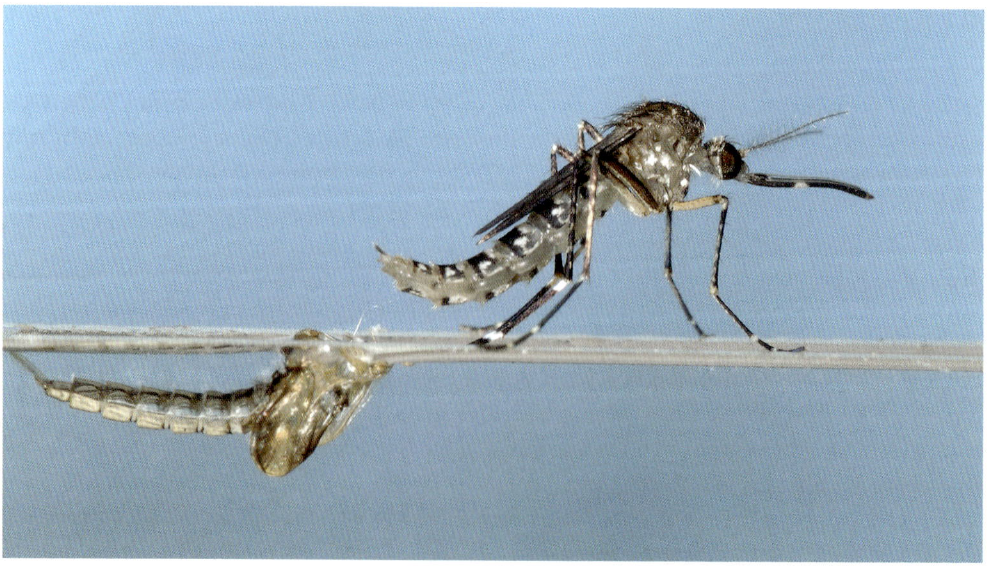

Figure 4.10. A salt marsh mosquito, *Aedes taeniorhynchus*, just after emerging as an adult. The empty pupal case can be seen to the left, just below the water surface.

Cultural control

Turnout

Avoid turning horses out at times of the day when mosquitoes are biting is one of the few methods of mosquito control that does depend on species. Some species feed in daylight such as *Aedes albopictus*, others during the night (many *Anopheles* and some *Culex* species), and many around dawn and dusk. If mosquitoes are identified as a pest, turning horses out at a different time of the day may reduce contact with feeding mosquitoes. Often, however, horse owners are unaware of the biting pressure that their horses sustain during the night hours when horses are unattended. It is recommended to spend time observing pastured horses at dusk or dawn, during the day, and during the night several times in the spring and summer to evaluate if horses are reacting to pests or if the observer is being bitten at different times of the day and avoid horses being pastured at those times if necessary.

Standing water elimination

Mosquitoes can develop in shallow water sources, especially if the water is rich in organic debris like mud, vegetation, or leaf litter. Even small wet areas like a hoofprint holding water can provide habitats for developing mosquitoes (Figure 4.11). Because it can take as few as 4 days for immature mosquitoes to develop into adults, anything that can hold water such as flower pots, water buckets and bowls, barrels, unused birdbaths, wheelbarrows, tarps, plants, and other items should be checked for water and drained or removed. Standing water includes all sources of water that are not flowing, even water troughs that are refilled frequently. Water troughs should be emptied and cleaned at least weekly (Figure 4.12). An alternative to standing water in troughs is using automatic waterers. Water stays cooler, cleaner, and has a more frequent turnover in automatic waterers making them less suitable for mosquito development.

If possible, drill small holes at the bottom of items found to hold water, like tyres and pots, to prevent standing water from accumulating. If items cannot be modified, empty standing water at least weekly. Options for handling excessive water in areas include French drain lines, water bars, grassy swales, and dry wells. Mosquitoes may develop in potholes in driveways, so levelling and maintaining these areas can reduce potential development sites.

Figure 4.11. Mosquitoes can develop in even small amounts of water like water caught on tarps (A) or even in hoof prints (B).

Figure 4.12. Water troughs should be emptied weekly and cleaned to prevent algal growth (A) and larval mosquito (circled) development (B).

Vegetation management

Vegetation management may help to minimize mosquito survival. Mowing vegetation and weeds around buildings and paddocks, and thinning or removing rushes and cattails near ditches or low-lying areas can reduce resting sites close to hosts and help predators hunt mosquitoes. Planting native vegetation in ditches or drainage areas can help quickly absorb excess water (Figure 4.13).

Pasture management

Pasture management can be an effective way to minimize development sites for mosquitoes. Improve pasture drainage in areas that are soft enough to permit standing water. Horses should be rotated among pastures when forage is at 7.5 to 10.0 cm (3 to 4 in) to allow pastures to dry and grow between grazing or irrigation. When grasses are dormant in the winter, pasturing horses in a dry lot or sacrifice paddock will prevent overgrazing (Figure 4.14). Keep horses off dormant grasses to prevent mud development, which can harbour mosquitoes. Preventing soil and forage damage in

Figure 4.13. Planting native grasses and other vegetation in ditches can help prevent standing water in these areas.

these ways protects against mosquito development by eliminating potential development sites and also by protecting the roots of pasture grasses, which allows for better water penetration.

Facility management

The configuration of the property or adjacent lands can be changed to prevent mosquito development sites. In many facilities, horses spend quite a bit of time in the stable area. Using landscaping and grading to ensure good drainage away from the stable helps remove wet areas where mosquito development occurs (Figure 4.15). Installing gutters and downspouts on all buildings to divert rainwater away from stables and paddocks will greatly reduce the amount of mud, standing water, and mosquito habitat around the farm. Diverted water can be sent to a dry well, rain barrel, water tanks, or well-vegetated woods. The use of water-loving native shrubs around pastures and stables may help keep these areas even drier and reduce runoff.

Figure 4.14. Dry lots or sacrifice paddocks (in the front of this image) can be used to keep horses off dormant or growing grasses (in the background of this image) to protect the root systems and prevent mud from developing. Special attention should be paid to the dry lot to ensure proper drainage while horses are present.

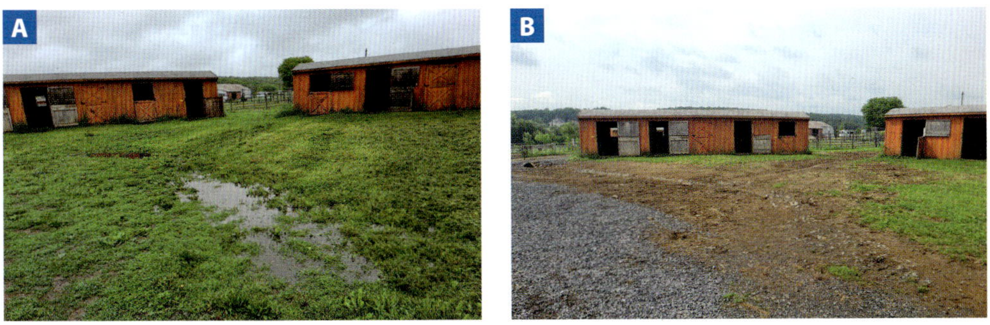

Figure 4.15. Areas of poor drainage can hold water (A), which can be development sites for mosquitoes. Improving drainage can keep development sites to a minimum and reduce the potential for mosquito bites (B).

Modification of footing in high-traffic areas of pastures and small paddocks can reduce mud. Seven to 15 cm (3 to 6 in) of chipped wood products, crushed rock, or sand can be applied to these areas to improve drainage and durability (Figure 4.16). Modifying large wet areas around facilities like wetlands may be possible, but in many places, large natural aquatic areas are protected. Before attempting any modification of these areas, the appropriate governmental entities should be consulted for any required permits or permissions.

Lighting modification

Changing lighting around stables can help reduce mosquito numbers around horses. Mosquitoes are attracted to light, so dimming or completely turning off stable lights at night can prevent the attraction of mosquitoes. Yellow incandescent or fluorescent lights are thought to be less attractive to mosquitoes than conventional lighting.

Feed additives

There is no evidence that additives to feed, including garlic or apple cider vinegar, help to repel mosquitoes. Garlic can potentially cause anaemia in horses (Pearson *et al.*, 2005), so these additives should be avoided unless recommended by a veterinarian.

Physical and mechanical control

Fly sheets, boots, and masks

Mesh fly sheets, boots, and masks can be used to reduce mosquito bites on horses (Figure 4.17). If the mesh is maintained (i.e. kept free of excessive soil) the material allows air to circulate over

Figure 4.16. High traffic areas can create mud (background), which can be a development site for some mosquitoes. Additions of crushed rock or other materials can improve drainage and durability of high traffic areas (foreground).

the skin, yet it blocks the entry of insects and some ultraviolet rays. Mesh leg boots cover from the animal's hooves to the knees or hocks; these are especially useful for horses that are sensitive to fly irritation on their legs. See Chapter 10 for more information on using fly sheets, boots, and masks.

Flight impediments

Mosquitoes are not strong fliers and do not handle moving air well, so installing fans in stalls or around smaller paddocks can help deter mosquitoes from landing (Figure 4.18). Be sure only to use fans rated for outdoor use, even when used in the stable to reduce the risk of fires started from dust and debris.

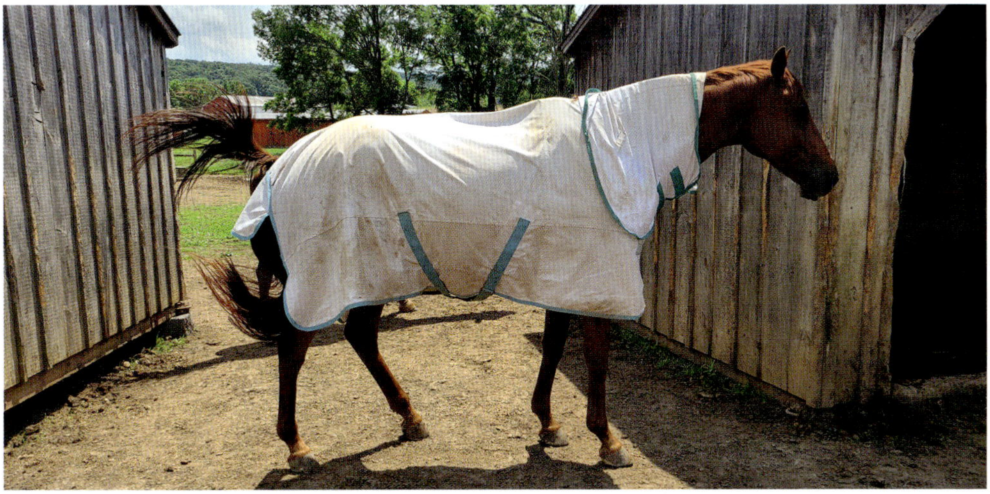

Figure 4.17. Mesh sheets can be used to reduce mosquito and other pest fly bites on horses by preventing them from landing directly on the body.

Figure 4.18. Fans can be mounted around stalls to circulate air, deterring mosquitoes from landing.

Trapping

Carbon dioxide traps can be used to draw mosquitoes away from horses and trap them. There are many trap models available depending on budget and power availability. Three traps that have been demonstrated to be effective at trapping mosquitoes are the Biogents BG-Sentinel trap, the Centers for Disease Control and Prevention (CDC) miniature light trap, and the Mosquito Magnet® (Figure 4.19) (Lühken *et al.*, 2014). Some of the traps have additional attractants, such as 1-octen-3-ol (often shortened to octenol and not to be confused with the fatty acid octanoic acid found in some repellents) or other host odours that can increase the numbers of certain mosquito species caught. It is important to make sure traps are not set up near areas where horses or people are active because these traps may bring more mosquitoes into those areas.

Autocidal gravid oviposition traps such as the Ovi-Catch™ and other lethal oviposition traps catch mosquitoes that lay their eggs in containers. Mosquitoes are attracted to the dark trap and the odour of organic material in the water. They enter the trap but cannot leave. Any mosquitoes emerging from the eggs that are laid are also captured. These traps are particularly effective for reducing the abundance of *Aedes aegypti* and *Ae. albopictus* (Barrera *et al.*, 2014).

Ultraviolet bug zappers are not recommended. In general, if used these traps should only be used for small spaces for pests indoors. Although these traps will kill mosquitoes, they can have the unintentional effect of attracting more mosquitoes and other unwanted pests to an area. They can also kill beneficial insect species that eat mosquitoes.

Figure 4.19. The Mosquito Magnet® is an example of a commercial carbon dioxide-baited trap in which the carbon dioxide is generated by the burning of propane.

Biological control

Predators

Mosquitoes have many natural predators. Dragonflies, bats, and birds all use mosquitoes as a food source. Bat and birdhouses can be installed to encourage these natural predators to establish themselves at stables (Figure 4.20). Be sure to set these houses far from the stable areas to reduce contact with wildlife. Mosquito-eating fish, like Koi, Shunbunkin, Gambusia, and Sarasa Comets will eat many mosquito larvae per day. Other fish species may be used as well, depending on location. However, before stocking ponds with fish, it is important to check local government agencies for regulations and restrictions.

Water additives

Mosquito larvae can be managed using dissolvable blocks containing mosquito-killing bacteria (Figure 4.21). These blocks are added to large areas of standing water, such as water troughs, ponds, and ditches. Adding these dissolvable blocks that contain the bacteria *Bacillus thuringiensis israelensis* (*BTi*) or *Bacillus sphaericus* (*BS*) control mosquito larvae effectively when applied weekly until a killing frost. These additives are safe for people, horses, other livestock and pets, and plants.

Chemical control

Area-wide use of foggers with an adulticide and large-scale applications of larvicides for mosquito control is an option, but vegetation and large pasture areas make their use complicated, expensive, and often ineffective.

Figure 4.20. Placing bat houses can help promote bat populations, which feed avidly on adult mosquitoes.

Figure 4.21. Dissolvable blocks of mosquito-killing bacteria, such as *Bacillus thuringiensis israelensis* are readily available, low risk, and effective methods for mosquito control on horse farms.

Insect growth regulators

Insect growth regulators can be used that target developing mosquitoes. These chemicals target critical stages in mosquito development, preventing them from emerging as adults. Several types of products can be purchased or applied by a pesticide applicator that includes sprays or even bait stations for female mosquitoes to bring the insect growth regulator to the water in which she lays her eggs. Insect growth regulators are most effective in smaller pools of standing water like small garden ponds and areas that cannot be drained consistently.

Treated fly sheets, boots, and masks

Fly sheets, boots, and masks with permethrin bound to the fabric fibres can be purchased to repel mosquitoes. The repellent action of these products generally lasts 20 to 40 washes, depending on the product. The duration of effectiveness may change with exposure to sunlight and weather patterns, so observe horses periodically to make sure it is still providing the protection desired. Fly sheets purchased without pesticide treatment can also be sent to a company that specializes in binding permethrin to fabric for treatment. It is important to recognize that do-it-yourself treatment of fabrics with pesticides that do not list fabric treatment for horses on the label may be a violation of the law. In addition, horses are sensitive to chemical pesticides and off-label treatment may cause skin reactions.

Topical products

There are many repellents marketed for protecting horses from mosquitoes and other pests. Repellent products can be effective, but generally for a short duration (less than 24 hours). Geraniol is not effective as a mosquito repellent for horses (Reeves and Miller, 2010). Repellent sprays should be applied using a microfiber mitt, towel, or brush to ensure complete coverage of the horse and reduce chemicals from dispersing in the environment as well as potentially affecting the eyes or nostrils of the horse. If applying repellents in this way it is important to consider using disposable gloves to protect the applicator's skin from the products. Roll-on or cream pesticides can also be used.

Roll-on (Figure 4.22) or spot-on pesticides can be applied to more sensitive areas and may help keep products working longer, and prevent chemicals from accidentally being applied to easily irritated areas of the horse or being taken into the respiratory system. Shampoos containing the active ingredient permethrin are drying to the skin and the residual effect of the active ingredient is short-lived like other topical products.

There is no evidence that rubbing orange or lemon peel, vanilla, onion, or other spices or essential oils has any repellent effect against mosquitoes.

4.1.6 Veterinary concerns

Mosquitoes are global pests of humans and other animals that transmit pathogens that cause many of the world's deadliest vector-transmitted diseases, including malaria, yellow fever, and dengue. They have been called the most dangerous animals in the world because of the number of human and other animal deaths associated with mosquito-borne pathogens. For North

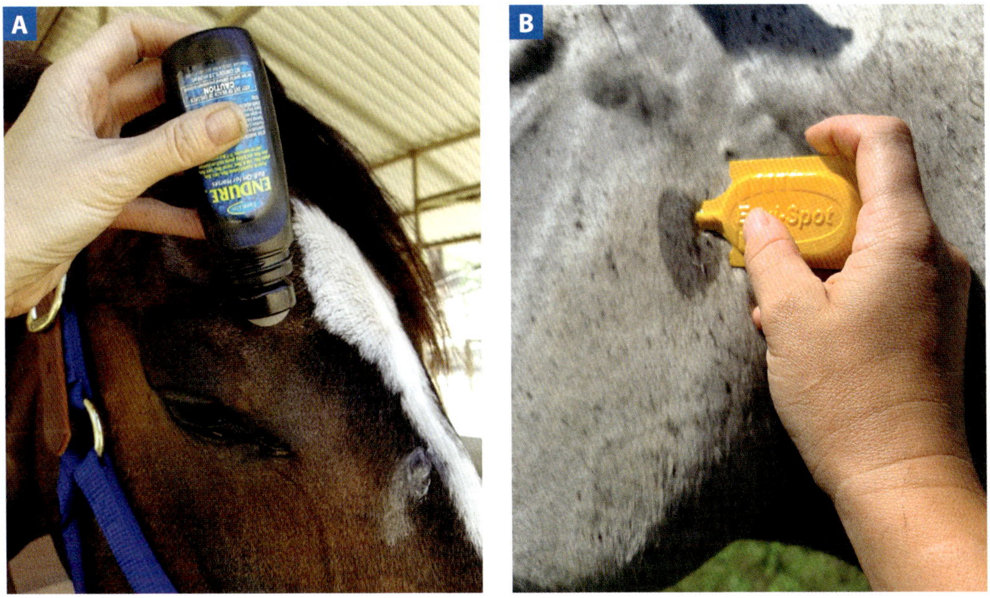

Figure 4.22. Roll-on (A) or spot-on (B) pesticide treatments are a simple way of getting short-term relief from mosquito pressure.

America, the most important disease concerns for equine owners are arboviruses known as the encephalitides, including eastern, western, Venezuelan, and St. Louis equine encephalitis and West Nile viruses (see Chapman *et al.*, 2018 in 'Continuing education' for more information about arboviruses).

For some horses, the simple bite of the mosquito can cause significant skin irritation. These present as welts ranging from 1 cm (0.4 in) to 8 to 10 cm (3 to 4 in) in size. Much like humans experiencing insect bites, these horses can be extremely itchy. If itching becomes excessive horses can be treated with steroids, or antihistamines to decrease the immune response.

Eastern equine encephalitis (EEE) and West Nile virus (WNV) are both endemic throughout the North American continent. Both viruses cause debilitating neurological diseases in affected horses. The initial symptom is often a fever, with EEE typically causing a higher fever that may reach up to 41 °C (106 °F). WNV generally presents with a low-grade fever. These fevers can be very transient so owners may not notice them. The horse then begins to become dull to its surroundings and has difficulty placing its feet. West Nile virus affected horses will often have fine tremors of the nose and face muscles. Both diseases usually progress for 24 to 48 hours before symptoms level off. Symptoms can progress to recumbency (laying down), seizures, and death in that time. Recumbency, even if only for a short time, is strongly associated with poor outcomes. Eastern equine encephalitis carries a 98% mortality rate in unvaccinated horses. West Nile virus has a lower mortality rate of approximately 33%. For most of North America, annual vaccination is recommended and is extremely effective in preventing disease. Vaccination two to three times per year is recommended in areas with mosquito activity year-round.

European cases of WNV have been more limited, but considering the severity of disease caused by West Nile virus vaccination should be considered essential for most European horses. In 2017, 2018, 2019, and 2020 there were 127, 285, 93, and 183, reported cases of WNV in horses throughout Europe, respectively (ECDC, 2017, 2018, 2019, 2021). In 2020, eight countries were affected with the majority of those cases occurring in Spain (ECDC, 2021).

Murray Valley encephalitis, Kunjin virus, and Ross Valley encephalitis are endemic encephalitides to Australia. All typically cause neurologic symptoms similar to West Nile virus. No vaccination is available for these viruses at this time. A veterinarian should be contacted as soon as symptoms are noticed so that treatment can be initiated early.

Venezuelan encephalitis has the potential to spread to the southern United States via Central America, where it is endemic. Its prevalence should be monitored, and vaccination discussed with a veterinarian, although in general vaccination it is not currently recommended.

4.2 Biting midges (genus *Culicoides*)

4.2.1 Pest summary (Table 4.2)

Biting midges (Figure 4.23) are flies in the insect family Ceratopogonidae. They are commonly called no-see-ums, punkies, and sand flies, although they are not true sand flies (Chapter 7). The family includes 123 genera and about 6,300 species, although most of the species that attack horses and other animals belong to the genus *Culicoides*. There are about 1000 biting midge species. They are small dark-coloured flies with long antennae and patterns on the wings that can be used to identify species with training. The females feed on blood to obtain protein to develop eggs. *Culicoides* bites can lead to a skin condition known as sweet itch, and they can transmit many important pathogens including viruses and nematodes throughout their global range.

4.2.2 Distribution and seasonality

Culicoides are found almost worldwide, the exceptions being New Zealand and the Antarctic, including the Antarctic regions of South America (Chile and Argentina). New Zealand practices a strict surveillance program with routine monitoring for the midges and the pathogens they transmit. Within their range, their distribution is influenced by host availability, development site availability, and climate. Although they develop in semi-aquatic areas, they can fly outside of these areas to feed. However, they typically do not travel more than a few kilometres (miles) from where they develop.

4.2.3 Description

Adult biting midges are tiny, 1.0 to 2.5 mm (0.4 to 0.1 in) in length, two-winged flies (Figure 4.23). They have long antennae and weakly-veined wings with dense hairs that form dark and light patterning (Figure 4.24). Eggs are an elongated oval shape, 0.25 mm (<0.01 in) in length, and brown to black. Larvae are worm-like (Figure 4.25) and 2 to 5 mm (0.08 to 0.20 in) in length. They have distinctive serpentine movements that resemble those of nematodes (worms).

Table 4.2. Pest summary of biting midges (genus *Culicoides*).

Distribution[1]	Australia: Yes Canada: Yes Europe: Yes New Zealand: No United States: Yes
Size	1.0-2.5 mm (0.4-0.1 in) in length
Key identification characteristics	Small, patterned wings, long beaded antennae
Mouthparts of adults	Cutting, sucking
Feed on	Blood
Development time[2]	14 days
Adult life span	1-3 weeks
Total eggs laid	30-450 per batch
May be mistaken for	Black flies (Section 4.4), dung midges (Scatopsidae) (Chapter 8), sand flies (Chapter 7)
Impact on horses	Annoyance, bites cause sweet itch skin conditions, vector of nematodes, and pathogens causing vesicular stomatitis and African horse sickness

[1] The pest distribution in the regions targeted by this publication are listed as of the date published. Areas of primary pest concern (if any) are in bold. Global distributions are listed in Section 4.2.2

[2] Development times are approximate and can change with environmental conditions including temperature, humidity, and availability of food resources.

Figure 4.23. Biting midge (*Culicoides* sp.) feeding on blood.

4.2.4 Life history

Biting midges have complete metamorphosis with four life stages: egg, larva, pupa, and adult. Larvae are semi-aquatic and live in the mud of both freshwater and saltwater wet areas. Larvae (Figure 4.25) feed on detritus and microbes or prey on other small organisms. Some species can tolerate habitats with high nutrient loading, such as the margins of manure settling lagoons that may be near equine facilities, or edges of wet manure piles or pit areas on the property (Figure 4.26) (e.g. *Culicoides sonorensis* and *Culicoides variipennis*). Other species, such as *Culicoides furens*, develop in salt marshes or mangrove swamps. The latter habitats are notorious for producing large swarms of flies that torment horses, livestock, and their owners in coastal areas.

Larval development times vary from 2 weeks to a year depending on the species and environmental conditions. The flies overwinter as larvae in temperate areas that are too cold for year-round development. Overwintering larvae pupate in the spring or early summer and emerge as adults soon after.

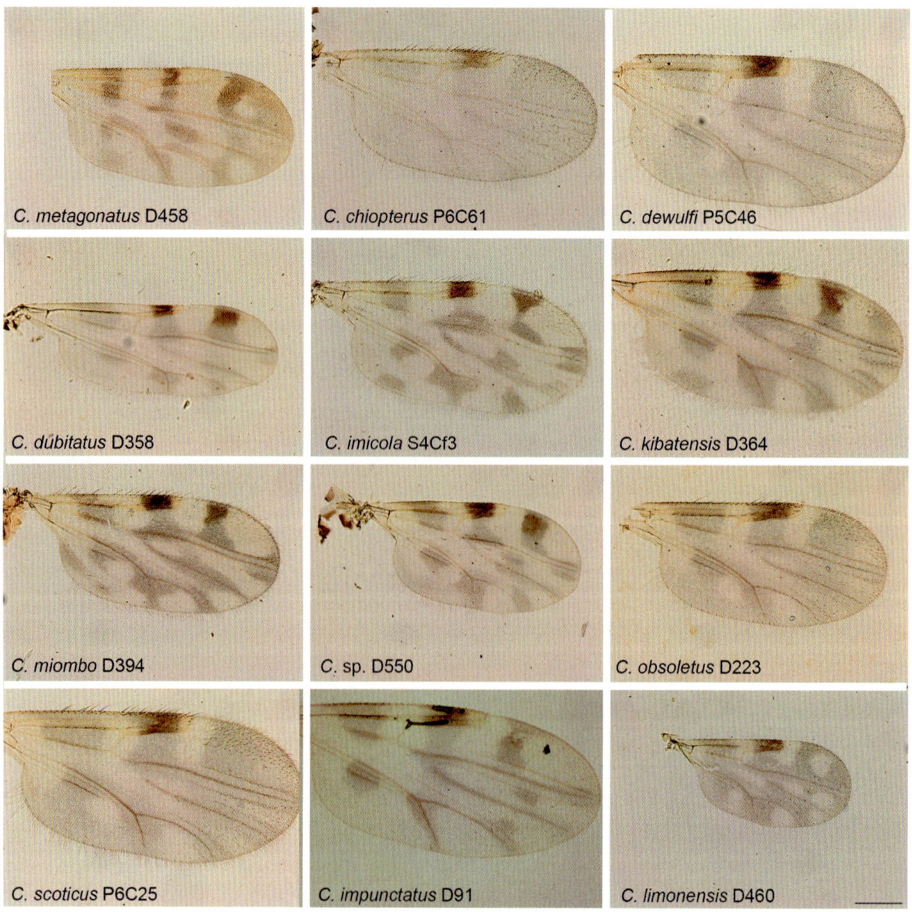

Figure 4.24. ***Culicoides*** **have distinctive markings on their wings, which can be helpful in identification.**

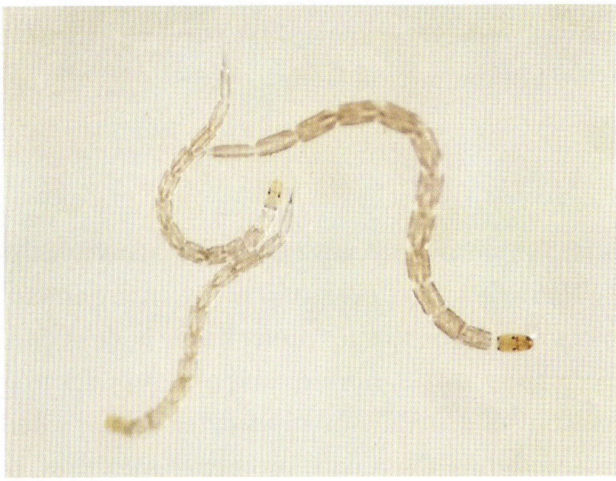

Figure 4.25. Biting midge (Ceratopogonidae) larvae.

Figure 4.26. Biting midges can develop in wet areas on the edges of manure piles or pits.

Both male and female biting midges feed on plant nectar for energy. Males form mating swarms, into which females fly to locate a suitable mate. Females require blood for egg development, with most species feeding around sunrise and sunset. They feed by lacerating the skin of the host and sucking up the resulting pool of blood.

Fed females move off the host to rest for several days for blood meal digestion and egg development. Females lay their eggs in batches ranging from 30 to 450 eggs. Females are generally ready to lay their first egg batch within a week of emerging, and the larvae hatch within a week. A second blood meal is required for the development of more eggs, but few flies live long enough to complete a second cycle. Biting midges can sometimes be confused for black flies (Section 4.4) and dung midges or minute black scavenger flies (Figure 4.27) (Chapter 8), the latter of which are harmless and assist with the breakdown of manure on farms.

4.2.5 Control

Control of biting midges is complicated because there are many species with often similar, but not identical, behaviours. These flies can develop in many environments with different levels of moisture, from damp to very wet. Some species prefer biting at dusk and dawn, and some during the day or night. Some species bite horses indoors, and others prefer to bite outdoors. Because of these different behaviours, control generally involves determining when horses are being bitten and then preventing flies from biting them with protective measures. In all cases, management of horses to avoid midge bites should be started before the midge season in the spring months to reduce bites and potential veterinary concerns like sweet itch (see Section 4.2.6). See Harrup *et al.*, 2016 in 'Continuing education' for more information about biting midge management.

Cultural control

Turnout

As with mosquitoes, simple adjustments to turnout schedules can be used to prevent high levels of contact between horses and biting midges. It is recommended to observe horses at different times of the morning, evening, and night for avoidance behaviours like head shaking, biting at the body or scratching to determine when or if horses are being affected by fly bites. Avoiding having horses turned out during these times may reduce contact with most biting midge species.

Habitat management

Habitat modification can be used to help reduce the density of immature biting midges. Biting midges are semi-aquatic, so modification methods involve either the removal or alteration of wet habitats where midges develop. However, the range of suitable habitat for biting midges is broad and not well understood, so managing midge development habitat should be considered along with other efforts to reduce contact with horses.

Figure 4.27. Dung midges or minute black scavenger flies, family Scatopsidae, are small, dark coloured flies that are often found within minutes of fresh dung landing on the ground. While these flies may look like biting midges or black flies, they are only found on the manure and are not pests of horses.

On equine farms, looking for and draining or drying areas suitable for immature development may reduce biting midge survival. Depending on the species, potential development areas including water troughs, leaking taps, hoses, runoff from rooflines, and standing water in low-lying pastures (Figure 4.28). Other sites for development are old manure storage areas, leaf litter, decaying vegetation (e.g. grass clippings), moist manure, mud, or tree holes. Biting midges also will develop in marshes and swamps where management options are limited or non-existent, and these areas may not be able to be addressed with landscape management, so other control methods may be necessary. However, removing as many of these development habitats as possible can reduce biting midge numbers.

Feed additives

As with mosquitoes, there is no evidence that additives to feed, including garlic or apple cider vinegar, help to repel biting midges. See Section 4.1.5 for more information on feed additives.

Physical and mechanical control

Physical barriers

Some biting midges will enter stables, but the number that enters can be reduced by excluding these flies with screening and other physical barriers (Meiswinkle *et al.*, 2000). Screening stalls with minimal gaps in screen can reduce biting midges by over 65% (Lincoln *et al.*, 2015). Because of the small size of biting midges, screen mesh size does make a difference, and the holes in mosquito netting are not small enough. Screens with mesh sizes of 1.6 mm^2 (0.06 in^2) reduce entry rates by just over 50%. Smaller mesh sizes of 0.9 mm^2 (0.04 in^2) reduce entry by 95% (Porter,

Figure 4.28. Standing water in pastures can be a development site for biting midges. Pasture management to limit standing water or increase drainage or drying can reduce the risk of midge development in these areas.

1959), but smaller mesh has the potential to also significantly reduce airflow and negatively impact the respiratory performance of the enclosed horses, so this should be considered when making screening decisions.

Fly sheets, boots, and masks

Fly sheets, boots, and masks are physical barriers that deter biting from multiple species of flies. These can be especially beneficial for horses that are more intolerant of biting midge bites. However, a good fly sheet to protect against biting midges may be different from what can be effective for other fly species. Midge protection sheets and other midge protection products have a much smaller mesh size than standard fly protection to prevent these small flies from reaching the horse's skin. A fitted sheet with a neck cover/hood and form-fitting belly part will be beneficial for horses that are especially sensitive to biting midge bites (Figure 4.29). Sheets should be breathable, especially for summer turnout. Sheet and boot use should be started before biting midge bites become a skin irritant or a potential vector of disease-causing pathogens. Sheet use starting when temperatures are consistently over 10 °C (50 °F) should reduce midge to horse contact, but protective covers like these can be used at any point if skin irritation is observed. See Chapter 10 for more information on using fly sheets, boots, and masks.

Fans

Ceiling fans in stables are not effective at preventing biting midge presence around horses in stables, but large fans installed around small pastures or in aisles can interrupt biting midge flight and reduce bites to horses (Lincoln *et al.*, 2015). Similarly, heavy-duty box fans rated for outdoor use can be installed in stalls to create air circulation that can prevent fly landing.

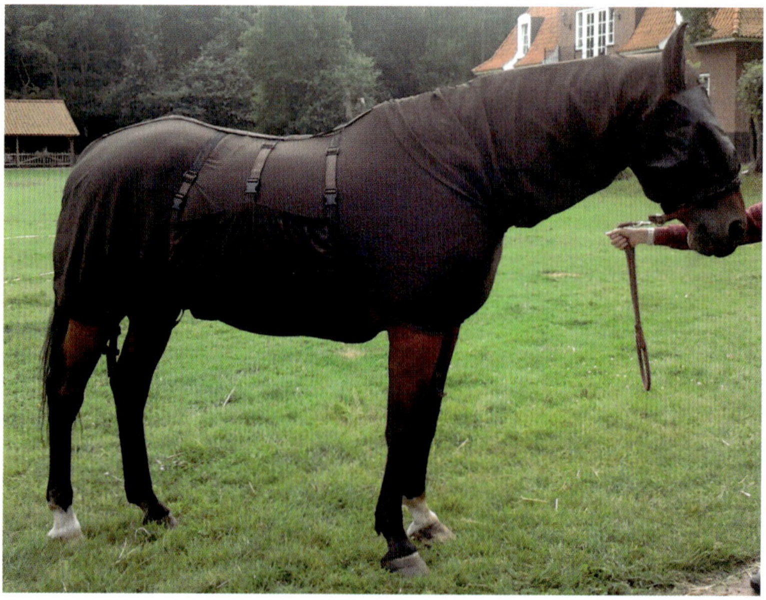

Figure 4.29. Fly sheets can mitigate the effects of biting midges on horses, but for biting midges fly sheets should be fitted to the horse with a neck cover and a form-fitting belly cover.

Chemical control

Chemical options for biting midge control are available, but few are very effective because of the different behaviours and development habitats of the various species. Broad-scale pesticide applications are unlikely to be effective for biting midges and even targeted applications of larvicides to development areas have the potential to harm other non-target aquatic animals. However, applications of pesticides labelled for use on animal facilities to areas in proximity to stables or pastures where biting midge females may rest, like fences, stable walls, or vegetation, may be useful. With many species of biting midge, each with different behaviour, pesticide treatments can not be expected to reach all individuals of all species. Repellents should be applied shortly before the time of day when midges come to horses to feed, usually just before dusk.

Topical products

Topical repellents do not appear to be effective against biting midges (Lincoln *et al.*, 2015). Applications of repellents may reduce biting midge feeding in the short term, but significant numbers of flies may still feed on horses.

Treated physical barriers

Protective mesh screens installed around stables can be treated with pesticides to improve midge blocking performance (Figure 4.30). A 0.2% synthetic pyrethroid can be applied on screens every 14 days reducing biting midges in the stables (Baker *et al.*, 2015). It is important to remember that screens, treated or untreated, can provide some protection from biting midges in stable environments but can not entirely exclude flies from entering stables. Pesticide-impregnated screening is also available from commercial sources.

Treated fly sheets, boots, and masks

As with mosquitoes, sheets, boots, and masks treated with permethrin can be used. These are described under chemical control in Section 4.1.5.

4.2.6 Veterinary concerns

Vesicular stomatitis

Pathogen transmission by biting midges to horses is minimal for both North America and Europe. In the Southwestern United States, biting midges can transmit the virus that causes vesicular stomatitis. Black flies are much more common vectors of vesicular stomatitis, but prevention of biting midge contact should be considered during an outbreak. See Section 4.4.6 for more information on vesicular stomatitis and additional resources.

Figure 4.30. Pesticide-impregnated screening can reduce biting pressure on horses in stables.

African horse sickness

The association of biting midges with the transmission of the viral pathogen causing African horse sickness is of global concern. This disease is endemic to central Africa and will be devastating if it establishes in Europe or North America. It carries a very high mortality rate, and effective treatment options do not exist currently. Diligent import procedures must be followed to keep this disease from spreading outside its range. Other viral diseases of concern are Peruvian horse sickness and equine encephalosis.

Onchocerciasis

Onchocerciasis is a form of dermatitis caused by the nematode *Onchocerca cervicalis.* These nematodes can be transmitted to horses by biting midges. When a horse is bitten by a midge carrying the immature nematodes, they migrate through the bloodstream and mature into microfilaria before moving to the skin. Inflammation in response to their presence causes fibrosis and mineralised nodules. Immature worms, or microfilariae, in the skin cause dermatitis in the ventral midline. *Onchocerca cervicalis* is extremely rare in horses who receive a macrolide deworming agent at least once yearly.

Sweet itch

Veterinary concerns with biting midges primarily involve self-trauma due to hypersensitivity to biting midge saliva, a condition called *Culicoides* hypersensitivity, Queensland itch, or sweet itch (Figure 4.31). While there is no official estimate of the prevalence of biting midge hypersensitivity, these authors would estimate it at a minimum of 10% of the population. One study performed in Israel in 2003 had a prevalence of 28% (Steinman *et al.*, 2003). Infected ears are a common issue. Horses with biting midge hypersensitivity will often cause severe wounds while attempting to itch affected areas. It is important to manage this condition early. Reducing exposure to midges as outlined in the previous section can reduce the risk of bites. Using electric fencing so horses do not have access to rubbing on fence lines can keep wounds to a minimum. Crushed flaxseed, when added to the diet, has shown some, albeit limited, benefits (O'Neill *et al.*, 2002).

Antihistamines are often used for treatment, with corticosteroids used for severe cases. These horses require frequent, extensive care to manage this hypersensitivity, and steps should be taken to avoid midge to horse contact at the start of the season. See Schaffartzik *et al.* (2012) in 'Continuing education' for more information about *Culicoides* hypersensitivity.

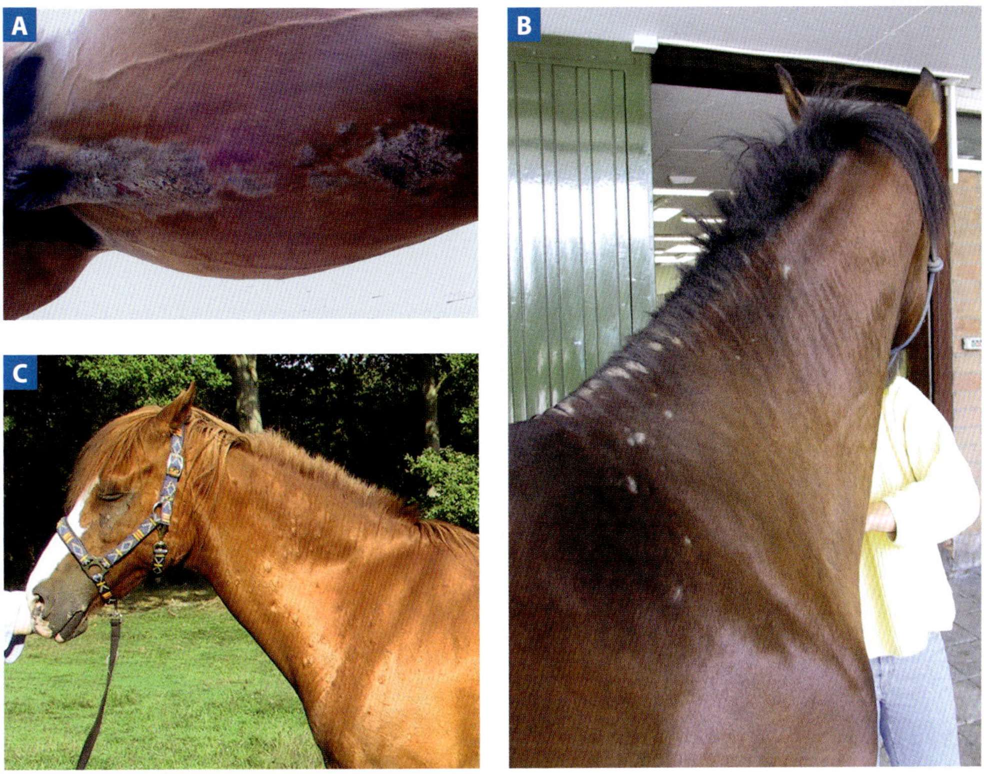

Figure 4.31. Sweet itch lesions can be found on nearly any area of the horse but are commonly are found on the belly (A), neck (B), and head (C).

4.3 Horse and deer fly/March fly (family Tabanidae)

4.3.1 Pest summary (Table 4.3)

Horse and deer flies (Figure 4.32) have many regional names including March flies (Australia), yellow flies, greenhead flies, gad flies, stouts, dun flies, moose flies, whame flies, buffalo flies, and elephant flies. These are large biting flies in the insect family Tabanidae. There are 4,300 species in the world and 335 in the USA. They are pests of horses (Figure 4.33), cattle, humans, and wildlife, as adult female flies feed on blood and can transmit pathogens, including equine infectious anaemia. The bites are painful due to their size and blade-like mouthparts.

4.3.2 Distribution and seasonality

Horse and deer flies have been reported to occur worldwide except for Hawaii, Greenland, and Iceland. Within their range, the larvae of most species are found in moist soil in marshes or bogs and close to freshwater sources. The adults are strong flyers and range over large areas to search for food, mates, resting sites, and suitable locations to lay eggs.

Table 4.3. Pest summary of horse and deer flies (family Tabanidae).

Distribution[1]	Australia: Yes Canada: Yes Europe: Yes New Zealand: Yes United States: Yes
Size	Horse flies 10-30 mm (0.4-1.2 in) in length, deer flies 6-11 mm (0.2-0.4 in) in length
Key identification characteristics	Large, multi coloured eyes, triangular body shape for deer flies, horse flies have larger, robust bodies
Mouthparts of adults	Cutting, sponging
Feed on	Blood
Development time[2]	1 year
Adult life span	30-60 days
Total eggs laid	100-1000 per batch
May be mistaken for	Hover flies
Impact on horses	Painful bites, annoyance, the transmission of the pathogen causing equine infectious anaemia

[1] The pest distribution in the regions targeted by this publication are listed as of the date published. Areas of primary pest concern (if any) are in bold. Global distributions are listed in Section 4.3.2.
[2] Development times are approximate and can change with environmental conditions including temperature, humidity, and availability of food resources.

Figure 4.32. An example of a horse fly, *Tabanus lineola* (A), and a deer fly, *Chrysops brunneus* (B).

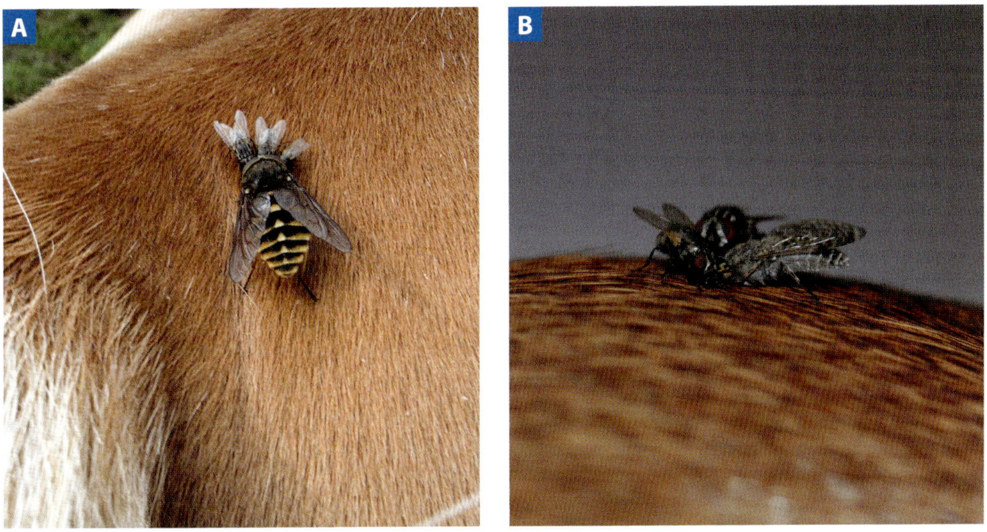

Figure 4.33. Horse flies (larger fly in photo A) are much larger than house flies (smaller flies shown in photo A). Smaller flies often gather at horse fly bite sites to drink blood from the wounds (B).

4.3.3 Description

Horse flies are large, 10 to 30 mm (0.4 to 1.2 in) in length, flies that generally belong to the genera *Tabanus* or *Hybomitra*, whereas deer flies are smaller, 6 to 11 mm (0.2 to 0.4 in) in length, flies that mostly belong to the genus *Chrysops*. Horse flies also are known as greenhead flies, clegs, March flies, and whame flies. Deer flies are often referred to as sheep flies and yellow flies. They are stout-bodied flies with large eyes that often are brightly coloured (Figure 4.34). Their large eyes allow the flies to distinguish movement and shapes that contrast against the background. Some species wait in forested edges adjacent to open areas and attack animals that move into or out of the open. Dark-coloured animals provide the flies with a clearer-contrasting visual target and often attract more flies than light-coloured horses in pastures.

Figure 4.34. The large eyes of horse and deer flies allow them to see contrasting colours and large shapes to find horses or other animal hosts for feeding.

Hover or flower flies in the family Syrphidae (Figure 4.35) sometimes can be confused for horse and deer flies, however, these flies are generally smaller, seem to hover in flight, and do not bite. Hover flies are usually found around flowering vegetation and will not land on horses.

The mouthparts of horse and deer flies are thick, pointed downward from the head, and designed for an unusual method of obtaining blood from hosts. Unlike stealthy blood-feeders such as mosquitoes, horse flies and deer flies have specialized mouthparts that first slash the skin to produce a pool of blood and then lap up the puddle with structures that function as a sponge and straw (Figure 4.36). Their attack style is aggressive, moving in quickly and attempting to slash, feed, and flee before the horse can use any defensive behaviours such as stomping, kicking, twitching, biting, and tail swishing. Only female flies feed on blood. Males primarily engage in efforts to locate females for mating, sometimes forming aggregations in areas where females are expected to visit.

Figure 4.35. Hover flies (A+B) can be mistaken for horse and deer flies, but these are generally smaller, have a distinctive hovering flight, and do not bite.

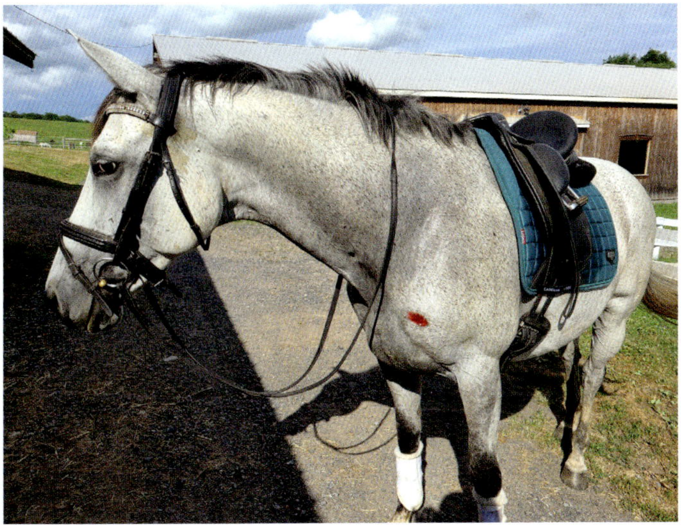

Figure 4.36. Horse and deer flies slash the skin to produce a pool of blood. It is recommended that repellent sprays are wiped on horses before they are turned out or ridden in areas with horse and deer flies.

4.3.4 Life history

Horse and deer flies have complete metamorphosis with four life stages: egg, larva, pupa, and adult. Female flies usually lay their eggs in masses of several hundred on vegetation adjacent to wet habitats such as marshes and next to streams and ponds (Figure 4.37). Larvae develop in mud in these semi-aquatic areas and prey on other insect larvae and other small invertebrates. Densities of developing horse and deer flies are generally low, usually fewer than 10 larvae per square metre or yard. Even at these low densities, populations can become high in areas with expansive habitats such as salt marshes or wooded areas that have open areas that serve as flight corridors (Figure 4.38). In cooler regions where they must survive winter conditions, they overwinter as larvae and pupae and emerge as adults the following year. Mature larvae move to drier areas in the substrate before pupating. Unlike many other insect pests, horse flies and deer flies only produce a single generation of flies each year. Each species has a typical seasonal pattern, resulting in different species being dominant during different times of the year.

4.3.5 Control

Control of horse and deer flies is challenging. Females generally spend only a few minutes feeding per day and can feed on horses, other domestic animals, or wildlife. In addition, they can lay eggs in many different aquatic sites and larval control in these areas would require large-scale environmental intervention that is not always practical. For example, if larvae occur at low densities in protected wetland habitats, there may be no environmentally sensitive way of controlling the flies before they emerge as adults. Therefore, control is usually focused on intercepting adult flight or protective measures for horses against adult bites.

Figure 4.37. Horse and deer flies lay eggs on vegetation near wet habitats like streams (A), ponds, and marshes (B) where the immature flies will develop.

Figure 4.38. Salt marshes are the source of some aggressive coastal horse fly species such as the greenhead fly in the north-eastern USA.

Cultural control

Pasture management

Management of grazing relative to horse and deer fly seasonal activity can limit the annoyance to horses. Horse and deer flies are most active around areas of pastures close to wooded areas. In general, activity decreases the further removed from pasture-wood edges. At about 50 meter (54 yards) into pastures, adult horse and deer fly activity decreases (Hansens, 1979). Horses pastured in large open fields away from wooded or marshy areas or at high elevations may have fewer bites. Increased pasture space allows horses to move naturally into areas with reduced horse and deer fly activity.

Horse and deer flies typically do not enter horse stables or other shelter areas, although this may be dependent on species and region. If they can, horses will move away from areas where they are being bitten by horse and deer flies. Adding run-in sheds or even shade areas from an overhang or canopy trees to pastures, provided they are located away from wooded edge areas, can provide a place for pastured horses to avoid horse and deer fly activity (Raymond and Rousseau, 1987) (Figure 4.39).

Turnout

Horse and deer flies are daytime feeders. If horses cannot be moved to pastures that are further from areas with high horse and deer fly populations, changing turnout times can reduce bites. Turning horses out during the night in the summer months can avoid horse and deer fly active periods.

Often, horse and deer flies are territorial in that they will not fly far out of a defined range. Bites can be avoided if horses are turned out in a different paddock further away from horse and deer fly activity during the time of the year the nuisance species is present.

Feed additives

As with other flies, there is no evidence that additives to feed, including garlic or apple cider vinegar, help to repel horse and deer flies. See Section 4.1.5 for more information on feed additives.

Physical and mechanical control

Trapping

Horse and deer flies are sight feeders, meaning they find their hosts primarily by colour and shape. Traps for horse and deer flies use the vision of these pests against them. The current commercially

Figure 4.39. Shade trees (A) and shelters (B) that are well away from wooded areas can provide refuge from biting horse and deer flies

available horse and deer fly traps use this principle to maximise attraction and trapping (Figure 4.40). Research has demonstrated differences overall in trap captures among traps, but with considerable variations depending on the location and species of flies present (Kline *et al.*, 2018; Mihok *et al.*, 2006). Where it is practical, trap catches can be increased by the addition of an odour, either carbon dioxide or 1-octen-3-ol (octenol), a chemical that is a constituent of ox breath (French and Kline, 1989).

The Nzi trap was originally developed for collecting tsetse flies in Africa but was discovered to be attractive to other biting flies, including horse and deer flies. The trap consists of blue cloth or painted wood panels mounted on the ground that channel the flies into an interior area. Once the flies are inside the trap, they climb up a cone towards the light and are collected in jars (Figure 4.40). The traps are effective but require some work to set up and are difficult to move. Instructions are available for making do-it-yourself Nzi traps out of fabric or wood (see Resources). Commercial versions are also available.

Figure 4.40. Traps like the Nzi (A), Horse Pal® (B), Epps Biting Fly Trap® (C), and BiteLite® H-trap (D) can be used to intercept adult horse and deer flies before they reach pastured horses. Except for the Epps trap, flies are collected in bottles or jars at the top that must be emptied frequently.

The Horse Pal® trap uses a large suspended black ball to represent the hind end of a large animal. This ball is covered by a canopy that encourages the natural upward flight of flies attracted to the swaying ball. Flies are trapped in a clear jar at the top of the canopy. The principle is similar for the Bite-lite® H-trap, a black ball suspended under a canopy with a trapping jar at the top of the canopy, but the final design is somewhat different. Instead of four legs and a square frame, the Bite-lite® H-trap has a single post.

The Epps Biting Fly Trap® uses large dark coloured sheets of polypropylene tarp with open areas towards the bottom that, to a fly, may appear like the side of a horse. Instead of trapping flies, the Epps Biting Fly Trap® deflects horse flies that land on the panels into trays of soapy water where they will drown. This particular trap requires regular maintenance to remove flies from the trays often, especially when populations are high, to prevent new captures from landing in the trays and using previously drowned flies as life rafts for escape.

Box traps use a conceptually simple design. A black box with an open bottom is mounted on four legs. The top of the box is covered with screening. Flies enter the underside of the box and are channelled upwards by screens or baffles to the upper part of the box, where they die. Instructions for making a box trap can be found under 'Resources' at the end of this chapter. A commercial box trap, the FlyCatch-r®, is available as well.

Trap placement is likely the most important component of operating traps for control of horse and deer flies. Traps should not be placed where they can be interfered with by horses, but ideally should be placed where the flies are a problem (e.g. small or shelter-less pastures where horses cannot escape bites, around riding arenas where horses and people are bitten). They should be located in sunny areas where they can be seen clearly from wooded and marshy areas and away from buildings or other obstructions. If initial captures are low, shifting traps even a few metres (yards) may increase trap captures because of the flight behaviour of these pests. The number of traps required depends on property size and design, but generally, one trap can cover 0.8 to 1.2 hectares (2 to 3 acres).

It is important to note that traps marketed for house flies and stable flies, such as jug traps, scatter baits, and other visual targets like Bite-Free™ sticky traps, will not be effective for horse and deer fly control.

Fly sheets, boots, and masks

Fly sheets, boots, and masks are physical barriers that deter biting from multiple species of flies. These can be especially beneficial for horses that are intolerant of fly bites. Zebra patterns on fly sheets may increase effectiveness. Research has demonstrated that fewer horse and deer flies land on zebras and horses with zebra print sheets than horses without zebra prints per unit of time (Caro *et al.*, 2019), concluding that striped surfaces prevent flies from landing successfully on zebras or horses with zebra print markings (Figure 4.41). See Chapter 10 for more information on using fly sheets, boots, and masks.

Figure 4.41. Fly sheets, boots, and masks can provide protective barriers for horses. Zebra print has been demonstrated to be particularly effective in reducing horse and deer fly bites.

Vegetation management

Horse and deer flies are ambush feeders in or near wooded areas. Reduction of trees and other vegetation has been associated with a reduction in deer fly numbers. Vegetation control around ponds, marshy sites, or other wet areas can also reduce fly numbers as it removes development areas for some species, but be sure to consult with any government regulatory institutions before modifying any bodies of water that may be protected.

Flight barriers

Adding barriers may also influence horse and deer fly flight. Roberts (1972) found that horse and deer flies would not fly over 2.4 m (8 ft) fences, supporting that this pest may fly around barriers, instead of over them. Morgan and Lee (1977) trapped higher numbers of horse and deer flies at gaps in a barrier between marshy areas and recreation areas, suggesting that barriers, like landscape trees, could be used strategically to direct horse and deer flies towards traps or away from horse areas (Figure 4.42).

Chemical control

Topical products

While there is limited information on chemical resistance in horse and deer fly species, synthetic pyrethroids appear to be effective for horse and deer flies as repellents. Other repellents, such as essential oil and fatty acid blends have not been evaluated with horse and deer flies. Horse and deer flies prefer landing on the belly, legs, and neck of horses, but will land on any part of the body so the full-body application of repellents will provide the greatest protection. Because flies are

Figure 4.42. Natural borders to wet areas, or strategically placed shrubs and landscape trees, can redirect horse and deer flies away from pastures or riding arenas and towards traps.

only on the horse for a short time, greater amounts of topical repellents applied to all areas of the horse may be necessary within the label limits. Horse and deer flies will bite people or horses, so it is important for the horse handler to also apply repellents to themselves as well, but be sure to not use repellents labelled for horses as these can cause skin reactions in humans. Repellents used against ticks and mosquitoes can be effective. Applications of product wiped on the horse's body with a microfiber mitt, towel, or brush, are the most effective, and repellents should be applied before turnout or riding in areas where horse and deer flies are a problem.

Treated fly sheets, boots, and masks

Fly sheets, boots, and masks treated with permethrin may add another layer of protection. These are described under chemical control in Section 4.1.5.

4.3.6 Veterinary concerns

Equine infectious anaemia

Horse and deer flies are the most effective vector of the virus responsible for equine infectious anaemia. Equine infectious anaemia is caused by a lentivirus, which can be present in high levels in the blood of infected horses. Horse and deer flies take large blood meals from their hosts, and, more importantly, have blood on their external mouthparts after a meal. This is exacerbated if the fly is interrupted during feeding. If this blood remains moist by the time the next horse is bitten, equine infectious anaemia can be transmitted. Horses infected with equine infectious anaemia display a biphasic disease process. Febrile episodes and anaemia accompany the initial infection. The immune system will often begin to recognize the virus and cause a decrease in viral load, leading to the cessation of symptoms. This happens repeatedly until most horses enter an inapparent carrier status. Inapparent carriers remain able to transmit infection via tabanid flies to other horses. These inapparent carriers are important reservoirs of disease as these horses can show no outward symptoms.

Tularemia

Tularemia has been reported to be transmissible by horse and deer flies in North America. However, ticks are likely a far more important vector in both North America and Europe. In general, instances of the disease are not common.

Injury and annoyance

Another veterinary concern is injury secondary to annoyance from horse and deer flies. Due to the large size, and relatively loud flight sounds of tabanids, some horses will display evasive flight behaviour (e.g. fence walking, running, bucking, etc.) when they hear a fly nearby. Horses can also experience raised welts (Figure 4.43) and, rarely, secondary bacterial infections of horse and deer fly bites. This is typically only seen in horses who are compromised either by advanced age or poor nutrition.

4.4 Black flies (genus *Simulium*)

4.4.1 Pest summary (Table 4.4)

Black flies (Figure 4.44) are biting flies in the insect family Simuliidae. There are approximately 2,200 species throughout their global distribution, although only a small percentage feed on horses or humans. About 80% of the species belong to the genus *Simulium*. They are small black or grey flies that can become an issue due to their blood-feeding and swarming behaviours. Female adult

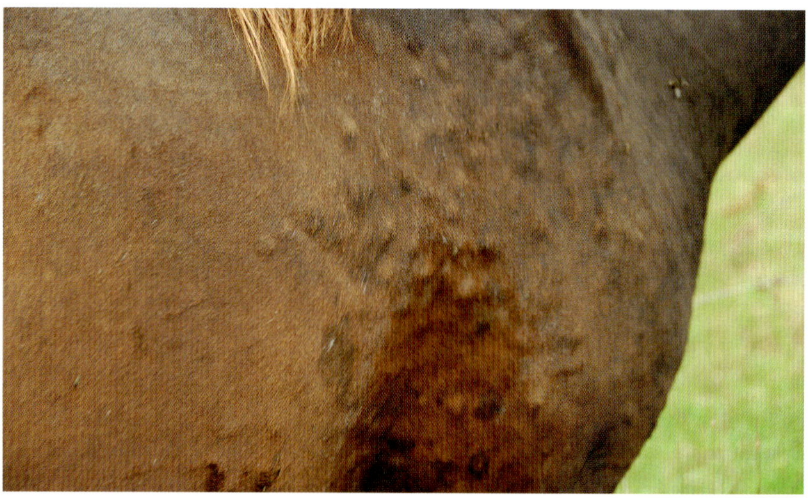

Figure 4.43. Horse and deer fly biting can lead to raised welts and, in rare cases, secondary infection.

black flies feed on animal blood for nourishment and reproduction, and the males feed mostly on nectar. As a family, they feed on a wide range of animal host species, but each fly species may prefer a specific host species. They bite outdoors during daylight hours and typically only fly in low wind conditions.

Table 4.4. Pest summary of black flies, genus *Simulium*.

Distribution[1]	Australia: Yes Canada: Yes Europe: Yes New Zealand: Yes United States: Yes
Size	1.0-5.5 mm (0.04-0.2 in) in length
Key identification characteristics	Humped back, short antennae, absence of scales or wing patterns, often found in swarms
Mouthparts of adults	Cutting, sucking
Feed on	Blood
Development time[2]	7-14 days
Adult life span	>28 days
Total eggs laid	150-500 per batch
May be mistaken for	Biting midges (Section 4.2), dung midges (Scatopsidae) (Chapter 8)
Impact on horses	Restless agitation, suffocation, exsanguination, toxaemia, vector of nematodes and pathogens causing vesicular stomatitis and auricular papillomavirus

[1] The pest distribution in the regions targeted by this publication are listed as of the date published. Areas of primary pest concern (if any) are in bold. Global distributions are listed in Section 4.4.2.

[2] Development times are approximate and can change with environmental conditions including temperature, humidity, and availability of food resources.

Figure 4.44. Black fly adult, *Simulium* sp., feeding on human skin.

4.4.2 Distribution and seasonality

Black flies have a global distribution and are found in any areas with freshwater rivers and streams even into Arctic regions, except for the Hawaiian Islands, Samoan Islands, Tristan da Cunha, and Kerguelen. They are also absent from islands that lack their necessary habitats, such as the Bahamas, Bermuda, and the Cayman Islands, and the coral islands and atolls of the Pacific and Indian Oceans.

The immature stages develop in aquatic areas with flowing water, which can vary from a temporary stream to a large river. They are not found in the still freshwater of ponds or lakes. Although they are particularly numerous in temperate wooded areas, they are not restricted to shaded or humid habitats and will often fly many miles from their aquatic development sites to locate an animal for a blood meal. Immature black flies are sensitive to environmental pollutants and can be useful indicators of water quality. Because of their requirement for clean water, an unintended consequence of reducing pollution in riverine systems can be the rebounding of black fly populations.

4.4.3 Description

Adult flies in this family can be recognized by their small size, 1.0 to 5.5 mm (0.04 to 0.20 in) in length, distinctly humpbacked appearance, short antennae, and the absence of scales or pronounced patterning on the wings (Figure 4.44). The larvae usually have a distinctive shape characterized by a somewhat bulbous expansion of the posterior end (Figure 4.45). The head of the larva is heavily sclerotized (hardened) and has fans that help direct small food particles into the mouth. Black fly larvae are found attached to the undersides of rocks and submerged vegetation in moving water.

Figure 4.45. Black fly larvae have a bulbous posterior and a well-sclerotized head.

4.4.4 Life history

Black flies have complete metamorphosis with four life stages: egg, larva, pupa, and adult. Female flies seeking blood meals often congregate in large swarms and are attracted to vulnerable body regions such as the ears, muzzle, belly, and exposed skin, particularly in areas where the hair parts. Heavy black fly harassment can make horses restless and agitated, and there have been reports of animals drowning or jumping from cliff tops to escape fly pressure. In some cases, animals have died due to suffocation from inhaling the flies, exsanguination, and toxaemia resulting from exposure to massive amounts of saliva during blood-feeding (Adler and McCreadie, 2019).

Unlike mosquitoes, which develop in still water, black fly larvae are found exclusively in moving water (Figure 4.46). Female flies will lay their eggs in batches of 200 to 300 either directly on the surface of streaming water or on objects in or near the water. Larvae produce silk that is used to anchor them to the undersides of rocks by the use of abdominal hooks (Figure 4.47). Silken threads allow the larvae to move into and out of the water stream, where they feed on particulate matter and microorganisms. Fan-like filters positioned near the mouth direct food particles into the mouth. Older larvae spin a silken cocoon attached to the rocks underwater in which they pupate. Emerging adult flies either climb or float to the surface in a bubble of air. Adult mating usually occurs near the larval development site, and female flies seek a blood meal after they have mated and fed on nectar for energy.

Figure 4.46. Unlike mosquitoes, biting midges, and horse and deer flies, black flies only develop in fast-moving streams and rivers.

Figure 4.47. Black fly larvae attach themselves with silk threads to rocks and submerged vegetation in moving water.

4.4.5 Control

Black flies are strong fliers that can fly up to 16 km (10 miles) from larval development sites to feed on animals. Control is difficult due to their small size, their development in aquatic systems where pesticide use is usually restricted, and because they often emerge as adults in large numbers at the same time.

Cultural control

Turnout

Black flies feed during the day and prefer not to enter buildings, so adjusting turnout schedules to stable horses indoors during daytime hours when black flies are a problem may help reduce bites.

Habitat management

Black flies develop in running water, so alterations to the water flow may reduce populations. However, even small-scale manipulation of streams or other bodies of water (e.g. damming) may require consultation and/or permitting with conservation organizations or other regulatory entities. Removal of leaf litter and other dead vegetation from the banks of running streams or rivers may help to reduce suitable development sites. Unlike other pest flies that develop in water, drying and managing standing water areas in pastures or containers, leaks, and other wet areas on a farm will not impact black fly development because of their requirement of moving water.

Feed additives

There is no evidence that additives to feed, including garlic or apple cider vinegar, help to repel black flies. See Section 4.1.5 for more information on feed additives.

Physical and mechanical control

Fly sheets, boots, and masks

The use of sheets, boots, and masks may help reduce black fly bites on horses, but black flies, like biting midges, are small-bodied flies that can pass through normal protective fly sheets. Fly sheets should be fitted with a hood or neck cover and a belly protector like those used for biting midges (Section 4.2.5, Figure 4.29). Although the mesh must be of small size, all protective gear should be breathable and light-coloured. Black flies use dark colours as visual indications of hosts, so white or tan sheets may provide better protection. See Chapter 10 for more information on using fly sheets, boots, and masks.

Black flies do not only bite the bodies of horses, but are often also found around the eyes, ears, and nose. Using masks with long nose guards and mesh around the ears may help prevent bites (Figure 4.48). If masks are not well tolerated by horses, application of petroleum jelly inside the ears (5 g per ear) can help reduce fly bites for up to 3 days (Figure 4.49) (Townsend and Turner, 1976).

Figure 4.48. Fly masks with long nose guards can help protect horses from black fly bites on the sensitive areas of the nose and mouth.

Figure 4.49. Petroleum jelly can be used in the ears if fly masks are not well tolerated to reduce black fly bites for up to 3 days. In this image, the left ear is treated and the right ear is not treated.

Fans

Black flies are strong fliers, but as with biting midges, large fans rated for outdoor use installed around small pastures may interrupt black fly flight and reduce bites to horses.

Trapping

Like mosquitoes, black flies are attracted to carbon dioxide. Mosquito traps that use carbon dioxide as an attractant may be effective at reducing local black fly populations if enough are used and they are maintained. The US Centers for Disease Control and Prevention miniature light trap can be used to trap black flies (Figure 4.50). The Mosquito Magnet® Executive™ trap can be used to monitor and potentially partially control black flies (López-Peña *et al.,* 2021). Attractive traps should not be placed near horses (e.g. stables, pastures) as they may have the unwanted effect of drawing black flies towards horses. However, if placed near streams or on the edges of pastures, they may be helpful for drawing flies away from horses or intercepting flight.

Biological control

Biological control of developing black flies using *Bacillus thuringiensis israelensis* (*BTi*) is the focus of most black fly control. However, not only does this method need to be community-based because of the extent of running water systems, it nearly always requires the involvement of government agencies for proper timing, dosages, application, and permitting.

Chemical control

Chemical control of adult black flies is ineffective with currently available products. Fogging using ultra-low volume pesticides may provide temporary relief. However, fogging has to be conducted during the day to ensure contact of these day-flying insects with the chemical, which is when fogging is least effective due to air currents. Humans, horses, and other animals will need to be

Figure 4.50. A US Centers for Disease Control and Prevention miniature light trap can be used to capture most small-bodied blood-feeding flies (e.g. mosquitoes, biting midges, and black flies). The trap comprises a rain guard, fan, collection chamber, and optional light or carbon dioxide. The trap in the photo has a thermos containing dry ice that produces carbon dioxide to bait the trap and attract more insects.

protected from the chemicals if fogging is used. Because of these limitations, fogging will not significantly reduce populations and may harm non-target beneficial insects. Using pesticides with residual activity to fly resting areas like fences has been suggested, but many of these pesticides have repellent effects, and black flies may move to alternative locations, like untreated areas or horses.

Treated fly sheets, boots, and masks

Fly sheets, boots, and masks treated with permethrin can help deter black flies. These are described under chemical control in Section 4.1.5.

Topical products

Repellents and other topical products for use on horses to reduce black fly bites have not been evaluated. Synthetic pyrethroid products likely provide short-term relief but will need to be reapplied daily or multiple times a day, depending on the need. There is some evidence that products containing geraniol can be effective for repelling black flies (Butler and Hogsette, 2020).

4.4.6 Veterinary concerns

Sweet itch

While biting midges (Section 4.2) are primarily responsible for causing sweet itch, black flies, and the species *Simulium equinum*, in particular, can also cause this allergic response. See Section 4.2.6 for more information on sweet itch.

River blindness

Black flies are among the most important arthropod groups with regard to pathogen transmission to horses. They are best known from a human epidemiological standpoint for their role in transmitting the nematode responsible for river blindness in Africa, *Onchocerca volvulus*. In horses, black flies may play a role in the transmission of the related filarial worm *Onchocerca cervicalis* which can also be transmitted by biting midges (see Section 4.2.6).

Vesicular stomatitis

Black flies are the insect vector for vesicular stomatitis. Vesicular stomatitis causes blistering at the junction of mucous membranes and skin (see Pelzel-McCluskey, 2016 in 'Continuing education' for more information about vesicular stomatitis). These blisters can be painful and will cause horses to decrease food and water consumption. The virus itself is not life-threatening but decreased food and water consumption result in malnutrition and dehydration. Vesicular stomatitis is indistinguishable from foot and mouth disease, and as such is cause for close regulatory monitoring. There are periodic outbreaks of the virus in North America, particularly in the desert Southwest. During these outbreaks, close monitoring of horses for symptoms and good fly control is paramount to management.

Auricular papillomas

Auricular papillomas (Figure 4.51) manifest as white plaques in the ears of affected horses. These can be caused by at least four different papillomaviruses that are transmitted by black flies. For some horses, they can lead to evasive behaviour issues when placing the bridle or halter. There are treatment options available, but they can be difficult to administer due to the painful nature of the lesions especially due to enhanced sensitivity during the treatment process (see Torres, 2013 in 'Continuing education' for more information about aural plaques). Treatment with imiquimod several times a week every other week is recommended, as are frequent applications of repellents and other cultural control methods (like stabling) to reduce contact with black flies and reduce discomfort.

Injury and suffocation

Large swarms of black flies have been observed to suffocate animals by flying into the nostrils. Horses also will run with little awareness of their surroundings when confronted by a swarm.

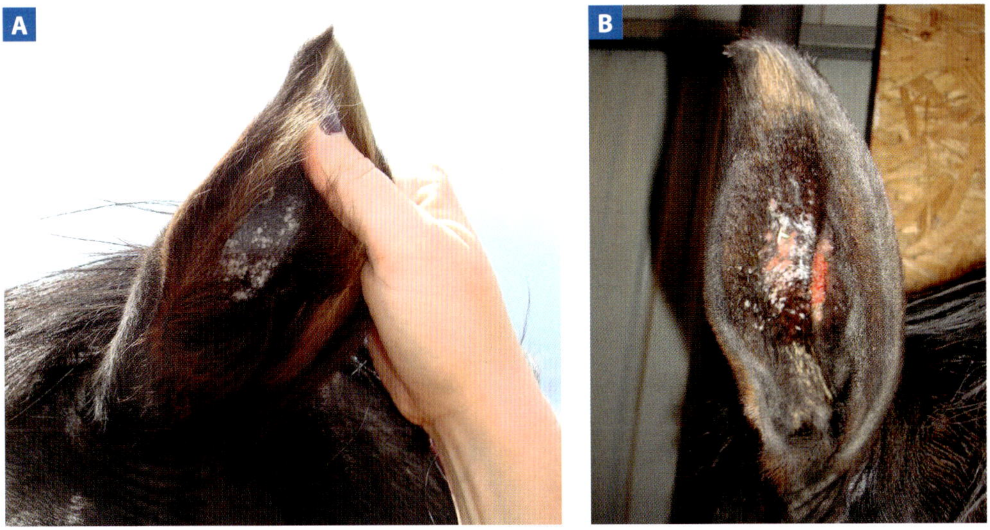

Figure 4.51. Auricular papillomas, or aural plaques, can be caused by papillomaviruses that can be transmitted by black flies. The plaques appear as white spots in the ear and can be crusty and painful (A+B).

This can lead to injuries when objects, such as fence lines, are encountered. Toxins in the saliva can lead to death from cardiovascular collapse. In rare instances, blood loss due to large numbers of flies feeding per horse can lead to death.

Continuing education

The challenges posed by equine arboviruses

Chapman, G.E., Baylis, M., Archer, D. and Daly, J.M., 2018.
https://doi.org/10.1111/evj.12829

- Arboviruses are viruses transmitted by biting arthropods including mosquitoes, black flies, and other aquatic flies. These are becoming an increasing risk for horses worldwide. This review discusses important arboviruses affecting horses, the challenges associated with surveillance, prevention, and control of these viruses and the problem of emerging viruses in new regions of the world.

Equine insect bite hypersensitivity: What do we know?

Schaffartzik, A., Hamza, E., Janda, J., Crameri, R., Marti, E. and Rhyner, C., 2012.
https://doi.org/10.1016/j.vetimm.2012.03.017

- Insect bite hypersensitivity is an allergic reaction of the skin of the horse caused by bites of *Culicoides*. This review evaluates the knowledge of insect bite hypersensitivity, including the mechanisms and treatment, and focuses on the allergens.

Advances in control techniques for *Culicoides* and future prospects

Harrup, L.E., Miranda, M.A. and Carpenter, S., 2016.
https://doi.org/10.12834/VetIt.741.3602.3

- In this review, the available options for *Culicoides* control techniques and their potential in reducing pathogen transmission are discussed. Future prospects and challenges facing *Culicoides* control are examined.

Vesicular stomatitis in horses

Pelzel-McCluskey, A., 2016.
https://aaep.org/horsehealth/vesicular-stomatitis-horses

- Vesicular stomatitis is a contagious disease that afflicts horses, livestock, wildlife, and even humans. The disease is caused by a virus that can be transmitted by insects. When contracted, blister-like lesions can usually be seen on the tongue, mouth, nose, or lips. In this article, diagnosis, treatment, and recovery of horses affected by vesticular stomatitis are discussed.

Equine aural plaques

Torres, S.M.F., 2013.
https://www.merckvetmanual.com/ear-disorders/diseases-of-the-pinna/equine-aural-plaques

- Equine aural plaques are caused by papillomavirus. This virus can be transmitted by black flies. In this article, the diagnosis and treatment of equine aural plaques are discussed.

Resources

Resources are listed in alphabetical order and may not be inclusive of all companies, materials, or products available.

CDC Miniature Light Trap. https://www.johnwhock.com/ or https://www.bioquip.com/
Biogents Mosquito Trap. www.biogents.com
BiteLite® H-Trap Professional Horse Fly Control System. https://www.h-trap.com/
Boett® sweet itch blanket. http://www.itchyhorse.co.uk/
Box trap instructions for horse flies. http://www.holgatetaxpayers.org/docs/greenhead-howto.pdf
Bucas sweet itch rug. https://bucas.com/ Busse Fly Rug Complete Plus. https://www.busse-reitsport.de
Epps Biting Fly Trap®. https://horselineproducts.com/
Felix Bühler sweet itch rug. http://www.kramer.co.uk
Flycatch-r® horse and deer fly trap. https://flycatch-r.com/
Horse Pal® Trap. http://www.bitingflies.com/
Insect Shield® Insect Repellent Fabric Treatment. https://www.insectshield.com/
Mosquito Magnet®. https://www.mosquitomagnet.com/
Nzi traps (commercial). https://www.indiamart.com/proddetail/nzi-trap-20920161930.html; https://www.rinconvitova.com/fly%20trap%20nzi.htm.
Nzi trap (fabric) instructions. https://www.tsetse.org/bait-technologies/how-to-assemble-an-nzi-trap
Nzi traps (wooden) instructions. https://projects.sare.org/wp-content/uploads/Nzi-Trap-Plywood-Web-Instructions-PDF.pdf

Ovi-Catch™. https://catchmasterpro.com/product/ovi-catch-ago-mosquito-trap/
SnuggyHoods. https://snuggyhoods.com/

References

Adler, P.H. and McCreadie, J.W., 2019. Black flies (Simuliidae). In: Mullen, G.R. and Durden, L.A. (eds.) Medical and veterinary entomology, 3rd ed. Academic Press, Cambridge, MA, USA, pp. 237-259.

Armstrong, P.M. and Andreadis, T.G., 2010. Eastern equine encephalitis virus in mosquitoes and their role as bridge vectors. Emerging Infectious Diseases 16: 1869-1874. https://doi.org/10.3201/eid1612.100640.

Augot, D., Mathieu, B., Hadj-Henni, L., Barriel, V., Mena, S.Z., Smolis, S., Slama, D., Randrianambinintsoa, F.J., Trueba, G., Kaltenbach, M. and Rahola, N., 2017. Molecular phylogeny of 42 species of *Culicoides* (Diptera, Ceratopogonidae) from three continents. Parasite 24: 23. https://doi.org/10.1051/parasite/2017020

Baker, T., Carpenter, S., Gubbins, S., Newton, R., Lo Lacono, G., Wood, J. and Harrup, L.E., 2015. Can insecticide-treated netting provide protection for equids from *Culicoides* biting midges in the United Kingdom? Parasites & Vectors 8: 604. https://doi.org/10.1186/s13071-015-1182-x

Barrera, R., Amador, M., Acevedo, V., Caban, B., Felix, G. and Mackay, L.J., 2014. Use of the CDC autocidal gravid ovitrap to control and prevent outbreaks of *Aedes aegypti* (Diptera: Culicidae). Journal of Medical Entomology 51: 145-154.

Butler, J.F. and Hogsette, J.A., 2020. Black flies, *Simulium* spp. (Insecta: Diptera: Simuliidae). UF/IFAS Featured Creatures EENY-30. Entomology and Nematology Department, Gainesville, FL, USA. https://entnemdept.ufl.edu/creatures/livestock/bfly.htm

Caro, T., Argueta, Y., Briolat, E.S., Bruggink, J., Kasprowsky, M., Lake, J., Mitchell, M.J., Richardson, S. and How, M., 2019. Benefits of zebra stripes: behaviour of tabanid flies around zebras and horses. PLoS One 14: e0210831. https://doi.org/10.1371/journal.pone.0210831

ECDC (European Center for Disease Prevention and Control). 2017. Epidemiological update: West Nile virus transmission season in Europe, 2017. https://www.ecdc.europa.eu/en/news-events/epidemiological-update-west-nile-virus-transmission-season-europe-2017

ECDC (European Center for Disease Prevention and Control). 2018. Epidemiological update: West Nile virus transmission season in Europe, 2018. https://www.ecdc.europa.eu/en/news-events/epidemiological-update-west-nile-virus-transmission-season-europe-2018

ECDC (European Center for Disease Prevention and Control). 2019. Epidemiological update: West Nile virus transmission season in Europe, 2019. https://www.ecdc.europa.eu/en/news-events/epidemiological-update-west-nile-virus-transmission-season-europe-2019

ECDC (European Center for Disease Prevention and Control). 2021. Epidemiological update: West Nile virus transmission season in Europe, 2020. https://www.ecdc.europa.eu/en/news-events/epidemiological-update-west-nile-virus-transmission-season-europe-2020.

French, F.E. and Kline, D.L., 1989. l-Octen-3-ol, an effective attractant for Tabanidae (Diptera). Journal of Medical Entomology, 26: 459-461.

Hansens, E.J., 1979. Review: Tabanidae of the east coast as an economic problem. Journal of the New York Entomological Society 87: 312-318.

Kline, D.L., Hogsette, J.A. and Rutz, D.A., 2018. A comparison of the Nzi, Horse Pal® and Bite-Lite® H-traps and selected baits for the collection of adult Tabanidae in Florida and North Carolina. Journal of Vector Ecology 43: 63-70.

Lincoln, V.J., Page, P.C., Koppa, C., Mathis, A., von Niederhäusern, R., Burger, D. and Herholz, C., 2015. Protection of horses against *Culicoides* biting midges in different housing systems in Switzerland. Veterinary Parasitology 210: 206-214.

López-Peña, D., Hawkes, F.M., Gibson G.I., Johnston, C., Vaux, A.G.C., Lis-Cantín, Á., Medlock, J.M. and Cheke, R.A., 2021. Mosquito Magnet® traps as a potential means of monitoring blackflies of medical and veterinary importance. Medical and Veterinary Entomology 35: 646-651. https://doi.org/10.1111/mve.12530

Lühken, R., Pfitzner, W.P., Börstler, J., Garms, R., Huber, K., Schork, N., Steinke, S., Kiel, E., Becker, N., Tannnich, E., and Krüger, A., 2014. Field evaluation of four widely used mosquito traps in Central Europe. Parasites & Vectors 7: 268. https://doi.org/10.1186/1756-3305-7-268

Meiswinkel, R., Baylis, M. and Labuschagne K., 2000. Stabling and the protection of horses from *Culicoides bolitinos* (Diptera: Ceratopogonidae), a recently identified vector of African horse sickness. Bulletin of Entomological Research 90: 509-515.

Mihok, S., Carlson, D.A., Krafsur, E.S. and Foil, L.D., 2006. Performance of the Nzi and other traps for biting flies in North America. Bulletin of Entomological Research 96: 387-397.

Morgan, N.O. and Lee, R.P., 1977. Vegetative barriers influence flight direction of saltmarsh greenheads. Mosquito News 37: 263-267.

O'Neill, W., McKee, S. and Clarke, A.F., 2002. Flaxseed (*Linum usitatissimum*) supplementation associated with reduced skin test lesional area in horses with *Culicoides* hypersensitivity. The Canadian Journal of Veterinary Research 6: 272-277.

Pearson, W., Boermans, H.J., Bettger, W.J., McBride, B.W. and Lindinger, M.I., 2005. Association of maximum voluntary dietary intake of freeze-dried garlic with Heinz body anemia in horses. American Journal of Veterinary Research 66: 457-465.

Porter, J.F., 1959. Some effects of screens in retarding entry of the common salt marsh sandfly *Culicoides furens* (Poey) (Diptera: Heleidae). Mosquito News 19: 159-163.

Raymond, H.L. and Rousseau, F., 1987. Abundance of horse flies (Diptera: Tabanidae) and reactions of cattle in a traditional farm of French Guiana. Acta Oecologica 8: 125-134.

Reeves, W.K. and Miller, M.M., 2010. Aqueous 2% geraniol as a mosquito repellent failed against *Aedes aegypti* on ponies. Journal of the American Mosquito Control Association 26: 340-341.

Roberts, R.H., 1972. The effectiveness of several types of Malaise traps for the collection of Tabanidae and Culicidae. Mosquito News 32: 542-547.

Townsend, L.H. Jr. and Turner, E.C. Jr., 1976. Field evaluation of several chemicals against ear-feeding black fly (*Simulium vittatum*) pests of horses in Virginia. Mosquito News 36: 182-186.

Steinman, A., Peer, G. and Klement, E., 2003. Epidemiological study of *Culicoides* hypersensitivity in horses in Israel. Veterinary Record 152: 748-751.

Photo credits

Titel page	Erika Machtinger, Pennsylvania State University.
Figure 4.1.	CDC/Dr Gary Alpert – Urban Pests – Integrated Pest Management (IPM), Public domain https://phil.cdc.gov/Details.aspx?pid=15992.
Figure 4.2.	Lary Reeves, Florida Medical Entomology Laboratory, University of Florida.
Figure 4.3.	Judy Gallagher, Flickr, license CC BY 2.0 https://www.flickr.com/photos/52450054@N04/49247115743/in/photolist-2i2NkLt-fKDRae.

Figure 4.4. Rob Mitchell, Flickr, public domain, https://www.flickr.com/photos/swallowedtail/18216934109/in/photostream/.

Figure 4.5. (A) E. Dronkert, Flickr CC BY 2.0 https://www.flickr.com/photos/dnet/5373280458/in/photolist-9bPt5w; (B) USDA NRCS Montana Public Domain https://www.flickr.com/photos/160831427@N06/25003792548; (C) Erika Machtinger, Pennsylvania State University.

Figure 4.6. United States Fish and Wildlife Service, Public Domain https://www.flickr.com/photos/160831427@N06/25003792548.

Figure 4.7. Seth Irish, Centers for Disease Control and Prevention.

Figure 4.8. Louisiana State University Agricultural Center.

Figure 4.9. Mick Talbot, Flickr, license CC BY 2.0 https://www.flickr.com/photos/micks-wildlife-macros/3470380619.

Figure 4.10. Jim Newman, Florida Medical Entomology Laboratory, University of Florida.

Figure 4.11. Erika Machtinger, Pennsylvania State University.

Figure 4.12. Erika Machtinger, Pennsylvania State University.

Figure 4.13. Erika Machtinger, Pennsylvania State University.

Figure 4.14. Erika Machtinger, Pennsylvania State University.

Figure 4.15. Erika Machtinger, Pennsylvania State University.

Figure 4.16. Erika Machtinger, Pennsylvania State University.

Figure 4.17. Erika Machtinger, Pennsylvania State University.

Figure 4.18. Erika Machtinger, Pennsylvania State University.

Figure 4.19. Dan Kline, USDA, ARS.

Figure 4.20. Jeffesp0, Flickr, CC BY 2.0 https://www.flickr.com/photos/36676878@N02/4449854125.

Figure 4.21. Erika Machtinger, Pennsylvania State University.

Figure 4.22. (A) Susan Friedland-Smith; (B) Erika Machtinger, Pennsylvania State University.

Figure 4.23. Lary Reeves, Florida Medical Entomology Laboratory.

Figure 4.24. Augot *et al.* (2017), license CC BY 4.O.

Figure 4.25. Dinesh Erram, Florida Medical Entomology Laboratory.

Figure 4.26. Thomas H Photo, Flickr CC BY 2.0-ND https://www.flickr.com/photos/15574096@N00/17361770661/in/photolist-sscBxt.

Figure 4.27. Lyle Buss, University of Florida.

Figure 4.28. E. Dronkert, Flickr, license CC BY 2.0, https://www.flickr.com/photos/dnet/5373280458/in/photolist-9bPt5w-9xYmJ2.

Figure 4.29. Courtesy Professor Marianne Sloet, Utrecht University.

Figure 4.30. Jerry Hogsette, USDA, ARS.

Figure 4.31. Courtesy Professor Marianne Sloet, Utrecht University.

Figure 4.32. (A) Judy Gallagher, Wikimedia Commons, license CC-BY-2.0 https://www.flickr.com/photos/52450054@N04/8726062826/in/photolist-ei6mt5; (B) https://www.flickr.com/photos/52450054@N04/18624478370/in/photolist-unMjy3.

Figure 4.33. (A) Andy Nowack, Dreamstime no. 53433751; (B) Andrei Mihalca, University of Agricultural Sciences and Veterinary Medicine.

Figure 4.34. United States Geological Survey Bee Inventory and Monitoring Lab, Flickr, public domain.

Figure 4.35. (A) John Tann, Flickr, license CC BY-2.0 https://www.flickr.com/photos/31031835@N08/5190703095; (B) https://www.flickr.com/photos/31031835@N08/5693864911.

Figure 4.36. Erika Machtinger, Pennsylvania State University.

Figure 4.37. (A) Elliott Brown CC BY 2.0 https://flickr.com/photos/ell-r-brown/5791706618/in/album-72157626746543853/; (B), Shiloh Gibson, Shutterstock No.1680684334.

Figure 4.38. Ken Schulze, Shutterstock no. 1021190296.

Figure 4.39. (A) Hans Klamm, Dreamstime no. 814506; (B) Erika Machtinger, Pennsylvania State University.

Figure 4.40. (A) Dave Carlson; (B,C) Wes Watson, North Carolina State University; (D) John-Fs-Pic, Shutterstock.

Figure 4.41. Erika Machtinger, Pennsylvania State University.

Figure 4.42. Erika Machtinger, Pennsylvania State University.

Figure 4.43. Andrei Mihalca, University of Agricultural Sciences and Veterinary Medicine.

Figure 4.44. Alice Abela, Flickr, all rights reserved.

Figure 4.45. Tom Murray, Massachusetts.

Figure 4.46. Jaminnbenji, Shutterstock No. 299813333.

Figure 4.47. Glacier National Park Service, Flickr, Public domain, https://www.flickr.com/photos/glaciernps/42501180000.

Figure 4.48. Virginia Hill, Flickr CC BY 2.0 ND https://www.flickr.com/photos/valeehill/6092852713/in/photolist-ahpsQH-ahseEb-ahpsyi-ahsdUA-ahpsXr-ahseoh-ahsdzy-ahpt8v-ahsdP5-ahsevU.

Figure 4.49. Erika Machtinger, Pennsylvania State University.

Figure 4.50. Emma Weeks, University of Florida.

Figure 4.51. Heather Maples.

Chapter 5. Pests that develop and live on horses (ectoparasites)

Chapter overview

This chapter covers arthropods that are ectoparasitic, meaning that they mainly live in or on the skin or hair of animals. These include eight-legged arachnids like ticks (Section 5.1) and mites (Section 5.2), six-legged insects such as lice (Section 5.3), and keds (or louse flies) (Section 5.4). These arthropods feed on blood in the case of ticks, sucking lice, and keds, and skin cells and other tissues in the case of mites and chewing lice. Mites vary from those that infest the follicles and skin of horses throughout their life cycle such as the horse follicle mite and equine body mite to those that feed opportunistically on horses such as the straw itch mite and chiggers. Two lice species affect equines throughout the world, the horse blood-sucking louse and the horse biting louse. Both can be highly irritating to the horse, causing intense itching that can lead to injury. The New Forest fly is the ked that is most commonly found on horses throughout Europe, where it causes irritation due to its itchy bites and relentless feeding. The deer ked is the most common ked found on horses in the USA.

Mites, lice, and keds generally only cause mild irritation associated with reactions to feeding that in many cases are self-limiting. In contrast, ticks can transmit pathogens that cause severe diseases such as the bacteria that cause anaplasmosis, Lyme borreliosis, and tularemia, the parasite that causes equine piroplasmosis, and the virus that causes louping ill. Control of ticks is typically limited to protecting the horse from bites, removing ticks that attach, and cultural techniques to reduce tick numbers in and around pastures. Mites and lice need careful management. Quarantine of infested animals and sanitation are key to prevent infestation of previously uninfested horses. Chiggers, straw itch mites, and keds are more challenging to control because horses are accidental hosts (animals which are not typical hosts of a parasite).

5.1 Ticks (families Ixodidae and Argasidae)

5.1.1 Pest summary (Table 5.1)

Ticks are small crawling (wingless) arthropods in the arachnid order Ixodida and are related to spiders and mites. The two main families of ticks are the hard ticks (Ixodidae) and the soft ticks (Argasidae). Globally, there are about 700 species of hard ticks and 200 species of soft ticks. Not all ticks are pests of horses, but those that are will attach to a horse and feed on blood for several days. Attachment is irritating especially when there are large numbers of ticks (Figure 5.1). The fluid transfer between the tick and horse after the tick has bitten can result in the transmission of pathogens that can cause disease. The wounds left from tick attachment also can become infected with other secondary pathogens or parasites.

5.1.2 Distribution and seasonality

Ticks are widely distributed throughout the world and are particularly common in warm humid climates. The risk of disease caused by pathogens transmitted by ticks depends on the species. To evaluate the regionality and distribution for common species of ticks, see detailed maps provided by the U.S. Centers for Disease Control and Prevention (2021), European Centre for Disease Control and Prevention (2021), Northern Territory Government (2021 , Australia), and the Australian Museum (2020) in 'Continuing education'.

Ticks are typically more active during warm and humid weather. The major limiting factors for whether ticks are present are climate conditions and the abundance of hosts on which they feed. The distribution of a tick species typically follows closely the distribution of the preferred hosts of the particular tick species. It is important to be aware of the tick species that are regionally abundant to understand the risks associated with tick bites. In many cases, there have been regional shifts in tick populations with species becoming more common in areas where they were previously rare and some species invading news areas they had not previously been reported. There is strong evidence that some of these population shifts are due to climate change and further changes in the distribution of many tick species may continue as the climate warms (Dumic and Severnini, 2018; Ostfeld and Brunner, 2015).

Table 5.1. Pest summary of ticks, families Ixodidae and Argasidae.

Distribution[1]	Australia: Yes Canada: Yes Europe: Yes New Zealand: Yes United States: Yes
Size	Larvae are not typically found on horses, nymphs 1-2 mm (0.04-0.08 in), adults 3-5 mm (0.12-0.20 in)
Key identification characteristics	Flattened, round shape, eight legs, no wings or antennae, can be grape-like when engorged (filled) with blood
Mouthparts	Piercing, sucking
Feed on	Blood
Development time[2]	Varies from several months to 2 years or more depending on the species
Adult life span	Variable, some species and life stages can live over a year without a host, so the total life span can be 2-4 years
Total eggs laid	Variable by species, generally several thousand but may be tens of thousands
May be mistaken for	Keds/louse flies (Section 5.4)
Impact on horses	Irritation, secondary infection at the site of the bite, Lyme borreliosis, anaplasmosis, equine piroplasmosis, tularemia, tick paralysis, louping ill virus

[1] The pest distribution in the regions targeted by this publication are listed as of the date published. Areas of primary pest concern (if any) are in bold. Global distributions are listed in Section 5.1.2.
[2] Development times are approximate and can change with environmental conditions including temperature, humidity, and availability of food resources.

Many ticks are found at the highest densities in areas that maintain high humidity like leaf litter, brushy vegetation, and grasses, but some species can also be found in pastures and the edge of trails.

5.1.3 Description

Hard ticks have a hard shield (scutum) on their backs and their mouthparts are visible from above protruding from the top of the shield. The patterns on their shields often help to distinguish the species. Soft ticks do not have a hard shield, have a wrinkled leathery appearance, and their mouthparts are positioned underneath and not visible from above. Ticks are not insects, they are arachnids like spiders, so the nymphs and adult stages have eight legs. The very small larvae have six legs, but this stage is not usually found on horses. It is very unlikely to find tick eggs as they are typically laid in protected areas such as soil or leaf litter. Eggs are small, usually brown and laid as a large mass of thousands of eggs (Figure 5.2).

Figure 5.1. Tick (*Anocenter nitens*) attaching to horse ear.

Figure 5.2. A female lone star tick with egg mass. Eggs are usually laid in the undergrowth away from view. A typical egg mass comprises many thousands of eggs.

The size of the tick depends on the species, but larvae are approximately 1 mm (0.04 in) and the nymphs are approximately 1 to 2 mm (0.04 to 0.08 in), and both look like small adults in shape although the final adult colour patterns may not be obvious (Figure 5.3). Adults have round, oval, or pear-shaped bodies and are typically about 3 to 5 mm (0.12 to 0.20 in) in length. When they feed on blood they swell over a few days and the females and immature stages, in particular, become much larger and almost grape-like in shape (Figure 5.4).

Many tick species can be found on horses causing annoyance, skin conditions, and in some cases transmitting pathogens that cause disease. The major tick pests of horses are in the genera *Ixodes, Dermacentor, Rhipicephalus, Hyalomma, Amblyomma, Haemaphysalis,* and *Otobius* (see Wall, 2016 in 'Continuing education' for assistance with tick identification). A few of the most common or relevant ticks found on horses are highlighted below.

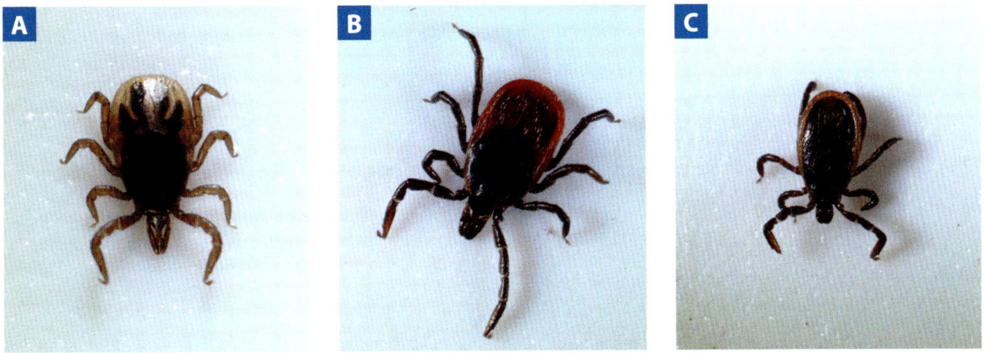

Figure 5.3. Hard ticks in the genus *Ixodes* transmit the pathogens that cause Lyme borreliosis and equine anaplasmosis. Pictured is a nymph (A), adult female (B), and male (C) *Ixodes scapularis*, the vector of these pathogens in the USA. Note the long mouthparts which can help distinguish this species from other ticks.

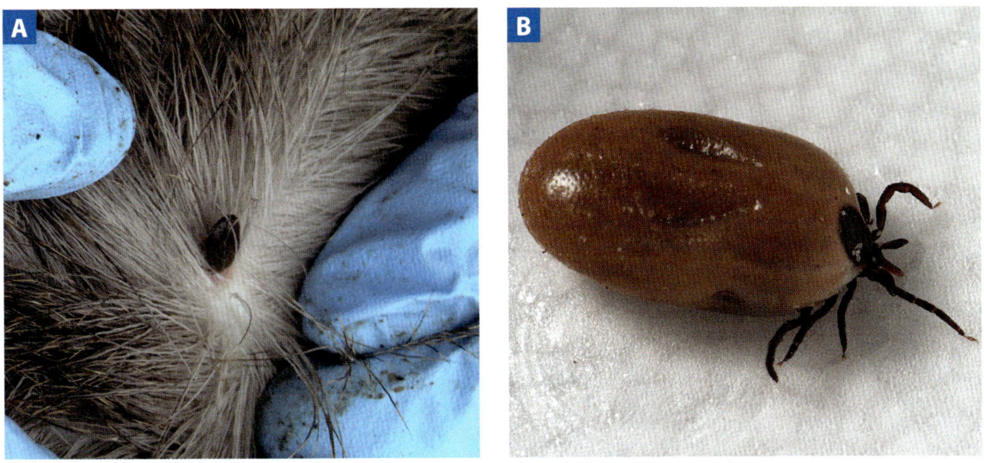

Figure 5.4. When ticks feed on blood they swell over a period of days and become much larger like the examples pictured (A+B).

Ticks of the genus *Ixodes* tend to be more oval than round and have long mouthparts. Hard ticks in the genus *Ixodes* transmit the pathogens that cause Lyme borreliosis and equine anaplasmosis (Figure 5.3 and 5.5). The pathogens are acquired by the larvae feeding on rodents and are transmitted to larger mammals such as horses, dogs, and humans by the nymphs and adults. In general, these species are common in the spring in temperate areas. The blacklegged tick, *Ixodes scapularis,* is the vector of these pathogens in the eastern and central USA (Figure 5.3), *I. pacificus* in the western USA, and in Europe the sheep/castor bean/deer tick *Ixodes ricinus* is responsible. *Ixodes ricinus* also transmits the virus that causes louping ill in the UK and *Ixodes holocyclus* causes tick paralysis in Australia (Figure 5.5).

The bacteria that causes tularemia is transmitted by the American dog tick, *Dermacentor variabilis* (Figure 5.6), which also can cause tick paralysis in the USA. Tularemia-causing bacteria also can be vectored by *Dermacentor andersoni*, the Rocky Mountain wood tick (Figure 5.6), and the lone star tick, *Amblyomma americanum* (Figure 5.7). *Amblyomma* have eyes and long mouthparts, and *Amblyomma americanum* females have a characteristic white spot on their shield (Figure 5.7). *Dermacentor* species are typically large ticks with eyes, mouthparts wider than they are long and decorative white or cream patterns on their shield.

Equine piroplasmosis or equine babesiosis can be transmitted by 33 species in six tick genera. While species in the *Dermacentor, Rhipicephalus,* and *Hyalomma* genera are confirmed vectors, transmission also has been demonstrated by *Ixodes, Amblyomma* (Figure 5.7), and *Haemaphysalis* species.

Haemaphysalis longicornis (Figure 5.8) is called the Asian longhorned tick, New Zealand cattle tick, or bush tick. It is native to Asia and is a pest of hooved animals, primarily cattle. However, this tick is a recent invader to other areas of the world including Oceania and North America and has been found on horses. Tropical species of *Hyalomma marginatum* complex (*H. rufipes* and *H. marginatum*) (Figure 5.9) have been recently found on horses in mainland Europe (Chitimia-Doblera *et al.,* 2016) and in the UK. (Hansford *et al.,* 2019), suggesting these ticks may have become established in some areas. These shifts in tick populations are likely related to climate change and are not limited to *Hyalomma* species (Fernández-Ruiz *et al.,* 2021).

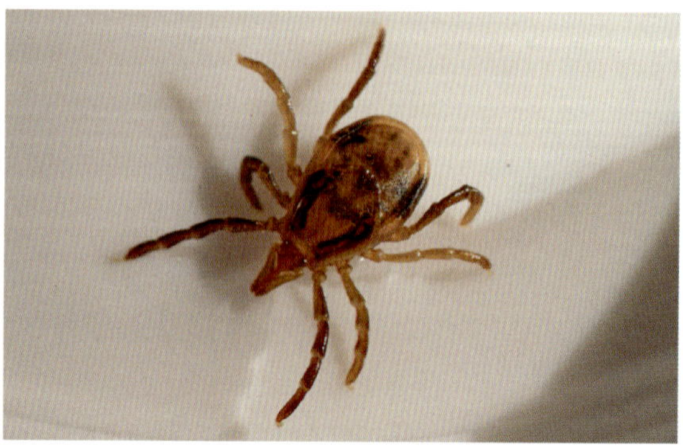

Figure 5.5. Adult female of the Australian paralysis tick, *Ixodes holocyclus*.

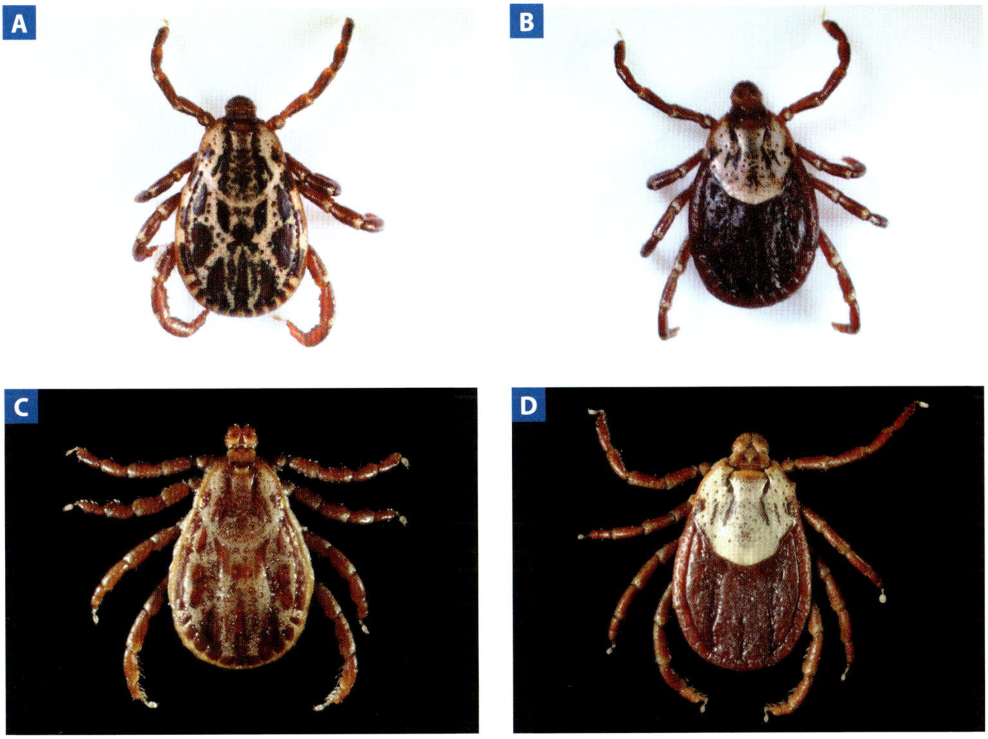

Figure 5.6. *Dermacentor* species transmit the pathogens that cause tularemia in the USA. *Dermacentor variabilis* bites were also the cause of the few isolated cases of tick paralysis in the USA. Pictured are *Dermacentor variabilis* male (A), *Dermacentor variabilis* female (B), *Dermacentor andersoni male* (C) and *Dermacentor andersoni female* (D). Note the colourful shield (scutum) in both sexes of the two species.

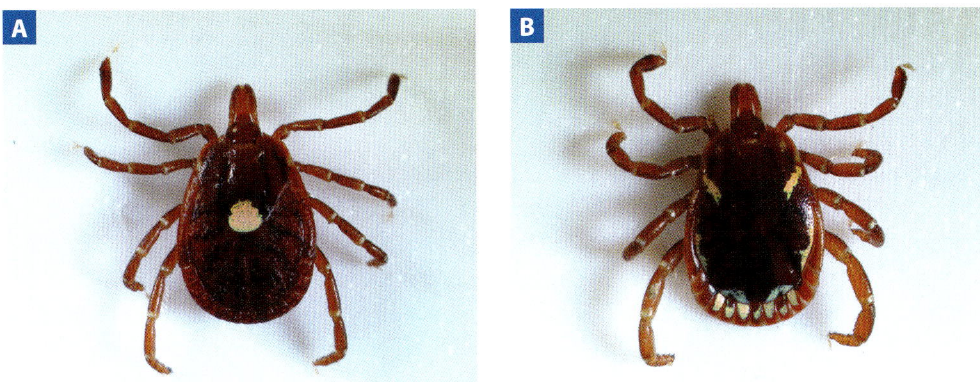

Figure 5.7. The female lone star tick, *Amblyomma americanum*, has a distinctive white dot on the shield (A) but the male does not (B).

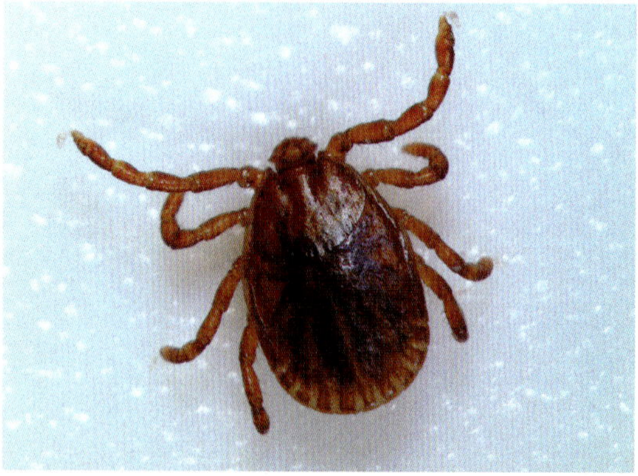

Figure 5.8. The Asian longhorned tick, *Haemaphysalis longicornis* (adult female pictured), originated in Asia but is a recent invader to other areas of the world including Oceania and North America.

Figure 5.9. Tropical *Hyalomma* species have been found on horses in Europe where they were previously not established.

The larvae and nymphs of one soft tick species, the spinose ear tick (*Otobius megnini*) (Figure 5.10), attach inside the ears of their equine and other hosts. This tick has been documented in western Canada (British Columbia) throughout the USA into Mexico, Central and South America, Africa, and India. They are particularly common in dry areas of the western USA and Canada. Adults are not parasitic and live off the host and do not feed. When off the host, the nymphs and adults are typically found in the environment in cracks and crevices, under logs, in fence posts, and similar areas. The immature stages are greyish to red-brown and they have a spiny exoskeleton. This tick is not known to transmit pathogens causing disease, but bites can be very painful.

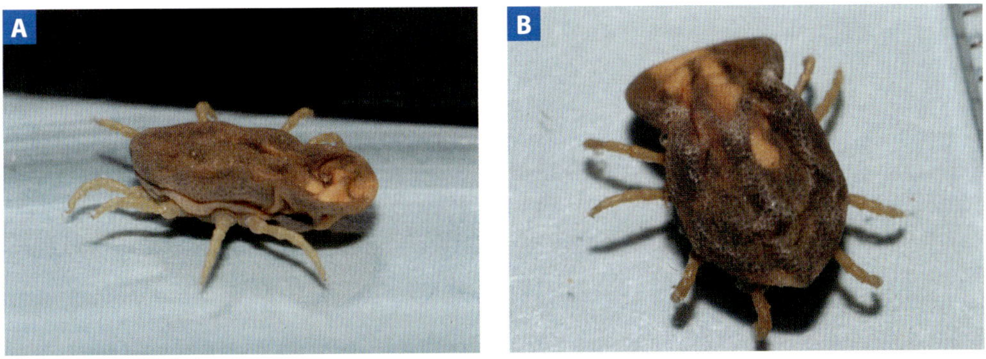

Figure 5.10. The spinose ear tick (*Otobius megnini*, A+B) is a soft tick species that attaches inside the ears of horses causing them to shake their heads.

5.1.4 Life history

Hard and soft ticks have four life stages: egg, larva, nymph, and adult (Figure 5.11). Hard tick species are often referred to as being either one-, two-, or three-host ticks, which refers to the number of hosts from which the tick takes a blood meal and drops off. For example, in a three-host tick, the larva hatches from the egg, locates a host, feeds, then drops off the host to digest its blood meal and moult. The resulting nymph then locates a new host and feeds. The nymph drops off after feeding, digests its blood meal and moults to the adult stage. The adult then locates a host, feeds, then the female drops off the host into the environment to digest her blood meal and lay eggs. In contrast, the winter tick (*Dermacentor albipictus*) is a one-host species that only leaves the host animal (e.g. moose, deer, elk, and occasionally horses) to lay eggs. Some tick species mate on the host and some mate in the environment. Soft ticks have up to seven nymphal stages with each needing to feed on blood. For this reason, soft ticks are usually not found on horses and instead are often nest parasites of mammals and birds while repeatedly feeding on the same individual animal.

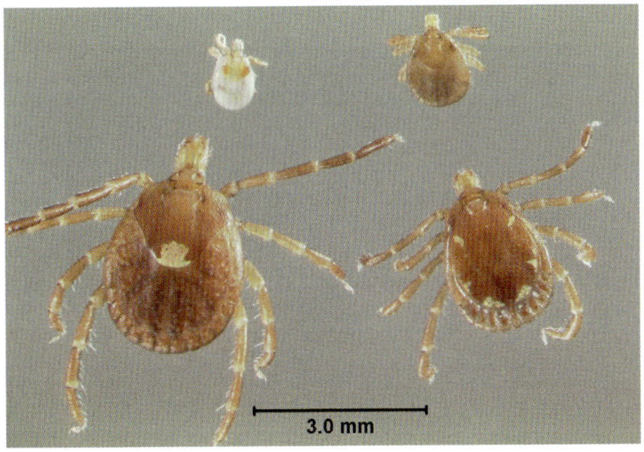

Figure 5.11. Life cycle of the lone star tick, *Amblyomma americanum*, from top left and clockwise larva, nymph, adult male, and adult female. Note the long mouthparts and characteristic white spot on the shield of the adult female.

Most ticks affecting horses are host generalists, meaning they do not have strong preferences for the species of the host that they bite. Tick species have different methods for locating a host. Some are nest parasites whereas most ticks affecting horses search for their hosts by moving towards host odour, heat, and vibration or use a more passive approach called questing (Figure 5.12). Questing is a behaviour unique to ticks. The ticks crawl up vegetation where they then stretch out their two front legs. These legs have sensory organs that allow them to detect an approaching host. The questing tick will use its legs to grasp onto the host as it passes. Once on the host, the tick will move to a preferred location on the body to attach and blood feed, which varies by animal host and tick species. Ticks attach by using their mouthparts to make a hole through which they insert their mouthparts and then use special substances in their saliva to make an anchor known as a cement cone that holds them in place. This is why it is hard to remove a tick without leaving the mouthparts behind. They will stay attached until they have finished feeding, which may be several days later.

5.1.5 Control

It is important to identify the species of tick most likely to be a pest. There are many species of ticks that can infest horses globally, and the biology and behaviour of these pests will influence which control methods are most likely to be effective.

Most of the life of a multiple host tick is spent off its host in the environment. Therefore, one of the best methods of tick control is to reduce the suitability of the environment for tick survival. There are some ticks that have adapted to challenging environments where landscape or environmental management is not effective. However, for many ticks, cultural control methods like trimming back vegetation and removing areas or plants where tick hosts occur off the host can be effective and provide long-term results but require substantial time and labour costs. Although there are a few options, such as wild animal host-targeted treatments, many chemical control methods that work well for residential areas are challenging to implement around pastures and large acreages. Preventative actions like using repellents and conducting tick checks to remove ticks on horses can reduce the risk of ticks getting on horses or transmitting pathogens.

Figure 5.12. Questing is a unique tick behaviour that enables them to get on a moving host. Ticks crawl up vegetation, stretch out their two front legs and then use their legs to grasp onto the host as it passes.

Cultural control

Landscape management

Many tick species of veterinary concern to horses are more likely to be found at the edge of pastures where they adjoin wooded habitats than in the middle of the pasture as they need moist or humid areas such as leaf litter or protected plants and undergrowth to prevent them from drying out and dying. Removing leaf litter and undergrowth around the edge of the wooded area can help to reduce tick survival and the chances of them getting onto pastured horses.

Horses should be prevented from grazing in wooded areas by putting up a fence, ideally at least 3 m (9 ft) from the wooded edge and keeping the vegetation between the wooded space and fence cut short (Figure 5.13). Removal of any woody debris like fallen logs and branches around the pastures and remove stone or rock walls if possible (Figure 5.14). This reduces the suitability of a nest site for rodents such as mice that not only feed ticks but also serve as reservoir hosts for many of the pathogens that feeding ticks can transmit to horses.

Figure 5.13. Moving fence lines that are close to wooded brushy areas (A) at least 3 m (9 ft) or more away from the wooded edge (B) can reduce the ability of some tick species to locate horses and reduce tick bites and possible pathogen transmission.

Figure 5.14. Tick infested rodents can live in stone or rock walls close to horses in pastures.

Physical and mechanical control

Wildlife exclusion

Preventing the access of wildlife, especially deer or other large animals, into pastures is important as these animals can move ticks from wooded habitats into grassy areas. However, this can be a challenge. Fencing that is at least 1.8 m (6 ft), and preferably 2.4 m (8 ft), tall is necessary to keep deer out of pastures, and this may not be feasible for many situations. Deer are less likely to jump over or crawl under electric fencing than traditional wood or similar fencing. For more information on fencing to exclude deer, see McPeake *et al.* (2019) in 'Continuing education'.

Habitat manipulation

In the pasture, keep the grass mowed short and trim any branches hanging over fences or the edges of pastures to allow for more sunlight, which will dry wet areas (Figure 5.15). Long grass (Figure 5.15) can provide a suitable habitat for some tick species and shorter grass can aid in drying areas, making them unsuitable for ticks. This is most important in areas where there are high numbers of ticks or near wooded edges and not necessarily as important in open pasture.

Tick checks and tick removal

Tick abundance in the environment and the risk of tick bites to horses depends on the species of tick, life stage of the tick, and local conditions. For many ticks, during the spring, summer, and autumn, daily tick checks to find and remove ticks are the most important ways to reduce the risk for tick-transmitted pathogens that may cause disease. Tick checks are very important if horses are turned out or ridden through wooded habitat, brushy areas, or grassy areas. The sooner a tick is removed the less risk of pathogen transmission.

Figure 5.15. Long grass promotes tick survival and makes it easier for questing ticks to get on the horses. Horses may be more at risk from ticks in areas of pasture with long grass near wooded areas.

Tick checks are conducted by running a hand over the horse. It may be difficult to see the ticks, so using a hand to feel for small bumps is most effective (Figure 5.16). These may be small or larger if the tick has been feeding for several days. It is important to check all areas of the horse, but ticks are most likely to be found in protected warm places. On horses, ticks are most commonly found under the jaw (Figure 5.17), under the mane, around the ears and eyes (Figure 5.17), on the underside of the neck and chest, and belly, behind the elbow, in the groin area, and next to or under the tail, but it may depend on species. If the horse is within the regional distribution of spinose ear ticks, be sure to check inside the ear (Figure 5.18); a tell-tale sign of infestation may be observed when horses excessively move or shake their heads from one side to the other. Tick checks should be conducted daily during seasons when local ticks are active as well as after riding in areas where ticks may be present.

Figure 5.16. Check for ticks by feeling along areas represented by the red colour for the presence of small bumps.

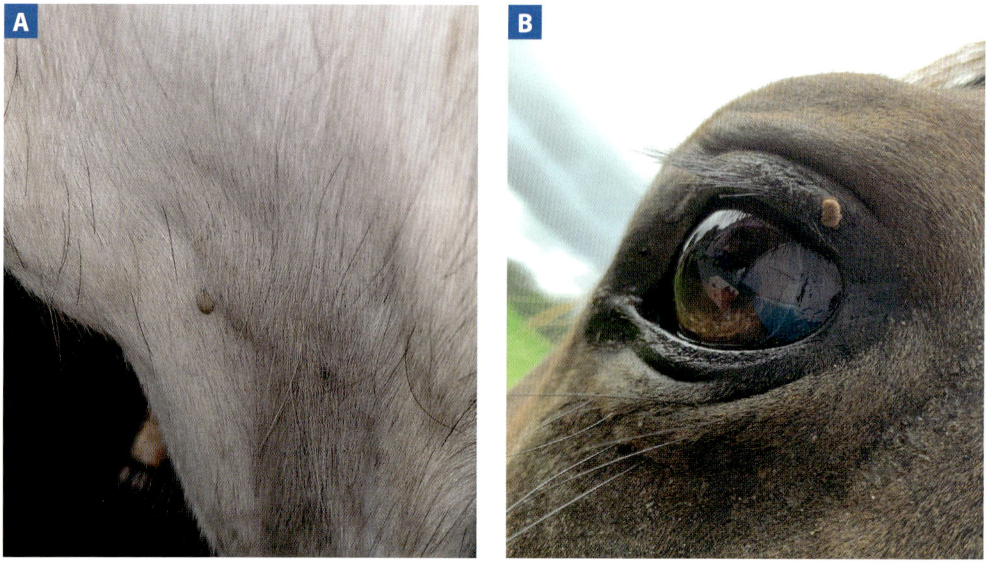

Figure 5.17. Different species of ticks can be found on different areas of the horse, but often ticks can be found on the head including under the jaw (A), eyes (B), or ears. Check all protected warm places on the horse during a tick check.

Figure 5.18. The spinose ear tick is common in dry areas of the western USA and Canada. In areas where this species is present examine the ears for infestations during routine tick checks.

To safely remove a tick, use a pair of fine-tipped tweezers or forceps or a specialised tick removal tool, such as a Tick Key™ or Tick Nipper™ (Figure 5.19). These specialised tools can be helpful if hair gets in the way of grasping the tick. Grasp the tick as close to the skin as possible. Pull the tick out from the skin in a firm steady movement (Figure 5.20). It is important to not quickly rip the tick out or grasp the end of the tick. Rapidly pulling a tick out of the skin makes it more likely the mouthparts will be left embedded in the horse, which can increase the risk of an allergic reaction, secondary infection, and pathogen transmission. In addition, do not use other methods, such as covering the tick with petroleum jelly, burning the tick, ice, or pinching the tick with fingers. All of these methods will cause the tick to regurgitate into the horse, which can increase the risk of pathogen transmission.

Hair can make tick checks a challenge. To improve the ability to locate a tick with fingers, winter guard hairs can be clipped short in common areas of tick attachment, like under the jaw. Body clipping in the spring and late autumn when some adult ticks are active, especially species that can transmit the agents of Lyme borreliosis, can also help with tick location and will allow for repellents to be more effective.

Once removed, dispose of the tick. If the tick is a species that is known to transmit pathogens causing disease, the tick can be placed in a container or bag with rubbing alcohol or hand sanitiser in a freezer. Label the container if it will be kept with the date, name of the horse, and location of the tick bite. Keep the tick for at least 2 months from the date of removal so that they can be tested at a later date to assist with a clinical diagnosis if the horse should present tick-borne disease symptoms. Regardless of species, clean the site of tick attachment with an antiseptic solution. A tick removal kit (Figure 5.21) can be made to add to a standard equine and human first aid kit or a grooming kit including fine-tipped tweezers or other removal tools, magnifying glass, local

Figure 5.19. Ticks can be removed with fine forceps or by specialised tools designed for this purpose. Some examples are shown here (A, B+C).

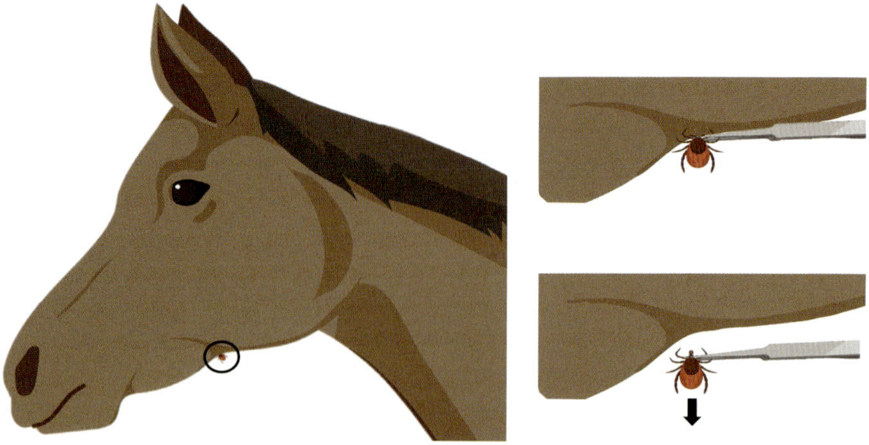

Figure 5.20. To remove a tick safely, use a pair of fine-tipped tweezers or forceps or a specialized tick removal tool. Grasp the tick as close to the skin as possible and pull the tick out from the skin in a firm, straight, and steady movement.

tick identification key, antiseptic wipe, premade vial or containers with alcohol or hand sanitiser or small bag to put the tick to be frozen, gloves, and a marker to label the vial or bag if it is to be saved for future reference or testing.

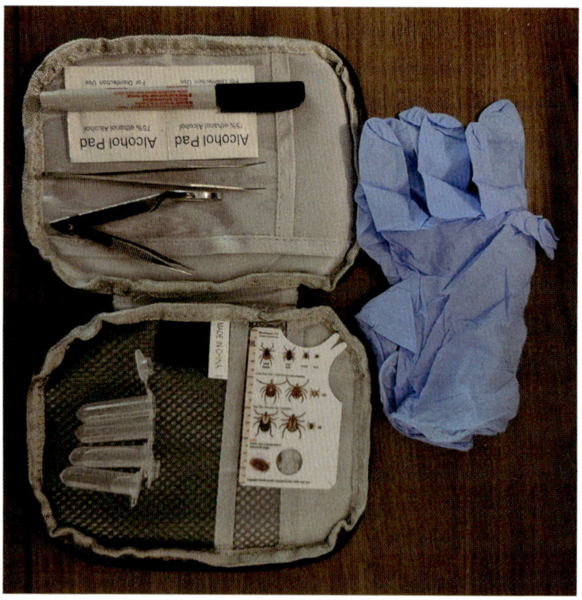

Figure 5.21. A tick removal kit should be part of a standard grooming kit for horses. It should include: fine-tipped tweezers or other removal tools, magnifying glass, local tick identification key, antiseptic wipe, premade vial or containers with alcohol or hand sanitiser or small bag to put the tick to be frozen, gloves, and a marker to label the vial or bag if it is to be saved for future reference or testing for pathogens.

In areas where Lyme borreliosis is endemic, horses located in pastures with oak trees are more likely to be positive for *B. burgdorferi* infection (Neely *et al.*, 2021). This is likely related to the effect of acorns on the survival success of wildlife, primarily rodents that eat the acorns and that carry the pathogen, and ticks biting those animals. While it is not recommended to remove oak trees, tick checks should be conducted thoroughly and regularly if oak trees are in horse pastures.

Biological control

Chickens and other fowl

Currently, there are limited options for biological control of ticks that are effective. Anecdotally, opossums, chickens, and Guinea fowl are believed to eat ticks. However, research studies have so far provided inconclusive results as to whether they reduce the number of ticks in the environment. Moreover, they can act as hosts for some tick species. Opossums have been found to eat high numbers of ticks in the laboratory, but recent studies have demonstrated that there is no evidence that this animal eats enough ticks in nature to control them (Hennessy and Hild, 2021). Chickens and Guinea fowl will eat ticks and other small arthropods (Figure 5.22), but studies that have evaluated these species have shown to have very little impact of these birds on tick numbers. In addition, spilt poultry feed may increase local rodent populations and their associated ticks, which may carry pathogens that can be transmitted to horses and humans.

Figure 5.22. Chickens will eat ticks but are unlikely to have much impact on total tick numbers in the environment.

Predators

Fire ants (and other ant species) (Chapter 6) are believed to reduce tick numbers (Castellanos *et al.*, 2016) but are not desirable to encourage in a horse pasture, due to the negative effects they have on humans, horses, and other animals.

Entomopathogenic fungi and nematodes

Entomopathogenic nematodes and fungi may be better biological control options for ticks. These organisms are found naturally in soil and fungi have been found to infect ticks. Research has shown that both nematodes and fungi can kill ticks but careful formulation is needed to protect the nematodes and spores in the environment (Figure 5.23) (Weeks *et al.*, 2018). There was one commercially available product in the USA that contained the fungus *Metarhizium brunneum* (also widely known as *M. anisopliae*). This product was labelled for use against ticks on turf in institutional and residential areas and was shown to be effective for managing tick populations, especially when combined with other control methods (Little *et al.*, 2021). However, the production of this product was terminated and it is unclear if it will become available in the future.

Chemical control

Topical products

Topical repellent sprays, roll-on, and wipes are available with a variety of active ingredients, percentages (or concentrations), and trade names. Topical repellents for ticks typically contain synthetic pyrethroids, such as permethrin and cypermethrin. These may or may not be effective depending on local pest tolerance of the active ingredient(s), and method of application. If applying repellents, it is important to consider using disposable gloves to protect the applicator's skin from the products (Fig 5.24).

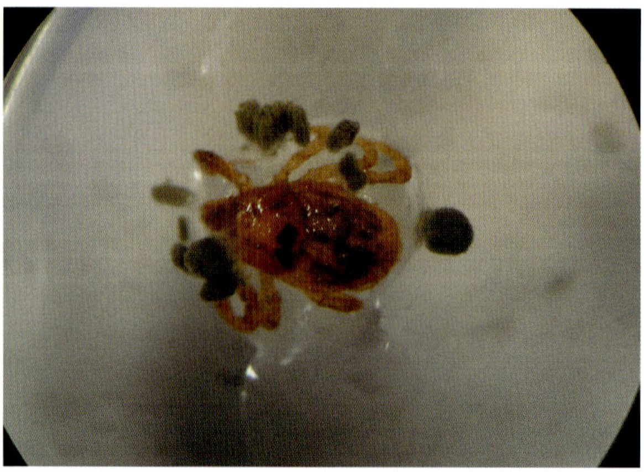

Figure 5.23. The entomopathogenic fungi, *Metarhizium brunneum* (*anisopliae*), has been demonstrated to kill ticks of various species. The image depicts a tick nymph with *Metarhizium* conidia.

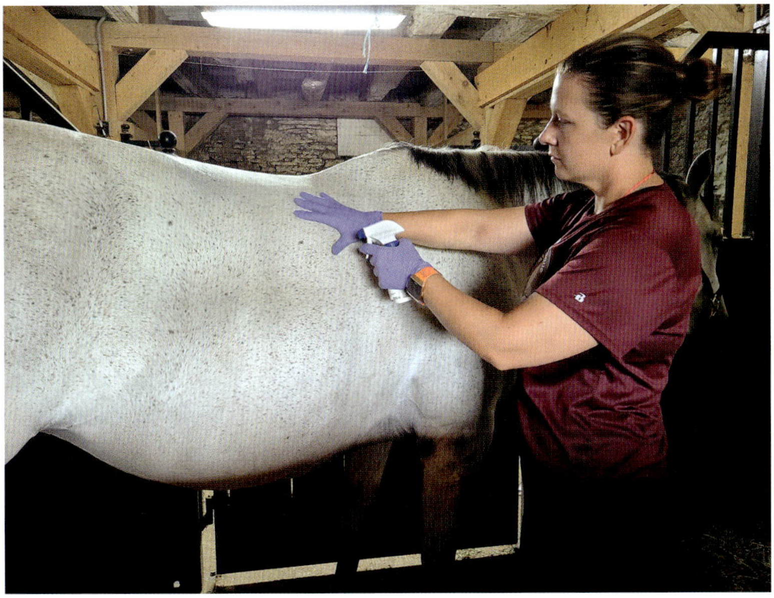

Figure 5.24. Human safety is as important as horse safety when applying repellents or other compounds. Disposable gloves should be worn if there is a risk for contact with the chemical. Read the product label carefully for other safety and application information.

Many formulations of topical repellents degrade quickly in ultraviolet light during turnout, wash off with sweat or rain, or can be hidden under layers of dirt after rolling. Some formulations may only retain an acceptable level of repellency for a few hours and others for about a day. Evidence suggests repellents with permethrin may only be effective for less than 24 hours against adult *Ixodes* ticks and exposure to rain or dew significantly reduces the duration of effectiveness (Machtinger, personal communication). Because of this, repellents should be applied before riding in tick habitat and reapplied daily before turning out during tick season. Shampoos containing synthetic pyrethroids have similar problems as topical repellents and will likely lose effectiveness quickly.

Pour-on repellents or pesticides can be applied to the back of horses to prevent tick bites. The active ingredient in pour-on pesticides is usually permethrin. Neem seed extracts are effective against *Ixodes* and *Rhipicephalus* ticks. In a laboratory study, ticks died when directly sprayed or exposed by walking on surfaces treated with the extracts in water at a 1:40 dilution and no skin irritation was observed even after the most concentrated application of 1:20 (Schmahl *et al.*, 2010).

Before using any chemical control methods, read the label directions carefully to understand the safe application. It is important to follow label application rates to ensure sufficient application amounts and full coverage over the body, head, and legs. The most effective applications are to clean horses and using a towel, applicator mitt, or brush to ensure coverage across the entire horse, but pay special attention to those areas where ticks are most likely to attach (Figure 5.17).

No repellent is 100% effective so regular tick checks are still necessary. Repellents are frequently applied to horses when fly populations are active because flies serve as a visual reminder of pest presence. However, it is important to remember to apply repellents to horses in the early spring and autumn as well when some ticks are active but fly populations may be low.

Pesticide resistance is less common in ticks than in other pests, such as flies, but may be a concern when chemicals are overused. When this happens, chemicals that were effective against ticks become less effective. Many products containing synthetic pyrethroids as active ingredients also contain piperonyl butoxide (PBO), which combines with synthetic pyrethroids to work more effectively and may help in situations where resistance has occurred. It is important to note that horses can have or develop a sensitivity to synthetic pyrethroids. This is most often seen with prolonged contact and presents as typical contact allergy symptoms such as hives and intense itching. These reactions underscore the need to only use products labelled for application to horses unless otherwise directed by a veterinarian. For more information on chemical active ingredients, resistance, and proper application and use of repellents see Chapter 10.

Repellents such as DEET, IR3535, picaridin, oil of lemon eucalyptus, para-menthane-diol, and undecane have proven efficacy against ticks on humans, but all except DEET are not labelled for use on horses. Other chemical compounds like fipronil and fluralaner that work against ticks have been developed for cattle and pets also are not labelled for horses. However, spray products containing fipronil have been used successfully to control ticks on horses and usually require applications every 30 days. Because these products are not currently labelled for use on horses, they should only be applied as directed by a veterinarian (Figure 5.24). While ivermectin and other macrocyclic lactones are available for oral administration for worm control in horses these products do not claim to control ticks on horses. It is important to note that ivermectin does not control the spinose ear tick. Another active ingredient in some pesticides for tick control is amitraz, which may be used in cattle and dogs, however, this active ingredient is toxic to horses and should not be applied.

Treated fly sheets, boots, and masks

Fly sheets, boots, and masks with permethrin bound to the fabric fibres can be purchased to repel biting arthropods. While this has not been evaluated specifically for ticks on horses, permethrin-treated clothing has been demonstrated to be very effective at repelling ticks on humans (Connally *et al.*, 2019). The repellent action of these products generally lasts 20 to 40 washes, depending on the product. The duration of effectiveness may change with exposure to sunlight and weather patterns, so observe horses periodically to make sure it is still providing the protection desired. Fly sheets purchased without pesticide treatment can also be sent to a company that specialises in binding permethrin to fabric for treatment. It is important to recognise that self-treating fly sheets or other items with pesticides that do not list fabric treatment for horses on the label may be a violation of the law. In addition, horses are sensitive to chemical pesticides and off-label treatment may cause skin reactions.

Landscape pesticides

While many chemical pesticides can be used on horses, some pesticides are used on vegetation surrounding the property or pasture. If tick numbers are high or if there are animals that are not able to have repellents or toxicants applied directly to them or checked frequently for ticks, these area sprays can be used. Landscape application of synthetic acaricides includes the use of sprays or granular preparations that vary in frequency of application, season targeted for treatment, or intensity of application.

Several products are labelled for use on a landscape level. The primary active ingredients are bifenthrin, carbaryl, cyfluthrin, deltamethrin, lambda-cyhalothrin, permethrin, and pyrethrins. Some products can be applied by the horse owner and some need to be applied by a licensed pesticide applicator, depending on local regulations and label directions. Single applications of acaricides are effective against several tick species in North America, resulting in short-term reductions in tick populations in some cases (reviewed in White and Gaff, 2018).

Before applying landscape pesticides, make sure that they are safe for use around horses by checking the product label. If equine facilities are not on the product label, the product may still be able to be used if horses are removed from the area. The number and timing of applications will depend on the tick species targeted for control. It is important to check with governmental regulations to ensure that the chemicals are permitted for use in the target area as some of the primary active ingredients have been banned in some parts of the world.

Repellents and landscape pesticides that are made from essential oils, garlic juice, or other more natural ingredients may be available and marketed for use on or around horses. However, many of these have either not been effective at reducing tick numbers or have not been tested against different tick species in rigorous studies, so their effectiveness against ticks and safety to horses and people is unknown. These types of chemicals may also degrade quickly, reducing their effectiveness. They also may be harmful to the environment by killing non-target species like pollinators (e.g. bees and butterflies). It is not recommended to make homemade products or use products that have not been tested for horse, human, or environmental safety or effectiveness against ticks on horses.

Wild host-targeted control

Pesticides can be used to interrupt the life cycle of ticks on their wild animal hosts. There are at least three wild animal host-targeted control methods that can be used as part of a tick control program for some species of ticks: 4-poster deer feeders, tick tubes, and rodent bait boxes. The use of these products will depend on regional approvals and availability, as well as the targeted tick species.

The 4-poster feeder was developed by the United States Department of Agriculture in the 1990s to assist with controlling ticks on white-tailed deer in the north-eastern USA (Figure 5.25). These are large plastic units with a central feed storage area that holds approximately 95 kg (210 lbs) of corn kernels. On either side of the feeder are two smaller troughs where the corn is dispensed in limited amounts to visiting deer. Two felt, wool or similar fabric rollers are held vertically and treated with a pesticide, usually a synthetic pyrethroid, which is then transferred to the deer as they rub on the rollers to get to the corn in the feeders. This kills the ticks on the deer. 4-poster feeders may be effective for any tick species that can be found on the head, ears, and necks of deer, but specifically for blacklegged ticks (*Ixodes scapularis*) and lone star ticks (*Amblyomma americanum*). 4-poster feeders may be useful in large areas with dense deer herds. Some studies have shown reductions in tick numbers using 4-poster feeders while others have not (summarised in White and Gaff, 2018).

There are some drawbacks to 4-poster feeders. Baiting wildlife always has some ecological consequences like increased survival and reproduction of tick hosts including deer and other animals that eat the corn (squirrels, raccoons, opossums, birds, etc.) and may not be permitted in some areas. The gathering of animals around a food source may also increase the risk of transmission of other wildlife diseases like chronic wasting disease in North America (Brown and Cooper, 2006). Non-target wildlife like bears and squirrels that are attracted to the corn food source can be a nuisance in areas where the 4-poster devices are implemented. In addition, these feeders can be expensive and time-consuming to maintain.

The other host-targeted tick control methods target ticks that are found on small rodents, primarily blacklegged and deer/sheep ticks (*Ixodes scapularis* and *Ixodes ricinus*). Many ticks that bite mice become infected with pathogens that are of veterinary concern. Tick tubes (Figure 5.26) are biodegradable cardboard tubes that are filled with cotton or cotton-like fabric that is treated with synthetic pyrethroids. Small rodents, especially mice, use this fabric in their nests. When they collect and bring the treated fabric back to their nest, they treat themselves and, if female, their offspring for ticks. These tubes are easy to purchase and apply in areas where their use is permitted,

Figure 5.25. Pesticide can be applied to rollers on 4-poster feeders. When deer come to eat from the feeders they get a dose of pesticide that kills ticks.

and do not need a professional pesticide company to use them. Another positive attribute of this method is that bait is not used, which may draw additional small mammals to an area. However, they do primarily target mice, so will not be useful to control tick species that do not bite small mammals. Moreover, they do not affect many other species of small and medium-sized animals that can also be hosts for ticks.

Tubes are deployed according to manufacturer recommendations during the spring, summer, and autumn months and should be replaced as they are depleted. Mice prefer to be protected when moving in the environment, so tubes are most effective if placed in areas with brushy understory in pastures (if wooded) or at the forest edge around pastures, or around rock walls or fallen branches and trunks where they like to make their nests (Figure 5.27).

Figure 5.26. Tick tubes contain cotton treated with an acaricide (a pesticide that targets acarids, e.g. ticks and mites) to treat small mammals that are hosts to ticks, especially mice. These small mammals collect the cotton for their nests and the pesticide is transferred to their coat, killing any ticks that they are carrying.

Figure 5.27. Small mammals like mice (A) are frequently found nesting in or travelling on fallen logs and branches, stacks of wood, or in dense brush (B). Tick control that targets mice should be placed in these types of habitats.

Rodent bait boxes, also called tick bait boxes, are used to control ticks on small mammals, primarily mice and chipmunks (Figure 5.28). These bait boxes provide food bait to small mammals that enter them. As they pass through the bait box to get to the bait, they go under a small wick that is treated with a low dose of the chemical fipronil. The small mammal becomes treated with fipronil, which subsequently kills any tick that attach while that animal is treated. Bait boxes were developed for the control of the blacklegged tick (*Ixodes scapularis*) in North America.

Bait boxes are highly effective (Williams *et al.*, 2018), but must be applied by a pesticide applicator in the areas where their use is permitted. Usually, boxes are installed in the spring and replaced once during the late summer, but manufacturer recommendations should be followed. Bait boxes are installed 3 m (about 10 ft) into the woods and placed about 10 m (30 ft) apart. If the wooded area extends more than 12 m (about 40 ft) another row of boxes is recommended. As with tick tubes, bait boxes should be placed where mouse and other small mammal activity is likely (Machtinger and Li, 2019).

There are a few disadvantages to bait boxes. Wooded areas surrounding pastures may be too extensive for full applications of bait boxes, so these may be best used in specific areas of the property. Also, like 4-poster feeders, these target only a limited number of hosts and ticks species, and providing a food bait to rodents could serve as a supplemental food source that may increase their survival or reproductive rate. The long-term ecological consequences of these bait boxes have not been evaluated.

Figure 5.28. Tick bait boxes are enclosed containers that contain bait that is attractive to small mammals. As the animal, such as a mouse, uses the bait, they walk under a wick treated with an acaricide (a pesticide that targets acarids, e.g. ticks and mites) that kills the ticks that are on the animal.

5.1.6 Veterinary concerns

Anaplasmosis and Lyme borreliosis

In North America, there are two primary concerns for tick-borne diseases in horses, *Anaplasma phagocytophilum* and *Borrelia burgdorferi*. *Anaplasma phagocytophilum*, the causative agent of anaplasmosis, causes acute onset high fevers, along with all the sequelae that come with high fevers: soreness, stiff movement, lethargy, and inappetance. There also can be jaundice and anaemia secondary to the destruction of red blood cells as the immune system tries to clear the organism. Symptoms appear 3 to 10 days after exposure. Testing can be performed, however, treatment with a tetracycline antibiotic is often started before testing in endemic areas. *Anaplasma phagocytophilum* infections can mimic infections with *Neorickettsiae risticii*, the causative agent of neorickettsiosis (Potomac horse fever). Fortunately, seasonalities are different so mistakes in diagnosis are unlikely and treatment is similar so if mistakes do occur the animal will receive effective treatment for the condition anyway. The similarities in early symptoms are reason enough to test because neorickettsiosis carries a higher incidence of complications, and requires a prolonged treatment duration compared to *A. phagocytophilum*.

B. burgdorferi is the causative agent of Lyme borreliosis. Symptoms of Lyme borreliosis are as frustrating in horses as they are in humans. Hypersensitivity to touch, shifting leg lameness, and subtle to profound neurological symptoms can all be signs of *B. burgdorferi* infection. These signs are often frustratingly vague. Testing also can be problematic as positive results are not always indicative of active infection, and active infections can fail to cause a testable immune response. Treatment with antibiotics can be curative, however, a small percentage of infected horses will have long-lasting effects.

Equine piroplasmosis

Equine piroplasmosis is a serious concern in South and Central America, the Caribbean, Africa, the Middle East, and Eastern and Southern Europe. Piroplasmosis can be caused by either *Babesia caballi* or *Theileria equi*. Both organisms are endemic in areas with significant amounts of the world's equine population. Horses can have different levels of infection: inapparent carrier, or peracute, acute, subacute, or chronic infections. Most horses develop inapparent carrier status. These horses have low levels of parasites and may display no symptoms. Mares can transmit the organism in utero, causing significant losses through abortion or severe compromise of foals. Horses who are inapparent carriers are also at risk for developing more severe infections if the immune system is compromised. Peracute infections are seen most commonly in foals, and naive horses brought to highly endemic areas. These horses develop profound anaemia secondary to the destruction of red blood cells, along with subsequent organ failure due to blockage of small capillaries by cellular debris. Acute and subacute horses experience fevers along with generalised swelling, and anaemia. Chronic horses are best characterised as poor doers. These horses have difficulty maintaining body condition scores, poor performance, and may experience varying levels of inappetence, and lethargy. Treatment success is highly variable and is often aimed at managing clinical signs. Treatment to eliminate carrier status so that import to the USA can occur has been attempted. However, it is unclear if this is successful. See de Waal, 2019 in 'Continuing education' for more information on equine piroplasmosis.

Tularemia

Tularemia is caused by the bacteria *Francisella tularensis*. Rabbits and hares are the most common source of the bacteria. Ticks moving from rabbits or hares to horses is the usual mode of infection. The disease is rare in horses and has only been identified in horses with severe tick infestations. Horses present with a high fever, swollen lymph nodes, and lethargy (Foley, 2020).

Tick paralysis

Tick paralysis has recently been described in horses in North America. In 2019 two miniature horses developed symptoms consistent with tick paralysis and had numerous *Dermacentor variabilis* (Trumpp *et al.*, 2019). Treatment to remove and kill the ticks resulted in the resolution of symptoms in 24 hours. Tick paralysis has been seen with increasing frequency in Australia due to *Ixodes holocyclus* and should be considered in any horse with even a few ticks and neurologic symptoms. Removal of the offending ticks is generally curative. See Padula, 2016 in 'Continuing education' for more information on tick paralysis in Australia.

Louping ill virus

Louping ill virus is a variant of tick-borne encephalitis virus (TBEV). While TBEV is found in mainland Europe and Asia and primarily affects humans, louping ill is confined to the UK, and Ireland, and primarily affects sheep. Louping ill virus is carried by *Ixodes ricinus*. There have been reports of louping ill in horses, although it is infrequent (Hyde *et al.*, 2007; Timoney *et al.*, 1976). Symptoms are a high fever followed by various neurologic symptoms depending on viral load. Treatment is supportive only and is aimed at hydration, reducing inflammation, and preventing decubital ulcers in down animals. Horses who recover often have residual neurologic symptoms.

Irritation

Otobius megnini will cause head tossing and shaking, ear twitching, and other similar signs of ear discomfort at typical infestation levels. Severe infestations may cause the horse to display extreme responses leading to injury. Additionally, foals who experience severe infestations may develop anaemia due to blood loss.

Nodules

Many horses will develop a nodule at the site of a tick bite (Figure 5.29). These can become mildly infected and are often extremely itchy. Mild infections with only some heat and tenderness can be treated with topical antibacterial ointments. The itching can be managed with topical steroid ointments. A veterinarian should be contacted for treatment if the nodule becomes extremely painful or continues to drain pus for longer than 3 days.

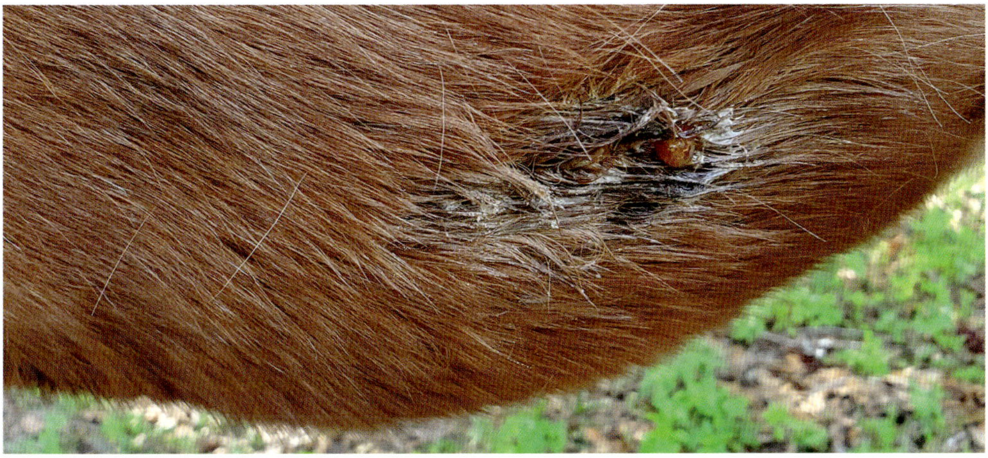

Figure 5.29. After the removal of a tick, a nodule can develop at the site of the bite.

5.2 Mites (subphylum Chelicerata)

5.2.1 Pest summary (Table 5.2)

Mites are tiny, mostly microscopic, arthropods that are closely related to ticks and are present throughout the environment (Figure 5.30). Most species are harmless to animals and provide ecological services like decomposing organic matter. A few parasitic mite species invade the skin of otherwise healthy animals and can cause problems for horses. Many of these mite infestations can be self-limiting and cause no issues, in fact, most animals have some mites present. However, when mites irritate the skin causing hair loss and inflammation the resulting relatively rare condition is known as mange. There are several types of mange caused by different mites, including sarcoptic mange or equine scabies, psoroptic mange or mane mange, chorioptic mange or leg mange, and demodectic mange. Trombiculid mites commonly known as harvest mites or chiggers (family Trombiculidae) are usually present on other animals, and straw itch mites/forage mites are usually found in straw or grain, but both of these groups of mites can also infest horses causing skin irritation and hives.

5.2.2 Distribution and seasonality

Mites are distributed globally, with different species being pests of horses in some areas and some groups being more common in regions than others. Sarcoptic, psoroptic, and chorioptic mange typically occur in the winter when horses are more likely to be confined indoors and in closer contact. The indoor humidity increases the speed of mite development. Horses are also more likely to be hungry in the winter from reduced forage intake and more likely to be stressed due to confinement. Both malnutrition and stress can have negative effects on the immune system and reduced immunity will increase the risk of mange. Demodectic mange is much less common than other forms of mange in horses.

Table 5.2. Pest summary of mites, subphylum Chelicerata.

Distribution[1]	Australia: Yes Canada: Yes Europe: Yes New Zealand: Yes United States: Yes
Size	0.2-0.6 mm (less than 0.1 in)
Key identification characteristics	Simple form, eight legs (nymph and adult)
Mouthparts	Chewing (itchy leg, equine body and chiggers), pin-like (follicle mite), plier-like (scabies mite), piercing (straw itch mite)
Feed on	Skin, fat, lymph, and oils
Development time[2]	Mange mites 2-3 weeks, straw itch mite 2-4 weeks, and chiggers 2-12 months
Adult life span	Follicle mite several weeks, itchy leg mite 4-6 weeks, chiggers 12 months, scabies 1-2 months
Total eggs laid	Straw itch mites 200-300, scabies 200, chiggers 15 per day, itchy leg mite 17, follicle mite 15-20
May be mistaken for	Ticks (Section 5.1), lice (Section 5.3), other non-parasitic mites
Impact on horses	Mange

[1] The pest distribution in the regions targeted by this publication are listed as of the date published. Areas of primary pest concern (if any) are in bold. Global distributions are listed in Section 5.2.2.
[2] Development times are approximate and can change with environmental conditions including temperature, humidity, and availability of food resources.

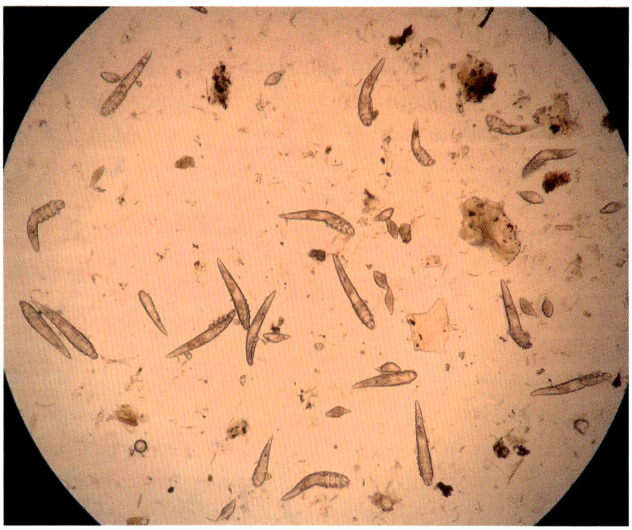

Figure 5.30. ***Demodex*** **species mites surrounded by debris that is typical in a skin scrape image when viewed down a microscope.**

5.2.3 Description

Mites are tiny arachnids in the subclass Acari; they are typically less than 1 mm (0.04 in) in length and have a simple form, meaning they look like they have one body region with attached legs. Like ticks, mite larvae have six legs and the nymphs and adults have eight legs. The structure of their legs varies depending on where they live. Some mite legs are adapted for grasping hairs and others for burrowing in the skin. There are four mite parasites of horses that live and feed exclusively on equines: *Chorioptes equi*, *Demodex equi*, *Psoroptes equi*, and *Sarcoptes scabiei* var *equis*.

Chorioptes equi, the itchy leg mite, is an oval microscopic mite, 0.4 to 0.6 mm (less than 0.01 in) in length, with two hairs near the end of the hind legs (Figure 5.31). *Demodex equi*, the horse follicle mite, has an elongated body shape to enable it to infest hair follicles (Figure 5.32). They are microscopic, typically approximately 0.25 mm (less than 0.01 in). *Psoroptes equi*, the scab mite or equine body mite, is typically oval and 0.5 to 0.6 mm (less than 0.01 in) in size (Figure 5.33). *Sarcoptes scabiei* var *equis*, the common mange mite, are spherical and typically 0.3 to 0.5 mm (less than 0.01 in) (Figure 5.34). They have two legs pointing forwards and two pointing backwards; the front legs have suckers.

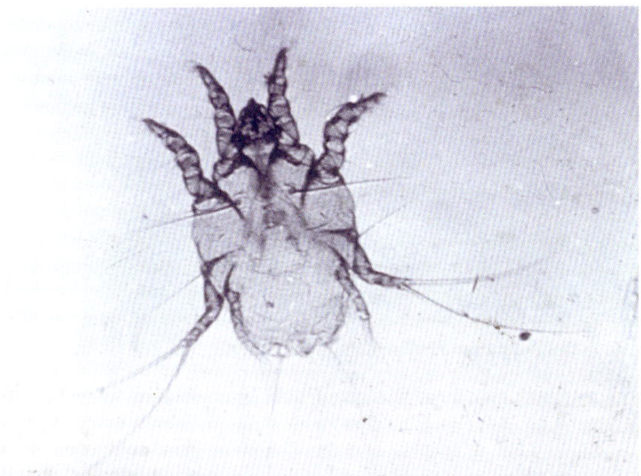

Figure 5.31. ***Chorioptes*** **mite infestations are typically not as severe as sarcoptic or psoroptic mange. The mites cause itching that can result in eczema and even crusty scabs if left untreated.**

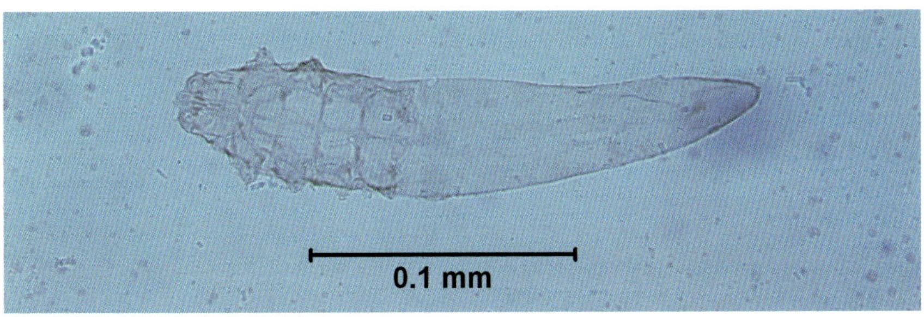

Figure 5.32. ***Demodex*** **species mite with the typical elongated body shape. May cause lesions on the forehead and around the eyes or shoulders.**

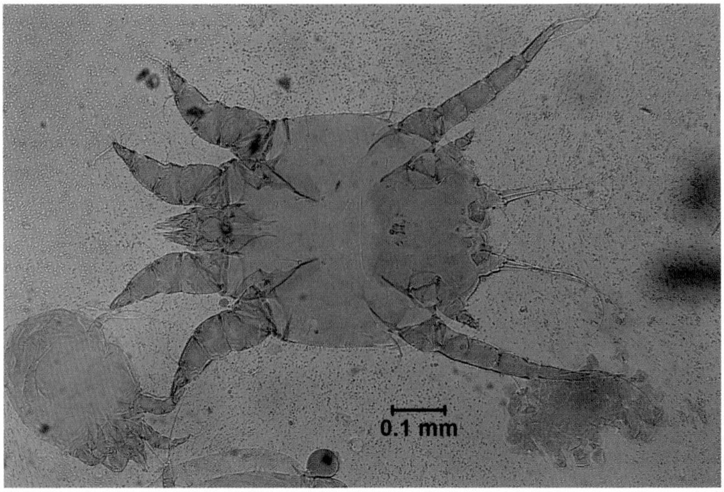

Figure 5.33. ***Psoroptes*** **species mites with the oval body shape. May cause scabs due to an allergic reaction to the faeces.**

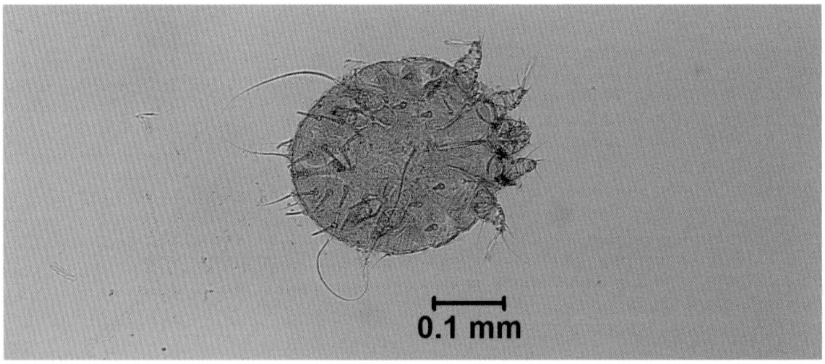

Figure 5.34. ***Sarcoptes scabiei*** **mites have a rounded body shape with suckers on the front legs.** ***Sarcoptes scabiei*** **are one of the mites that can cause the condition known as mange. These mites burrow in the skin and can cause lesions and extreme itching.**

Two other mite groups may accidentally or opportunistically bite horses: straw itch mites and chiggers. The straw itch or forage mites are in the genus *Pyemotes* (Figure 5.35). They are elongated oval in shape and approximately 0.2 mm (less than 0.01 in). Trombiculid mites, commonly known as chiggers, are tiny, 0.2 mm (less than 0.01 in), yellow to red organisms (Figure 5.36).

5.2.4 Life history

Most species of mites are harmless and even parasitic species rarely cause harm. Like ticks, mites have four life stages: egg, larva, nymph, and adult. Two types affect horses, those that are parasitic throughout their life cycle and those that are parasitic only in one life stage. Life cycles between the genera of mites that affect horses vary and the life histories of most important mite pests on horses are described below.

Figure 5.35. Straw itch mites can infest and attempt opportunistically to feed on animals that come into contact with infested feed, including horses and humans.

Chorioptes equi, the itchy leg mite, lives on the skin surface and feeds on skin, fat, and lymph with their chewing mouthparts. They preferentially infest the lower legs. Horses with feathers (long hair) on the fetlocks are especially at risk (Figure 5.36). *C. equi* infestation is associated with chronic progressive lymphedema. This species of mite cannot infest humans, but can survive off the host for more than 2 months.

Figure 5.36. Larvae of trombiculid mites or chiggers inject digestive enzymes into the skin to form a hole and then feed on internal skin tissue in this hole causing a delayed allergic response in the form of itching and swelling.

Demodex equi, the horse follicle mite, eats skin cells and oils within the follicle. Almost every animal in the world has a species-specific follicle mite, including humans, and they typically cause no issues. This species can be present on horses and cause no problems at all. When they do cause issues they can cause a scaly or pustular reaction. The pustular form may result in abscesses, especially on the neck area. Other areas where lesions can be found are on the forehead, around the eyes and shoulders. *D. equi* do not usually infest humans and can survive off the host for approximately 4 months.

Psoroptes equi, the scab mite or equine body mite, feed on the skin with chewing mouthparts like the itchy leg mite. This mite is less common on horses than other vertebrate hosts with little or no affiliation for horses. It is believed that an allergic reaction to their faeces causes the characteristic scabs. Infestations are typically in areas with a lot of hair such as the mane and tail. This species does not usually infest humans and can survive off the host for 15 to 20 days.

Sarcoptes scabiei var *equis*, the common mange mite, burrow in the skin of their host and use enzymes to digest skin tissue, feeding on the liquid. The females lay eggs in burrows in the skin, which cause allergic reactions. This species can infest humans and other animals, although specific types of this mite occur on certain host species. The active stages can live for just a few days off the host as the mites are susceptible to drying out. This mite is highly contagious, spreading easily from horse to horse. Skin lesions caused by mange mites can be found anywhere on the body, but they are most frequent on the head, neck, and ears.

The straw itch or forage mite (Figure 5.35) females feed on insects in hay, straw, and grain. They puncture the insect with their mouthparts and inject a toxin that paralyses the insect. They remain attached feeding on the insect. The feeding females swell in size up to 2 mm (0.08 in) and, unlike most other mite species, her offspring develop inside her and emerge from her birth canal as adults. The males are born first and then they assist and mate with the females as they are born. Straw itch mites can infest mammals opportunistically and attempt to feed on them, including horses and humans. The allergic reaction that follows is typically in the area that comes into contact with the feed. This is typically the head and neck on horses feeding from a manger whereas rashes on humans are commonly on the arms from carrying and handling the feed.

Approximately 30 species of trombiculid mites, commonly known as chiggers (Figure 5.37), feed on animal cells. Instead of biting the host directly, they inject digestive enzymes into the skin to form a hole and then feed on internal skin tissue in this hole. This feeding can cause a delayed allergic response in the form of itching and swelling. Only the larvae of these mites feed on animal cells. The nymphs and adults of this group are not parasitic and are free-living in the environment. Unlike the other mite groups discussed here, they actively leave the host and can move within the environment. Most of the time, the chigger larvae have already dropped off the animal by the time it begins to show irritation from their feeding.

Figure 5.37. Cobs (shown), draught breeds, and other horses with long feathers on the lower legs are at particular risk for *Chorioptes equi* infestation.

5.2.5 Control

Prior to attempting mite control, it is important to determine that the issue is due to mites or mite related allergies and not allergic reactions to other factors, behavioural disorders, or other irritants, and to also determine the mite species responsible. Control is different for different species and while on-animal treatments will be necessary to treat mites that are parasitic throughout their whole life cycle, prevention of exposure is more helpful to reduce chigger and straw itch mite bites.

Mites that are parasitic throughout their life cycle cannot travel great distances so transmission of the mites usually occurs through direct contact between an infested and uninfested horse or contact of an uninfested horse with a surface recently contacted by an infested horse. Some species can survive for several weeks off the host and may be moved from horse to horse by sharing saddlery or grooming equipment.

Cultural control

Risk factors associated with mite infestation include a higher density of horses, sharing blankets, brushes, or other equipment between horses, poor feed quality, gestational status, underlying health issues, longer body coat (like a winter coat or feathering), and season with infestations being more common in winter.

Periodically checking horses for mites or skin lesions during regular grooming sessions can help make infestations apparent before they become more challenging to control. Mites prefer thick hair as is found in ears, in the mane, tail, jaw area, and feathered horses (Figure 5.38), and this explains why they are more common in cooler months when horses grow winter coats.

Figure 5.38. Mites are most common around the ears, mane, tail, jaw, and in the feathers as indicated in red.

Because some mites can survive off the host for weeks to months it is important to practice good hygiene to reduce the risk of introducing infestations or transferring them between individuals. Second-hand or used horse rugs or blankets should be washed before use. Saddlery and grooming tools should be disinfected before use among different horses. Ideally, there should be separate labelled sets of grooming tools, saddlery, and blankets for each horse (Figure 5.39). Stables and sheds should be cleaned and disinfected before moving in a new horse.

Figure 5.39. Grooming kits should be kept separate and labelled corresponding to an individual horse to avoid inadvertently moving mites to an uninfested horse.

Keeping horses in good health and well-fed reduces the risk that they will suffer from mite infestations that require treatment. Sick, malnourished, or stressed animals have a compromised immune system and are at increased risk of parasites. Regular grooming and periodic bathing will help reduce mite numbers on an animal and also allow for time to examine horses carefully (Figure 5.40).

If horses need to be isolated due to a suspected mite infestation, any handlers should be careful to use gloves and other protective equipment before handling the infested horse to not spread the mites to new areas or other horses.

Mites can be spread easily by direct contact with other horses. Horses suspected of mite or lice infestations should not travel to equestrian events until proper treatment is completed. Conversely, keep uninfested horses away from other horses during competitions, exhibitions, or other travel.

Physical and mechanical control

Before the winter months, horses with heavy coats or long feathers can be clipped to reduce the risk of infestation (Figure 5.41). Clipping the hair on the legs helps by removing the hair so the skin is more easily accessed and it also exfoliates the skin crusts caused by hyperkeratosis (thickening of the skin), making treatment easier and more effective. Mites can be transferred from animal to animal and can survive in the environment for several weeks so clipped or cut hair and other debris should be carefully disposed of and equipment sanitised after using it on an infested animal.

Figure 5.40. Regular grooming and periodic bathing will help to reduce mite numbers on horses and provide an opportunity for close inspection.

Figure 5.41. Infested areas can be clipped and shampooed, but be careful to dispose of hair and debris and sanitise clippers afterwards to avoid transferring mites to other horses.

Horses with mites should be isolated if possible. If isolation is not possible, all horses that have contacted the affected individual should be treated as it is likely the mites have been distributed among the group. If the horse is diagnosed with sarcoptic mange, be aware that humans and other animals can become infested with this mite as well, although infestations are generally self-limiting in people.

Chemical control

Although chemicals can be used to remove mites from the skin of animals it is important to also remove them from the hair by removing hair and scales, crusts, and dead skin by clipping off the affected hair and shampooing the areas with selenium sulphide or keratolytic shampoos.

Hot lime sulphur spray or dip is labelled for treatment of *Psoroptes*, *Chorioptes*, and *Sarcoptes* mites (Reichard and Thomas, 2015). This yellow concentrate is made from calcium hydroxide, sulphur, and water, and can be an effective treatment for mites when repeated once a week for at least 4 weeks. The yellow colour does stain lighter coloured horses, so this will need to be taken into consideration.

Permethrin sprays can be used to treat mite infestations, but it is important that the animal is wet when treated and treatments must be repeated according to label directions. For *Sarcoptes* mites, organophosphate pesticides can be sprayed, dipped or sponged onto the animal for effective treatment. Multiple treatments, possibly up to four will be required every 12 to 14 days and the entire horse should be treated, not just the problem area. This is because the eggs and mites are mostly in burrows under the skin and are unlikely to be exposed to the active ingredient every application. Eprinomectin, fipronil, and doramectin have also been used to treat mite infestations but there are no products labelled for use on horses with these active ingredients so their use should be directed by a veterinarian. Treatment of demodectic mites is usually not necessary as incidences of demodectic mange are typically self-limiting.

For chiggers and straw itch mites, infestations are accidental and reproduction is not occurring on the host, so removing the remaining mites is typically sufficient. Use a spray product containing pyrethrin or synthetic pyrethroids to kill any feeding mites and then treat the symptoms.

Similarly to tick control, many chemicals are labelled for control of mites on livestock or pets, but are not labelled for horses and should not be used unless directed by a veterinarian.

5.2.6 Veterinary concerns

Mites are not known to be carriers for viruses or bacteria to horses at this time so veterinary concerns are largely a result of pruritus (itching) and other skin issues associated with mange or chronic progressive lymphedema. Mites create extreme skin irritation, which is a reaction to the mite saliva and faeces and anti-itch medications are not effective at relieving the irritation. In an attempt to relieve the itching, horses may cause lacerations or abrasions that require veterinary care. Superficial scrapes can be treated with topical antibacterials as the only therapy. More commonly, the wounds to the skin develop secondary skin infections that require diligent wound care along with antibiotics.

Mange

Mange is a generalised parasitic skin disorder caused by mites. As mites are microscopic usually the owner will not see the mites but may observe stamping feet, biting or chewing the back legs, rubbing of the body or legs on the fence or walls, rashes, scabs, flakey scales, thickened skin, or patchy hair loss. It is important to understand that during the warmer months, horses can also exhibit an allergy to the bites of *Culicoides* midges called sweet itch (Chapter 4), which may resemble mite infestations or may have a louse infestation, which presents similar symptoms.

To diagnose a mite infestation, a veterinarian will generally take a skin scrape of the affected area and identify the mites, if present, under a microscope (Figure 5.30). In most circumstances, a veterinarian will develop a treatment plan tailored to the specific horse and circumstance. *Psoroptes*, *Chorioptes*, and *Sarcoptes* mites have been effectively treated with topical and injectable monocyclic lactones, such as ivermectin and doramectin administered 10 to 14 days apart. However, these products are typically not labelled for use with mites on horses and their use must be directed by a veterinarian. Mobile stages of the mites (larvae, nymphs, adults) may be eliminated after one treatment, but additional treatments are usually necessary to treat mites that hatch from eggs after the first treatment, especially if there is more than one affected horse. Corticosteroids can be used to help relieve signs. Please see Karasek *et al.* (2020) in 'Continuing education' for a review of treatment options for mite infestations in horses. Please see Osman *et al.* (2006) in 'Continuing education' for more information on clinical and therapeutic studies on mange in horses.

Chronic progressive lymphedema

While not caused by mites, *Chorioptes equi* infestation can be associated with chronic progressive lymphedema (CPL). This is a disease of some breeds of draft horses, where infestations of this mite can exacerbate the condition resulting in excessive itchiness and lead to kicking, self-trauma, and dermatitis (Figure 5.42). If CPL is suspected, the clinical presentation will be categorised from mild to extreme based on several factors like hair and skin condition. CPL is incurable and will require intensive management to slow the progression. Management of CPL includes controlling the tissue oedema, providing skin care, and treating secondary infections. Please see de Keyser *et al.* (2015) in 'Continuing education' for more information on the diagnosis and treatment of chronic progressive lymphedema.

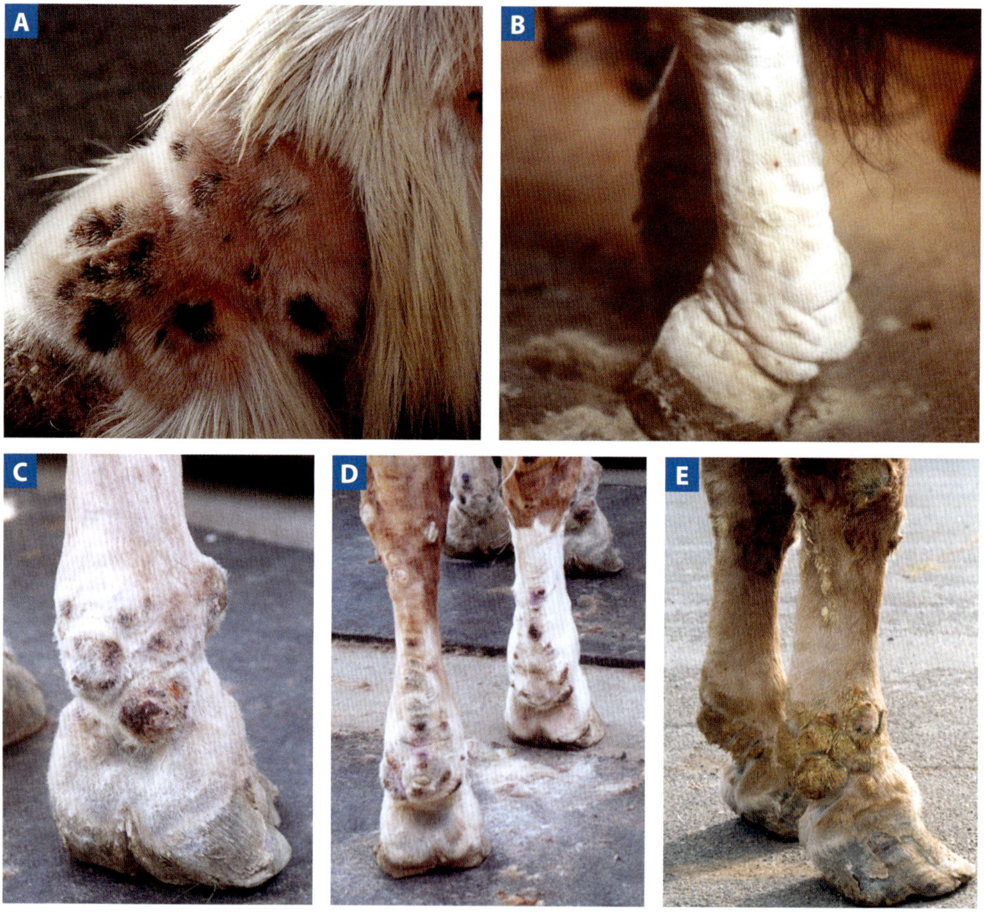

Figure 5.42. Examples of clinical presentation of chronic progressive lymphedema (A-E).

5.3 Lice (families Haematopinidae and Ischnocera)

5.3.1 Pest summary (Table 5.3)

Lice are small insects in the order Phthiraptera that live in the feathers and hair of animals. There are nearly 5,000 species, all of which are obligate parasites, living on warm-blooded hosts. Almost every species of bird and mammal has an associated lice species. Horses have two lice species: the horse blood-sucking louse (*Haematopinus asini*) and the horse biting louse (*Bovicola* [*Werneckiella*] *equi*) (Figure 5.43). When the lice on a horse begin to cause a problem the associated disease is known as pediculosis.

5.3.2 Distribution and seasonality

The two species of lice that affect horses are globally present. As obligate parasites, they are always found within the hair of their hosts. They do not disperse from the host unless by direct contact and they survive for a very short time off the host. Lice infestations are worse in the cooler months when horses are typically in poorer condition and have longer coats. These long coats create a warm humid microclimate for the lice to develop quickly and without risk of desiccation. However, louse infestations can occur at any time of the year and can spread through direct contact with infested horses.

5.3.3 Description

Sucking lice have an oval flat body that varies from 0.5 to 5.0 mm (less than 0.01 to 0.20 in). The colour of a louse species varies from beige to grey and is dependent upon the typical colour of its host species as they are often camouflaged. They have heads that are narrower than their bodies with piercing-sucking mouthparts, and short antennae. Chewing lice are a similar size but some

Figure 5.43. Sucking louse (*Haematopinus* sp.) eggs (nits) are attached to the hair of the host.

Table 5.3. Pest summary of lice, suborders Mallophaga and Anoplura.

Distribution[1]	Australia: Yes Canada: Yes Europe: Yes New Zealand: Yes United States: Yes
Size	Chewing lice 0.5-6.0 mm (less than 0.01-0.24 in), sucking lice 0.5-5.0 mm (less than 0.01-0.20 in)
Key identification characteristics	Small but visible to the naked eye, six legs, flat, sucking lice have wide oval bodies, narrow heads and enlarged claws, chewing lice have narrow oval bodies, wide heads and finer legs
Mouthparts	Chewing or biting
Feed on	Chewing lice feed on skin and other debris, sucking lice feed on blood and other fluids
Development time[2]	Biting louse 4-5 weeks, sucking louse 4-7 weeks
Adult life span	Up to 35 days
Total eggs laid	Up to 10 eggs/female/day
May be mistaken for	Mites (Section 5.2), ticks (Section 5.1)
Impact on horses	Itching, dermatitis, secondary infections

[1] The pest distribution in the regions targeted by this publication are listed as of the date published. Areas of primary pest concern (if any) are in bold. Global distributions are listed in Section 5.3.2.
[2] Development times are approximate and can change with environmental conditions including temperature, humidity, and availability of food resources.

species may be slightly larger, 0.5 to 6.0 mm (less than 0.01 to 0.24 in), they are also flat but the head is as wide as or wider than the body. The antennae are thin and vary in shape by species. The legs are short and have claws for grasping hairs. The nits are oval and pale, almost translucent (Figure 5.44). The adults and nymphs look very similar.

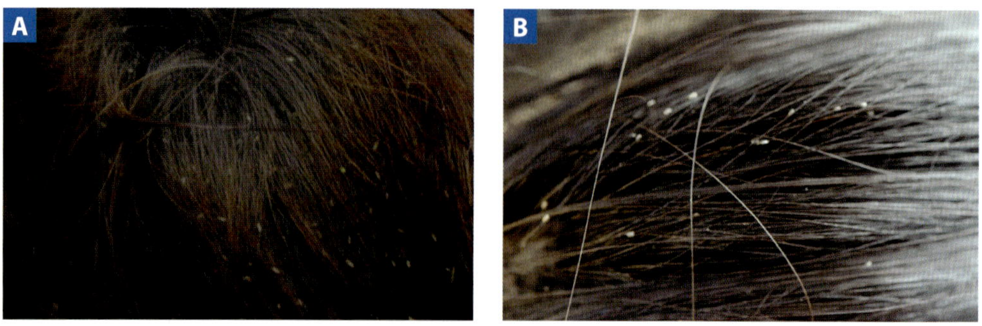

Figure 5.44. Louse infestations can be recognised by the presence of eggs (nits) glued to the hairs of the animal (A+B).

The horse biting louse (*Bovicola* [*Werneckiella*] *equi*) is a relatively small louse of approximately 1 to 2 mm (0.04 to 0.08 in) and brown with a yellow abdomen that has brown stripes (Figure 5.45). These lice are quite active and can be seen moving quickly through the hair.

The horse blood-sucking louse (*Haematopinus asini*) (Figure 5.46) is a larger grey to yellow-brown louse of approximately 3 to 5 mm (0.12 to 0.20 in). This species is slower and less mobile than the chewing louse and can often be seen with their mouthparts inserted.

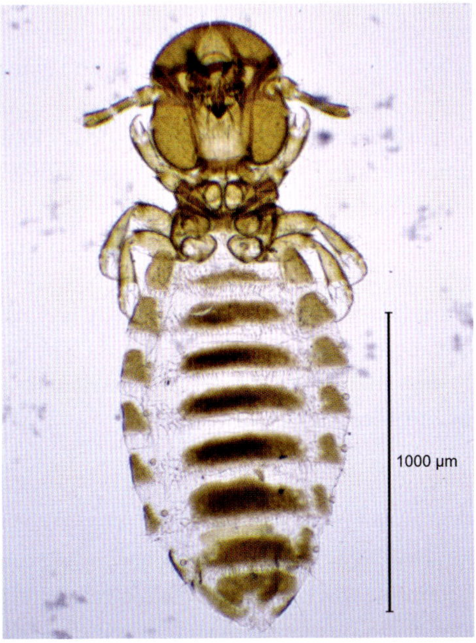

Figure 5.45. Horse biting louse (*Bovicola* [*Werneckiella*] *equi)* showing the characteristically broad head and chewing mouthparts of a chewing louse.

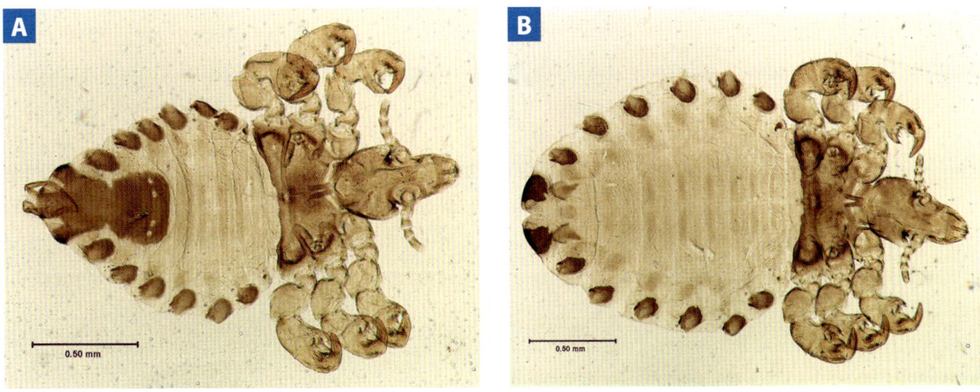

Figure 5.46. The horse blood-sucking louse (*Haematopinus asini*), showing the characteristic head that is narrower than the body of sucking lice. Images show male (A) and female (B).

5.3.4 Life history

Lice are small insects that are parasites of birds and mammals. There are two types of lice: sucking and chewing lice. Sucking lice pierce the skin and feed on blood and other fluids whereas chewing lice feed on skin and other debris. Almost every bird and mammal species has its own lice species. Mammals typically have one to three species of specialised lice. These lice species often have particular parts of the body where they infest and feed.

Lice have a hemimetabolous (incomplete) life cycle and have three stages: egg or nit, nymph, and adult. All stages take place on the host. The nymphs moult three times before becoming adults. The adults mate and the females lay their eggs, also known as nits, and glue them to the base of hairs.

The horse biting louse (*Bovicola* [*Werneckiella*] *equi*) prefers to lay its eggs on the finer hairs on the body of the horse, common sites to find eggs are the sides of the neck, the flank and at the base of the tail. Nymphs of the horse biting louse hatch from eggs after 5 to 10 days and develop into adults in 3 to 4 weeks.

The horse blood-sucking louse (*Haematopinus asini*) typically lays its eggs by the roots of the thicker hairs of the forelock, mane, and tail, or on the long hairs of the fetlocks (feathers). Nymphs of this species hatch 11 to 20 days after eggs are laid, immediately start sucking the blood of their host and mature to adults in 2 to 4 weeks.

5.3.5 Control

As with mites, lice are small and can be difficult to see but chewing lice can be seen moving through the hair. It is important to determine whether the problem is related to a louse infestation before attempting control. Symptoms of a louse infestation in horses include itching, skin inflammation, hair loss, scarring, biting at the skin, rubbing against walls and fences, depression, weakness, dull coat, matted hair, and raw skin. It is important to understand that during the warmer months, horses can also exhibit a *Culicoides* biting midge allergy called sweet itch, which may resemble louse infestations or may have a mite infestation, which presents similar symptoms. Although they are small, lice should be visible by the eye when parting the hair around the affected area. Nits (eggs) also may be visible, especially on a dark-coloured horse (Figure 5.44). In most circumstances, a veterinarian will develop a custom treatment plan that is most suited for the affected horse.

Lice are parasitic throughout their whole life cycle and cannot travel great distances or last for more than a couple of days off the host so transmission of lice between animals usually occurs through direct contact between an infested and uninfested animal or contact with an uninfested animal with a surface recently contacted by an infested animal such as a fence post or piece of equipment.

Cultural control

As with mites, risk factors associated with louse infestation include crowding of animals, sharing blankets, brushes, or other equipment between horses, poor feed quality, gestational status, underlying health issues, longer body coat (like a winter coat or feathering), and season; with infestations being more common in winter and early spring.

A louse check can be used to find infestations before they become a problem, or if an infestation is suspected. To do a louse check, part the hair in common lice infestation areas and look for nits (translucent eggs) or small, about 1 to 6 mm (0.04 to 0.24 in) in length, flat insects (Figure 5.47). The most common areas for lice infestations to check include around the forelock, mane, and tail for horse blood-sucking lice and on the neck and flank for horse biting lice (Figure 5.48).

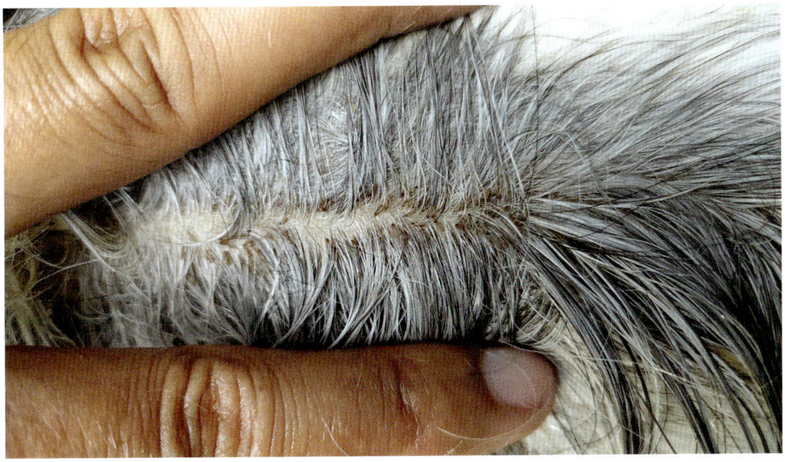

Figure 5.47. To check for lice, part the hair in areas either suspected of having lice or in areas where lice are common.

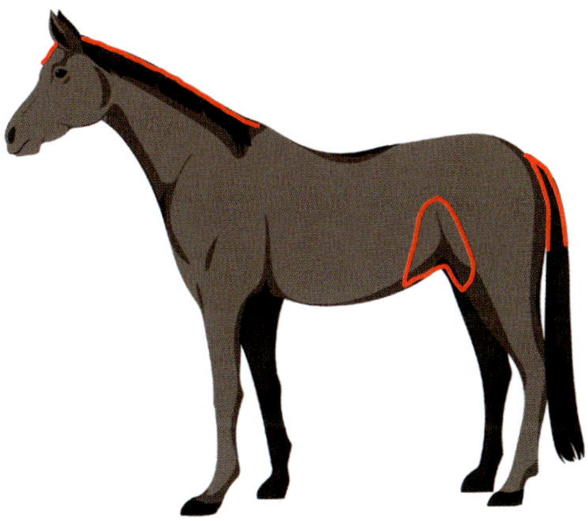

Figure 5.48. Check for lice around the forelock, mane, tail, and flank areas as shown in red.

When checking for lice, a magnifying glass and good light, like a headlamp or flashlight, is needed. A flea or louse comb can be useful when checking hair. Forceps might be useful to remove an insect suspected to be a louse under the magnifying glass, but tape can also be used to collect specimens off the comb for evaluation under a microscope or magnifying glass (Figure 5.49).

Grooming horses regularly, especially those with longer coats and in the winter months will disrupt lice and allow for a chance to check the coat for evidence of infestation. Another good time to check for lice if they are suspected is after exercising the horse because the warmth and sweat will bring the lice to the ends of the hair. Lice will not be found in areas of hair loss because they grip the hairs to stay in place on the horse.

Lice can survive off a host for a short time. Therefore, it is important to practice good hygiene to reduce the risk of introducing infestations or transferring them between individuals. If second-hand blankets/rugs or saddle pads are purchased or inherited, be sure to wash and dry them on high heat if possible (Figure 5.50). If not possible, store them for at least 14 days before using them to allow any remaining lice to die from starvation. Disinfect saddlery and grooming tools before use by soaking items in a bleach-water solution (1:20 ratio) and wiping with a pesticide and allowing them to dry (Figure 5.51). Keep separate grooming tools, saddlery and blankets for each horse and keep them well labelled. Stables should also be disinfected if there is a possibility of a lice infestation. Keep in mind that mites can live longer off the host than lice, so for routine hygiene, sterilisation is preferred to 14 days isolation.

Keeping horses in good health and well fed reduces the risk that they will suffer from a lice infestation that needs treatment. Sick, malnourished or stressed horses have reduced immunity and are at increased risk of lice. Regular grooming and periodic bathing can help reduce lice numbers on an animal, but also allows time to examine horses carefully.

Figure 5.49. After grooming areas with a flea comb that are suspected to have lice, tape can be used to collect specimens off the comb for evaluation under a microscope or magnifying glass.

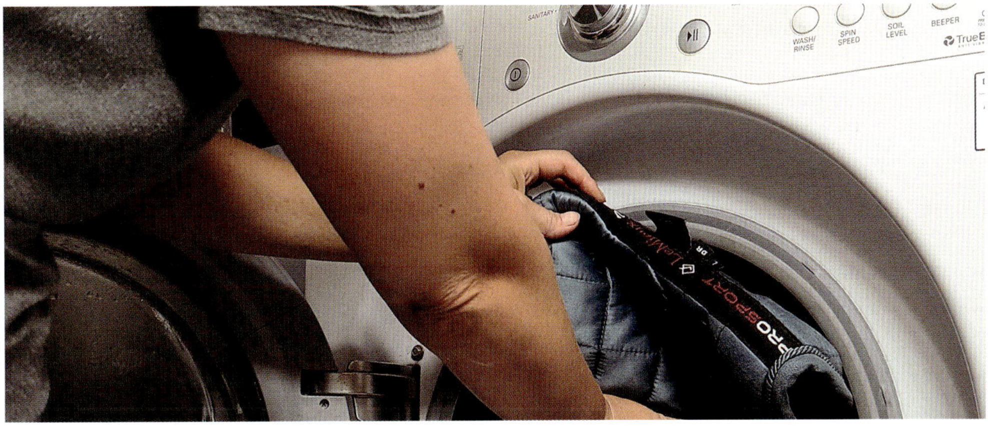

Figure 5.50. Saddle pads, blankets/rugs, and other equipment should be washed before use in between horses to prevent transferring lice among animals.

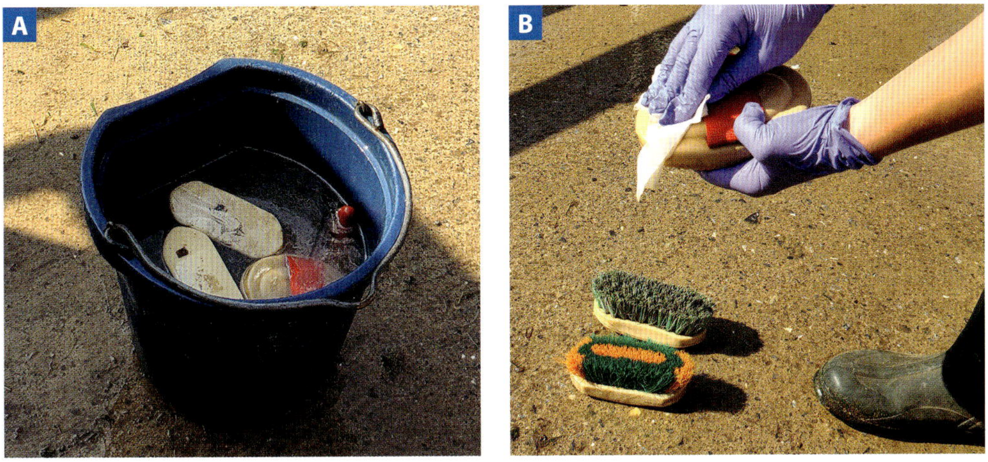

Figure 5.51. Disinfecting saddlery and grooming tools with a bleach solution (A) and then wiping with a pesticide (B) can help reduce the risk of moving lice to uninfested animals.

If new horses are going to be introduced to a herd, they should be checked for lice and treated as necessary. If horses need to be isolated due to a suspected louse infestation, any handlers should be careful to use gloves and other protective equipment before handling the infested horse to not spread the parasites.

Like mites, infestations of lice can be spread through direct contact with other horses. Horses suspected of mite or lice infestations should not travel to equestrian events until proper treatment is completed. When possible, keep uninfested horses away from other horses during competitions, exhibitions, or other travel. Louse checks are a good idea after travel for breeding or show, and especially if they have been housed with other animals.

Physical and mechanical control

Although chemicals can be used to kill lice it is important to also remove the eggs or nits and active stages from the hair by removing hair and skin debris, through clipping and shampooing the affected areas. As lice can be transferred from animal to animal and can survive in the environment for short periods, be sure to carefully dispose of hair and other debris and sanitise equipment.

If a horse is infested with lice it should be isolated if possible. If other horses have been in contact with the infested animal, they should be considered infested and treated as well.

Remove and dispose of bedding from stalls where infested horses have been and disinfect the stall and/or leave it unused for 14 days. If possible, remove the horses from pastures where infested horses have been for the same duration as the fence posts where horses have scratched may have lice.

Chemical control

Before chemical control, it is important to prepare the infested area to maximise chemical effectiveness. Clipping the hair in the affected area will help expose the area for treatment and remove some of the lice. After removing excess hair, clean the area with warm water and antibacterial soap.

Insecticidal shampoos, sprays, and wipe-on products are the most common but should only be used if labelled for horses as products designed for other animals may cause greater skin irritation. Suitable products typically contain pyrethrins, synthetic pyrethroids, imidacloprid, phoxim, selenium sulphide, triflumuron, or fipronil. Deltamethrin has been demonstrated to be effective and safe for the treatment of lice in horses (Castilla-Castaño *et al.*, 2017). Neem seed extract applied topically is effective on the horse biting louse. A study by Al-Quraishy *et al.* (2012) treated 300 horses with the extract diluted in water and found that the lice were killed even in those animals where infestations were not initially evident. Earlier studies that applied the extract to horses reported no skin irritation at concentrations of up to 1:20 (Schmahl *et al.*, 2010). Gloves and other protective equipment should be worn while treating the animal. It is important to note that none of the treatments available will kill the nits and so repeat treatments are essential to eradicate the infestation.

If using a shampoo, thoroughly wash the horse (Figure 5.52) and if using a spray or wipe-on product the horse must be thoroughly wetted. Repeat both types of treatment in 14 to 21 days to kill any newly hatched nymphs. If the weather is too cold to completely soak the animal then use a dust-based permethrin product. Use a brush to ensure that the dust gets into the coat; remember that lice live close to the skin. When treating for lice the entire animal should be treated, not just the affected areas. However, avoid sensitive areas like the eyes, nostrils, mouth, and mucosal membranes.

Horses that are regularly dewormed with monocyclic lactones, such as ivermectin, may be less likely to suffer from sucking lice infestations. However, these products are typically not labelled for use with lice on horses and effects on chewing lice will be minimal because they do not feed on blood.

Figure 5.52. Thoroughly wet the horse before using an insecticidal shampoo for the treatment of lice.

As with other parasites, many other products are labelled for louse control on cattle or sheep, but these should not be used unless also labelled for use on horses as an adverse reaction may occur. Amitraz is a chemical used in some mite control products for cattle and dogs but should not be used on horses as it is toxic.

If horses cannot be removed from pastures where infested horses have been, treat the fence posts, which may have lice, with a residual pesticide.

5.3.6 Veterinary concerns

The first signs of lice are the horse biting and rubbing the infested areas. The horse may be restless and hair or skin loss may occur. In areas where there are abundant lice, the hair may be matted. The main veterinary concern from lice is due to the extreme itching they cause. While lice do not carry viruses or bacteria, the skin damage the horse causes itching can be severe. Lacerations may require stitching, and severe abrasions are subject to secondary skin infections that require diligent wound care, and, potentially, antibiotics.

5.4 Keds/louse flies (family Hippoboscidae)

5.4.1 Pest summary (Table 5.4)

Keds (Figure 5.53), or louse flies (hereafter called keds), are true flies (Diptera) in the insect family Hippoboscidae. All of the 200 species in this family are obligate parasites of mammals and birds. Some ked species are winged, such as *Hippobosca equina*, the New Forest fly, which is an ectoparasite of horses and other mammals. Others, such as *Lipoptena cervi*, the deer ked

(Figure 5.53), have wings but lose them once they find a host animal. Both sexes of keds feed on the blood of their hosts. Each species of ked has a preferred host species and feeding on other hosts is unusual.

Table 5.4. Pest summary of deer ked/louse fly, family Hippoboscidae.

Distribution[1]	Australia: Yes Canada: Yes Europe: Yes New Zealand: No United States: Yes
Size	New Forest fly 6.0-8.5 mm (0.24-0.32 in), deer ked 5-7 mm (0.2-0.28 in)
Key identification characteristics	Flattened, strong forelegs, large eyes, brown or reddish-brown, move easily through the hair
Mouthparts	Piercing, sucking
Feed on	Blood
Development time[2]	6-7 weeks
Adult life span	4-5 months
Total eggs laid	Eggs develop internally, females produce 5-12 total offspring
May be mistaken for	Ticks (Section 5.1)
Impact on horses	Itching and irritation

[1] The pest distribution in the regions targeted by this publication are listed as of the date published. Areas of primary pest concern (if any) are in bold. Global distributions are listed in Section 5.4.2.

[2] Development times are approximate and can change with environmental conditions including temperature, humidity, and availability of food resources.

Figure 5.53. Deer ked (*Lipoptena cervi*) is shown here with wings, which will be lost when it finds a host. See Figure 5.56 for a ked that has lost its wings.

5.4.2 Distribution and seasonality

The New Forest fly, *Hippobosca equina* (Figure 5.54) is distributed throughout Europe and in parts of Africa and Asia. In the UK, the adults are most commonly seen from August to September. In the Americas, there are many louse fly species but none that are specific pests on horses. However, the deer ked, *Lipoptena cervi*, present in Europe and North America, may land on and bite horses, particularly if ridden or pastured in forested habitats where deer are present. Deer keds are most likely to be flying looking for hosts in the autumn.

5.4.3 Description

Keds are dorsoventrally flattened (like a pancake or plate). They have strong legs with claws, large eyes, and short antennae. Some species are wingless and others have long well-developed wings. While there are over 200 species of keds globally, the most important ectoparasites of horses are the New Forest fly and the deer ked.

The New Forest fly, *Hippobosca equina*, is pale red-brown with yellow spots on the abdomen (Figure 5.55). They have one pair of triangular wings that are 6.0 to 8.5 mm (0.24 to 0.32 in) in length. This species has a characteristic sideways movement through the hair.

The deer ked, *Lipoptena cervi*, is brown and 5 to 7 mm (0.20 to 0.28 in) in length (Figure 5.56). On the host, it will be wingless, although it emerges from the pupa with wings that it uses to find a host.

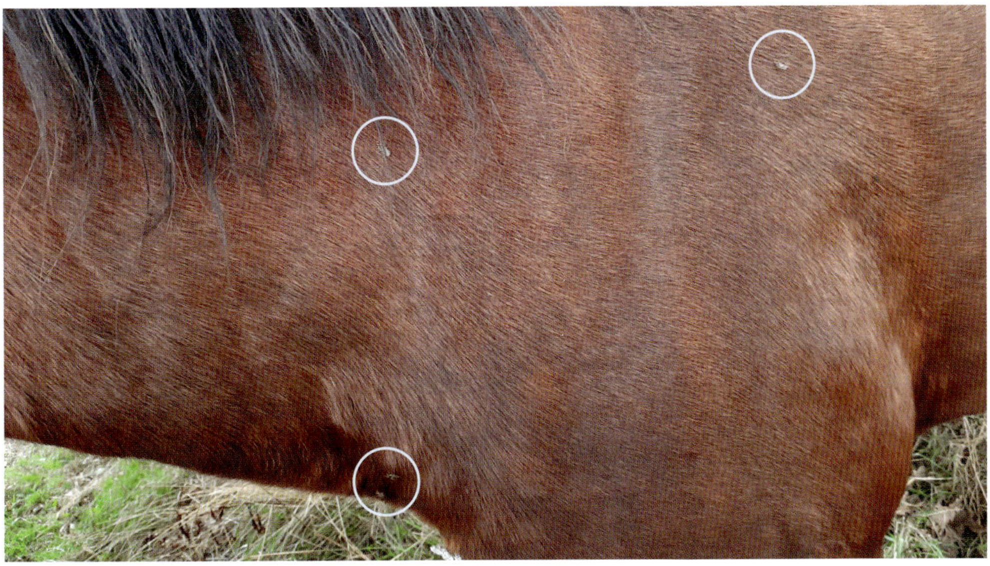

Figure 5.54. Some keds like *Hippobosca equina* (circled) can fly and bite horses while they are pastured or during rides.

Figure 5.55. The New Forest fly (*Hippobosca equina*) is distributed throughout Europe and in parts of Africa and Asia. This fly is pale red-brown with yellow spots on the abdomen.

Figure 5.56. The deer ked, *Lipoptena cervi*, may accidentally get on horses or people and bite, particularly if riding through forested areas. Deer keds lose their wings once they find a host, see Figure 5.53 for a deer ked with wings, which it uses to travel to a potential host.

5.4.4 Life history

Keds have complete metamorphosis with four life stages: egg, larva, pupa, and adult. However, unlike most flies, keds do not lay their eggs in the environment or on a host. Female keds develop one egg at a time in a modified uterus. The larva that hatches from the egg feeds from milk glands inside the female for approximately 3 weeks. The larva does not leave its mother until it is almost ready to pupate. The mother will deposit the pupa on the ground in a location where the emerging adult will easily find a new host. Adults emerge after 3 to 4 weeks and live for 4 to 5 months. A female will deposit five to 12 larvae in her lifetime. All keds are parasites of other animals and take blood meals from their host for nutrition.

New Forest flies do not lose their wings after landing on a host like some species do, so they can leave and return to the host if disturbed or to deposit larvae. They are most common in summer and autumn and attack areas with sensitive skin like around the anus and perineum, and the inner surface of the thighs in horses, as well as under the lower part of the external genital area in mares and fillies.

5.4.5 Control

Keds are rarely a major concern for horses so few control methods have been developed. Regular grooming, stable management, and pasture management are typically sufficient to keep ked numbers low or eliminate them on horses.

Cultural control

Keds are often mistaken for ticks due to their size and flat appearance but they will not be attached like a tick (Figure 5.57) and they will move very quickly across the skin and through the hair, unlike ticks that move more slowly (if they are not already attached). Regular grooming will disturb or remove louse flies. Keep in mind the common areas for infestation are between the hind legs and perineum area. Using a flea comb and forceps can help remove keds from the hair (Figure 5.58).

Regularly changing the bedding in stalls will remove pupae that may have dropped off the animal. Similarly, dragging pastures in the spring and summer may disrupt developing pupae before they can emerge as adults.

Because ked bites are more likely when riding through forested habitats, riding in open, windy areas or in the early morning when it is cooler and flies are not active can help avoid most bites.

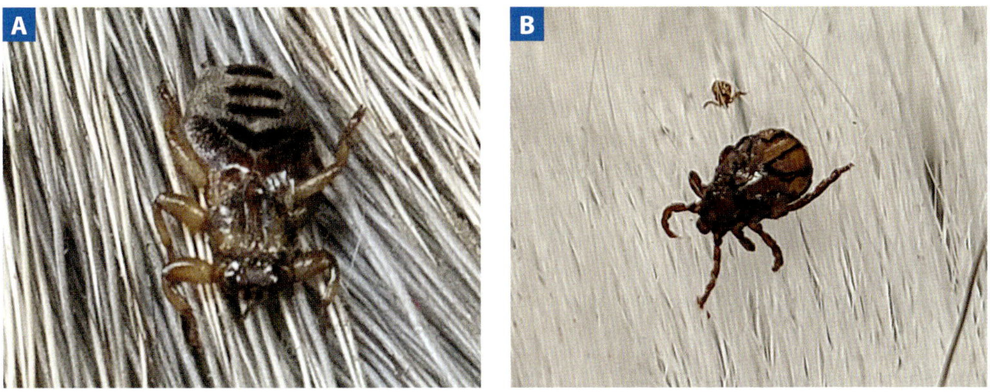

Figure 5.57. Keds (A) are often mistaken for ticks (B), but unlike ticks, they move very quickly through the hair.

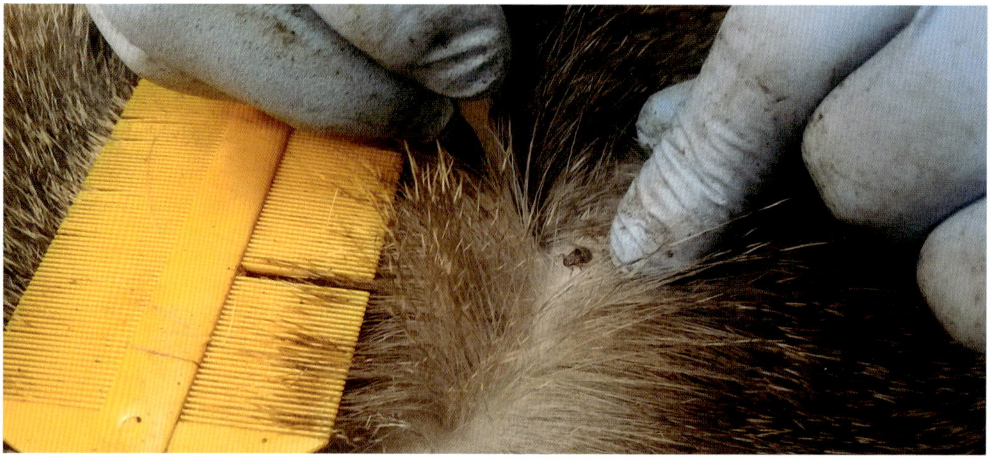

Figure 5.58. A flea comb and forceps can be used to help remove keds from the hair.

Chemical control

There has been little research into the management of keds as they are more of a nuisance pest in their native or invaded ranges. It is reasonable to assume that products for repelling or controlling keds or other insect pests on sheep or cattle will provide some degree of control on horses. For example, topical deltamethrin treatment has been shown to reduce ked numbers on horses for a month (Parashar *et al.,* 1991). However, currently, there are no products labelled for ked control on horses and recent research in the United States has demonstrated that keds are not repelled by common active ingredients used against ticks including clothes treated with permethrin and DEET (Michael Skvarla, personal communication).

5.4.6 Veterinary concerns

At low infestation levels, deer ked bites are mildly irritating to horses. In higher numbers, ked bites can cause extreme itching (pruritus) that can last several weeks. Itching can be controlled with typical anti-itch treatments. The New Forest fly has a very painful bite and horses infested with this species may rub against objects and exhibit nervous behaviours like tail swishing. In extreme cases, horses may act erratically to evade ked pressure, which may result in self-injury such as limb fractures (Sokół and Michalski, 2015).

Continuing education

Regions where ticks live

U.S. Centers for Disease Control and Prevention, 2021.
https://www.cdc.gov/ticks/geographic_distribution.html

- Detailed maps for common ticks of medical and veterinary importance in the United States.

Tick maps
European Centre for Disease Control and Prevention, 2021.
https://www.ecdc.europa.eu/en/disease-vectors/surveillance-and-disease-data/tick-maps
- Detailed maps for common ticks of medical and veterinary importance in Europe.

Cattle tick control
Northern Territory Government, 2021.
https://nt.gov.au/industry/agriculture/livestock/animal-health-and-diseases/cattle-tick-control
- Detailed maps for the cattle tick, *Rhipicephalus microplus*, in Australia.

Australian paralysis tick
Australian Museum, 2020.
https://australian.museum/learn/animals/insects/australian-paralysis-tick/
- Detailed maps for the paralysis tick, *Ixodes holocyclus*, in Australia.

Bristol University tick identification, online photographic guide to ticks: key to genera
Wall, R., 2016.
http://www.bristoluniversitytickid.uk/page/Key+to+Genera/6/#.YNl-6kwpCUl
- An easy to use tick identification key for hard ticks permitting identification to genera using photos and drawings.

Electric fence designs for deterring white-tailed deer
McPeake, B., Fryer, J., Simon, K. and Perez, A.P., 2019.
https://www.uaex.edu/publications/FSA9111.pdf
- Detailed instructions on installing fencing to exclude deer from pastures.

Tick paralysis of animals in Australia
Padula, A.M., 2018. In: Clinical toxicology in Europe, Australia, and the Americas. Springer, Berlin, Germany, pp. 279-304. ISBN 978-94-017-7438-3.
https://doi.org/10.1007/978-94-017-7438-3_65
- This is a chapter in the book *Clinical Toxicology in Europe, Australia, and the Americas*. This chapter is a review of all aspects of tick paralysis relevant to animals in Australia and discusses signs and treatment.

Equine piroplasmosis: a review
De Waal, D.T., 1992. British Veterinary Journal 148: 6-14.
https://doi.org/10.1016/0007-1935(92)90061-5
- This review focuses on equine piroplasmosis with specific reference to its distribution, diagnosis and clinical and pathological signs. The more commonly used drugs and treatments are discussed.

A review on the treatment and control of ectoparasite infestations in equids

Karasek, I., Butler, C., Baynes, R. and Werners, A., 2020. Journal of Veterinary Pharmacology and Therapeutics 43: 421-428. https://doi.org/10.1111/jvp.12874

- A limited number of products are approved for use of ectoparasite control in horses. This review discusses the currently labelled products as well as products that have been used off label under veterinary supervision.

Clinical and therapeutic studies on mange in horses

Osman, S.A., Hanafy, A. and Amer, S.E., 2006. Veterinary Parasitology 141: 191-195. https://doi.org/10.1016/j.vetpar.2006.04.039

- This research paper discusses clinical signs observed with mange infested horses including the common locations of different types of mange. Treatment of mange is evaluated and discussed.

Chronic progressive lymphoedema in draught horses

De Keyser, K., Janssens, S. and Buys, N., 2018. Equine Veterinary Journal 47: 260-266. https://doi.org/10.1111/evj.12256

- This review summarises and evaluates the current knowledge about chronic progressive lymphedema in draught horses.

Resources

Resources are listed in alphabetical order and may not be inclusive of all companies, materials, or products available.

4-poster deer feeder. https://crdaniels.com/shop/ols/categories/4-poster-deer-feeder
Damminix. https://ticktubes.com
Insect Shield® Insect Repellent Fabric Treatment. https://www.insectshield.com/
Tick boxes. http://www.tickboxtcs.com/
Tick Key. https://tickkey.com/product-catalog/
Tick Nipper. http://ticknipper.com/
Tick Tornado. https://zenpetusa.com/products/tick/tick-remover-tool/
Tick tubes. https://www.thermacell.com/products/tick-control-tubes

References

Al-Quraishy, S., Abdel-Ghaffar, F., Al-Rasheid, K.A.S., Mehlhorn, J. and Mehlhorn, H., 2012. Observations on effects of a neem seed extract (MiteStop®) on biting lice (mallophages) and bloodsucking insects parasitizing horses. Parasitology Research 110: 335-339. https://doi.org/10.1007/s00436-011-2495-0

Brown, R.D. and Cooper, S.M., 2006. In my opinion: the nutritional, ecological, and ethical arguments against baiting and feeding white-tailed deer. Wildlife Society Bulletin 34: 519-524.

Castellanos, A.A., Medeiros, M.C., Hamer, G.L., Morrow, M.E., Eubanks, M.D., Teel, P.D., Hamer, S.A. and Light, J.E., 2016. Decreased small mammal and on-host tick abundance in association with invasive red imported fire ants (*Solenopsis invicta*). Biology Letters 12: 20160463. https://doi.org/10.1098/rsbl.2016.0463

Castilla-Castaño, E., Vischi, A., Navarro, C., Lecru, L.A., Ribeiro, C., Pradier, S. and Cadiergues, M-C., 2017. Control of lice infestation in horses using a 10 mg/mL deltamethrin topical application. Irish Veterinary Journal 70: 22. https://doi.org/10.1186/s13620-017-0100-2

Chitimia-Dobler, L., Nava, S., Bestehorn, M., Dobler, G. and Wölfel, S., 2016. First detection of *Hyalomma rufipes* in Germany. Ticks and Tick-borne Diseases 7: 1135-1138.

Connally, N.P., Rose, D.A., Breuner, N.E., Prose, R., Fleshman, A.C., Thompson, K., Wolfe, L., Broeckling, C.D. and Eisen, L., 2019. Impact of wearing and washing/drying of permethrin-treated clothing on their contact irritancy and toxicity for nymphal *Ixodes scapularis* (Acari: Ixodidae) ticks. Journal of Medical Entomology 56: 199-214. https://doi.org/10.1093/jme/tjy138

Dumic, I. and Severnini, E., 2018. 'Ticking bomb': the impact of climate change on the incidence of Lyme disease. Canadian Journal of Infectious Diseases and Medical Microbiology 2018: 5719081 https://doi.org/10.1155/2018/5719081

Fernández-Ruiz, N. and Estrada-Peña, A., 2021. Towards new horizons: Climate trends in Europe increase the environmental suitability for permanent populations of *Hyalomma marginatum* (Ixodidae). Pathogens 10: 95. https://doi.org/10.3390/pathogens10020095

Foley, J.E., 2020 Tularemia in animals. Merck Veterinary Manual https://www.merckvetmanual.com/generalized-conditions/tularemia/tularemia-in-animals

Hansford, K.M., Carter, D., Gillingham, E.L., Hernandez-Triana, L.M., Chamberlain, J., Cull, B., McGinley, L., Phipps, L.P. and Medlock, J.M., 2019. *Hyalomma rufipes* on an untraveled horse: Is this the first evidence of *Hyalomma* nymphs successfully moulting in the United Kingdom? Ticks and Tick-borne Diseases 10: 704-708. https://doi.org/10.1016/j.ttbdis.2019.03.003

Hennessy, C. and Hild, K., 2021. Are Virginia opossums really ecological traps for ticks? Groundtruthing laboratory observations. Ticks and Tick-borne Diseases 12: 101780 https://doi.org/10.1016/j.ttbdis.2021.101780

Hyde, J., Nettleton, P., Marriott, L. and Willoughby, K., 2007. Louping ill in horses. Veterinary Record 160: 532.

Little, E.A., Williams, S.C., Stafford, K.C., Linske, M.A. and Molaei, G., 2020. Evaluating the effectiveness of an integrated tick management approach on multiple pathogen infection in *Ixodes scapularis* questing nymphs and larvae parasitizing white-footed mice. Experimental and Applied Acarology 80: 127-136.

Machtinger, E. and Li, A.Y., 2019. Tick control bait box use by *Peromyscus* spp. influenced by habitat placement but raises questions on disease ecology. Ecosphere 10: e02972. https://doi.org/10.1002/ecs2.2972

Neely, M., Arroyo, L.G., Jardine, C., Moore, A., Hazlett, M., Clow, K., Archer, H. and Weese, J.S., 2021. Seroprevalence and evaluation of risk factors associated with seropositivity for *Borrelia burgdorferi* in Ontario horses. Equine Veterinary Journal 53: 331-338.

Ostfeld, R.S. and Brunner, J.L., 2015. Climate change and *Ixodes* tick-borne diseases of humans. Philosophical Transactions of the Royal Society B: Biological Sciences 370: 20140051. https://doi.org/10.1098/rstb.2014.0051

Parashar, B.D., Gupta, G.P. and Rao, K.M., 1991. Control of the haematophagous fly *Hippobosca maculata*, a series pest of equines, by deltamethrin. Medical and Veterinary Entomology 5: 363-367. https://doi.org/10.1111/j.1365-2915.1991.tb00562.x

Reichard, M.V. and Thomas, J.E., 2015. Mange in horses. Merck manual, Veterinary manual. https://www.merckvetmanual.com/integumentary-system/mange/mange-in-horses#v3279848

Schmahl, G., Al-Rasheid, K.A.S., Abdel-Ghaffar, F., Klimpel, S. and Mehlhorn, H., 2010. The efficacy of neem seed extracts (Tre-san®, MiteStop®) on a broad spectrum of pests and parasites. Parasitology Research 107: 261-269. https://doi.org/10.1007/s00436-010-1915-x

Sokół, R. and Michalski, M.M., 2015. Occurrence of *Hippobosca equina* in Polish primitive horses during the grazing season. Annals of Parasitology 61: 119-124.

Timoney. P.J., Donnelly, W.J., Clements, L.O. and Fenlon, M., 1976. Encephalitis caused by louping ill virus in a group of horses in Ireland. Equine Veterinary Journal 8: 113-117. https://doi.org/10.1111/j.2042-3306.1976.tb03311.x

Trumpp, K.M., Parsley, A.L., Lewis, M.J., Camp, J.W. Jr. and Taylor S.D., 2019. Presumptive tick paralysis in two American miniature horses in the United States. Journal of Veterinary Internal Medicine 33: 1784-1788.

Weeks, E.N.I., Machtinger, E.T., Leemon, D. and Geden C.J., 2018. Biological control of livestock pests: Entomopathogens. In: Garros, C., Bouyer, J., Takken, W. and Smallegange, R.C. (eds.) Prevention and control of pests and vector-borne diseases in the livestock industry. Ecology and control of vector-borne diseases Volume 5. Wageningen Academic Publishers, Wageningen, the Netherlands, pp. 337-387. https://doi.org/10.3920/978-90-8686-863-6_12

White, A. and Gaff, H., 2018. Review: Application of tick control technologies for blacklegged, lone star, and American dog ticks. Journal of Integrated Pest Management 9: 12. https://doi.org/10.1093/jipm/pmy006

Williams, S.C., Stafford III, K.C., Molaei, G. and Linske, M.A., 2018. Integrated control of nymphal *Ixodes scapularis*: Effectiveness of white-tailed deer reduction, the entomopathogenic fungus *Metarhizium anisopliae*, and fipronil-based rodent bait boxes. Vector-Borne and Zoonotic Diseases 18: 55-64. https://doi.org/10.1089/vbz.2017.2146

Photo credits

Title page	C.E. Taylor
Figure 5.1.	Mogrzewalska, Shutterstock image no. 1572828457.
Figure 5.2.	Lyle Buss, University of Florida.
Figure 5.3.	Erika Machtinger, Pennsylvania State University.
Figure 5.4.	(A) Erika Machtinger, Pennsylvania State University; (B) Centers for Disease Control and Prevention, Dr Gary Alpert – Urban Pests – Integrated Pest Management, public access.
Figure 5.5.	Jude Black, Shutterstock_1259774959.
Figure 5.6.	(A,B) Erika Machtinger, Pennsylvania State University; (C,D) Centers for Disease Control and Prevention, Dr Christopher Paddock, Public Access.
Figure 5.7.	Erika Machtinger, Pennsylvania State University.
Figure 5.8.	Erika Machtinger, Pennsylvania State University.
Figure 5.9.	Armando Frazao, Shutterstock.
Figure 5.10.	David Kattes, Tarleton University.
Figure 5.11.	Chris Holderman, University of Florida.
Figure 5.12.	Jim Occi, BugPics, Bugwood.org.
Figure 5.13.	Erika Machtinger, Pennsylvania State University.
Figure 5.14.	Emma Weeks.
Figure 5.15.	Irina Orlova, Shutterstock.
Figure 5.16.	Erika Machtinger, Pennsylvania State Entomology.
Figure 5.17.	(A) Erika Machtinger, Pennsylvania State University; (B) Roxie White, USDA, ARS.
Figure 5.18.	Chulantha P. Diyes, Department of Zoology, University of Peradeniya.
Figure 5.19.	(A) American Red Cross; (B) Alexander62, Shutterstock no. 622966775; (C) Chris Geden, USDA, ARS.

Figure 5.20. Erika Machtinger, Pennsylvania State University.
Figure 5.21. Erika Machtinger, Pennsylvania State University.
Figure 5.22. Erika Machtinger, Pennsylvania State University.
Figure 5.23. Emma Weeks, University of Florida.
Figure 5.24. Erika Machtinger, Pennsylvania State University.
Figure 5.25. Erika Machtinger, USDA, ARS.
Figure 5.26. Erika Machtinger, Pennsylvania State University.
Figure 5.27. (A) Holly Gellerman, Island Conservation, National Science Foundation Multimedia Gallery https://nsf.gov/news/mmg/mmg_disp.jsp?med_id=72912&from=; (B) Erika Machtinger, Pennsylvania State University.
Figure 5.28. Erika Machtinger, USDA, ARS.
Figure 5.29. Erica Lacher, Springhill Equine Veterinary Clinic.
Figure 5.30. Wimala Namket, Shutterstock_1228617484.
Figure 5.31. Image provided courtesy of the National Center for Veterinary Parasitology, ncvetp.org.
Figure 5.32. Emma Weeks, University of Florida.
Figure 5.33. Emma Weeks, University of Florida.
Figure 5.34. Emma Weeks, University of Florida.
Figure 5.35. Eric Erbe, United States Department of Agriculture, Image Number K9077-26, Bugwood.org.
Figure 5.36. Erika Machtinger, Pennsylvania State University.
Figure 5.37. Susan Ellis, Bugwood.org.
Figure 5.38. Erika Machtinger, Pennsylvania State University.
Figure 5.39. Erika Machtinger, Pennsylvania State University.
Figure 5.40. Erika Machtinger, Pennsylvania State University.
Figure 5.41. Chelle129, Shutterstock no.1415304791.
Figure 5.42. University of California, Davis, School of Veterinary Medicine, Center for Equine Health.
Figure 5.43. Image provided courtesy of the National Center for Veterinary Parasitology, ncvetp.org.
Figure 5.44. (A) courtesy of Prof. Sameeh M. Abutarbush; (B) Khnunkorn, Shutterstock image no. 1467351707.
Figure 5.45. Bilal Dik, Selcuk University.
Figure 5.46. The Trustees of the Natural History Museum, London. Licensed under the Open Government Licence.
Figure 5.47. Berta Alviyanto, Shutterstock.
Figure 5.48. Erika Machtinger, Pennsylvania State University.
Figure 5.49. Erika Machtinger, Pennsylvania State University.
Figure 5.50. Erika Machtinger, Pennsylvania State University.
Figure 5.51. Erika Machtinger, Pennsylvania State University.
Figure 5.52. Erika Machtinger, Pennsylvania State University.
Figure 5.53. Erika Machtinger, Pennsylvania State University.
Figure 5.54. Emma Weeks.
Figure 5.55. Emma Weeks; identified by Steffen Rehbein, Boehringer Ingelheim.
Figure 5.56. Steffen Rehbein, Boehringer Ingelheim.
Figure 5.57. Erika Machtinger, Pennsylvania State University.
Figure 5.58. Erika Machtinger, Pennsylvania State University.

Chapter 6. Pests found in pastures and forage

Chapter overview

As grazing animals, most horses spend time in pastures. Filth flies, such as face flies and horn flies are common pests of pastured horses that have already been discussed in Chapter 3. Some arthropods can be accidental or incidental pests of horses in pastures or forages. Fire ants (Section 6.1) and Africanized bees (Section 6.2) are accidental pests of horses. If nests in a pasture are disturbed, stings may lead to severe allergic reactions. Eye gnats (Section 6.3) can be found in pastures and can cause annoyance and corneal eye ulcers. Some insects can cause major health problems if ingested. Blister beetles (Section 6.4) are a unique group of pests. They do not associate with horses directly but occur on the plants used to make hay, especially alfalfa. If ingested by a horse, cantharidin, the chemical associated with these beetles, can cause cantharidin poisoning which can result in death. Caterpillars with stinging hairs (called urticating caterpillars), like the tent caterpillar (Section 6.5.1) and the bag-shelter moth (Section 6.5.2), can be found in trees in pastures. Ingestion of these caterpillars can cause abortions. Many of these pasture and forage pests are regionally limited, so the first step of control is knowing whether that insect is present in that location. Further control depends on the species, but cultural control methods and prevention are the most important for pasture pests.

6.1 Fire ants (genus *Solenopsis*)

6.1.1 Pest summary (Table 6.1)

Ants are in the insect family Formicidae in the order Hymenoptera. In the genus *Solenopsis,* several species are known as fire ants because of their painful stings. Some species may also be called red ants due to their light brownish reddish colour (Figure 6.1). Many species of ants bite, and some cause irritation by spraying formic acid. However, fire ants bite using their mandibles (mouthparts) and then inject venom using their stinger. The sting is painful to large animals and can cause allergic reactions. For smaller animals, these stings can be deadly when large numbers of fire ants attack.

6.1.2 Distribution and seasonality

In general, fire ants are in the genus *Solenopsis* and are from the New World, but most have spread beyond their native ranges to other continents. For example, *Solenopsis invicta* (Figure 6.1), the red imported fire ant, is native to South America but has been introduced to the USA, Australia, and several Asian and Caribbean countries. Similarly, *Solenopsis geminata*, the tropical fire ant, is native to the New World (USA, South America, and the Caribbean) but has spread to

Table 6.1. Pest summary of fire ants, *Solenopsis* spp.

Distribution[1]	Australia: Yes Canada: No Europe: No New Zealand: Yes United States: Yes
Size	2-6 mm (0.08-0.25 in)
Key identification characteristics	Narrow waist with two lumps, antennae with 10 segments; usually identified based on aggressive behaviour and nest building
Mouthparts of adults	Chewing
Feed on	Omnivorous, feed on invertebrates, plants, and plant parts
Development time[2]	23-55 days
Adult life span	1-6 months as a worker or soldier, 2-7 years as a queen, a few days as a male drone
Total eggs laid	Up to 1,500 per day
May be mistaken for	Other ant species
Impact on horses	Local allergic reactions

[1] The pest distribution in the regions targeted by this publication are listed as of the date published. Areas of primary pest concern (if any) are in bold. Global distributions are listed in Section 6.1.2.
[2] Development times are approximate and can change with environmental conditions including temperature, humidity, and availability of food resources.

Figure 6.1. Red fire ant (*Solenopsis invicta*) worker caste.

Asia, Australia, New Zealand, and Africa. Fire ants are common throughout the southern USA and Mexico and appear to be limited by the winters in more northern states. Expansion of these cold-limited species may become more common with climate change. Fire ants form their nests in the soil and can be associated with moist areas, such as rivers or ponds, but they prefer open, sunny areas like pastures.

6.1.3 Description

Solenopsis species have three characteristics to distinguish them from other ants: a narrow waist (or petiole) with two lumps (or nodes), antennae with 10 segments ending in a two-segmented club shape, and a smooth propodeum (segment before the narrow waist) (Figure 6.2). However, several species of ants share these characteristics. Some key behaviours of fire ants can help with identification. First, fire ant workers seen around the nest are different sizes. Fire ant mounds are also larger than many other ant species without a central entrance hole. Fire ants also are easily identified by their aggressive behaviour. When a nest is disturbed, swarms of ants will exit the nest to defend it (see Canaday and Parkman, 2020 in 'Continuing education' for more information on identifying *Solenopsis* species from other common ant species in the USA).

Adult fire ants vary in size and colour depending on the type of adult, called a caste. There are four castes of adult fire ants: the queen, males (called drones), soldiers, and workers. Worker ants, which are the most commonly observed, may be 2 to 6 mm (0.08 to 0.25 in) in length and can be black to reddish in colouration (Figure 6.3). Queens and male drones can have wings. Queens lose their wings after they establish a nest. Drones will die outside the nest after mating.

Figure 6.2. *Solenopsis* species have a narrow waist (or petiole) (labelled A) with two lumps (or nodes) (labelled B).

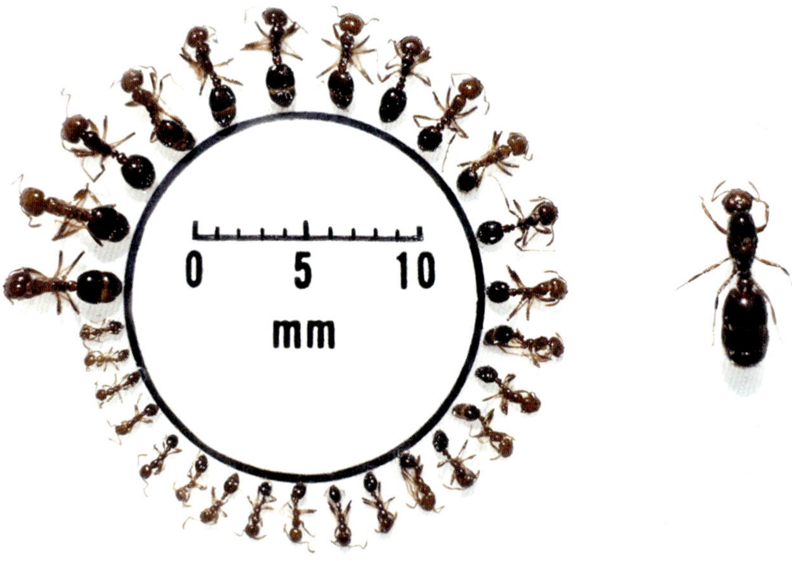

Figure 6.3. Fire ant workers (in the circle) vary in size and colour. The fire ant queen is on the right.

6.1.4 Life history

Fire ants have complete metamorphosis with four life stages: egg, larva, pupa, and adult, and live in colonies like honeybees. In pastures and other open areas, nests will appear as dome-shaped mounds of dirt up to 40 cm (16 in) tall (Figure 6.4) and may not have an obvious entrance hole. Fire ant nests may also be under wood, stones, or other objects and escape notice.

One or multiple fire ant queens will start a new nest site. Colonies with single queens are territorial, which limits populations to about 150 mounds with around 7 million ants per acre (approximately 60 mounds and 3 million ants per hectare). Multiple-queen colonies will share resources and are tolerant of closer mounds allowing for up to 300 mounds or 40 million ants per acre (approximately 120 mounds and 16 million ants per hectare) (National Parks Service, 2015).

Fire ant queens may live for 7 years and produce up to 1,500 eggs per day. Unfertilised eggs become female workers that can have many roles depending on their size. The largest ants have large mandibles that they use to chew and store food for the colony. The smaller workers care for the immature ants, clean the nest, and forage for food. Although they feed mostly on dead animals, seeds, and small plants, they may also attack and kill small animals. Fertilised eggs become male drones that mate with queens.

Figure 6.4. Fire ant nests in pastures and other open areas will appear as dome-shaped mounds without a distinct entrance hole (A-C).

6.1.5 Control

If the equine farm is located in a region where fire ants occur, ants will nest in the open in pastures or under logs, buckets, troughs, fence posts, or other areas of the pasture. Prevention can be difficult because of the size of colonies. However, populations can be reduced with some cultural, physical, and mechanical control methods. If more aggressive management is required, chemical control options can be used. In some areas of the world, like Australia, there are regional eradication programs in place that have restrictions on control mechanisms and require reporting of fire ant mounds. It is important to check with local governments to determine if these restrictions apply (see Queensland Government, 2018 in 'Continuing education' for more information on the fire ant eradication program in Australia).

Cultural control

Prevention of fire ants is difficult, particularly in the pasture environment. Like all pests, fire ants need food, water, and protected areas to live in, but colonies can usually find these resources. However, limiting access to additional human food or animal food waste prevents inadvertently supplementing the ants' food supply.

Physical and mechanical control

Mound treatment

Three gallons (11.5 litres) of boiling water poured onto the mound can be up to 60% effective at reducing ant numbers (Tschinkel and Howard, 1980). These treatments may kill some of the colony, but many repeated treatments will be necessary as this generally does not kill the entire colony, and it may encourage fire ants to spread to other areas. Treatments may be the most effective in the morning or when ants may be near the surface like after heavy rain. Ant colonies have a complex underground structure with many entrances and can easily relocate if the main mound is threatened. Studies have shown that several home remedies, including applying instant grits, molasses, aspartame, or club soda to ant mounds, do not work (Brown, 2014). Pouring chlorine, ammonia, gasoline, or diesel fuel on mounds not only do not work, but these methods can contaminate the soil and groundwater and are dangerous.

Mound flattening

Frequent mowing discourages the building of large mounds. Shallow disking, harrows (Figure 6.5), or dragging heavy objects like railroad ties or drags across pastures can temporarily flatten tall mounds and reduce fire ant numbers, although this will not eliminate fire ants and may spread colonies over a wider area. However, frequent manipulation can keep mounds lower if they are in riding areas or if large mounds present a tripping hazard.

Figure 6.5. Shallow disking or use of drag harrows (pictured) on pastures can provide temporary reductions in ant numbers.

Biological control

Other ant species are the main predators of fire ants. For this reason and environmental protection, it is important to apply targeted fire ant control treatments whenever possible to avoid harmful effects on potentially beneficial ant species.

Parasitoid flies

Ant-decapitating flies are tiny flies in the family Phoridae that are small hump-backed fly parasitoids of fire ants. The flies lay their eggs in the ant's thorax. After hatching, the fly larvae migrate to the head where they feed on various body tissues. Once they are ready to pupate they release enzymes that cause the ant's head to fall off, and the fly larvae then pupate in the detached head. These flies appear to provide fire ant control in their native ranges and have been introduced for biological control in the southern USA (Porter and Calcaterra 2013), but are currently not available commercially.

Entomopathogenic fungi

Beauveria bassiana and *Metarhizium brunneum* (*anisopliae*) are natural fungi and natural enemies of fire ants, but use in pastures was initially hampered by fungal sensitivity to extreme environmental conditions like ultraviolet (UV) light and high temperatures and humidity, which reduces the effectiveness of the fungi. Microencapsulated products being developed may reduce the harmful effects of UV radiation and might provide the answer to this problem (Bextine and Thorvilson, 2002). While there are no commercial products currently available, the use of these fungi for the control of other pests is increasing and products may be marketed in the future.

Chemical control

There are two options for chemical control: (1) targeted applications to mounds through either bait, dust, injection, or drench, and (2) broadcast applications of granules or baits applied over a large infested area. If the area has many mounds, the latter may be necessary, but the targeted application is preferred in limited infestations to reduce the toxic effects on beneficial insects.

Before chemical controls are applied, the cost of the product, labour, and equipment, and the time needed to apply the control must be considered. Baits and drenches should not be applied at the same time. Wait 7 to 10 days between applications. Ants will not actively forage if they encounter a toxin in the drench, which will make the baits ineffective. The best time of the day to apply chemical control is mid-morning when temperatures are not too hot, about 21 °C (70 °F).

Baits should not be applied if the ground is wet or if rain is expected within 2 days. Before applying bait, make sure that workers are actively foraging by placing a piece of snack food like a chip/crisp or a piece of hot dog on the ground close to, but not on, the mound (Figure 6.6). The ants should arrive at the food item within 5 to 10 minutes. Apply the bait around the mounds but not on the mound as ants will not forage on the mound itself. Do not disturb the mounds when applying. Make sure to read the label directions as some products may require reapplication. When applying bait over a larger area, a handheld spreader or a seed spreader mounted on a tractor or similar equipment can be used (Figure 6.7).

After bait is applied, worker ants will collect the bait as food for the colony, and then it will be chewed and fed to the queen and immature ants. The main active ingredients in these baits are insect growth regulators, such as pyriproxyfen, s-methoprene, and fenoxycarb. These products are not directly toxic to the ants, but they prevent individuals from becoming adults or adults

Figure 6.6. Fire ants are highly attracted to some human foods, such as chips/crisps (pictured) and they can be used to monitor for activity before mound treatment.

Figure 6.7. Fire ant bait can be applied with a hand spreader (shown here) or by using a seed spreader mounted to a tractor.

from producing offspring. Hydramethylnon is another active ingredient; it is a metabolic inhibitor preventing cellular respiration. However, death occurs slowly, allowing the individuals that contact the bait to return to the nest and feed it to the rest of the colony. Indoxacarb, an oxadiazine pesticide, can be applied in the same way as baits. However, its action is on the nervous system of insects.

It is necessary for the chemicals in drench applications to contact the ants, but if applied properly, the ants should be dead within a few hours. It is important to completely drench the mound and to apply the treatment first as a barrier around the outside of the mound to kill any escaping ants. Drench products, containing the active ingredients carbamate and carbaryl, are labelled for ant control when applied directly to mounds.

It is important to choose a product that is labelled for use in horse pastures. The application of chemicals that are not labelled for use with horses is illegal and potentially dangerous for horses and other animals. Several products have been labelled for fire ant control (See Billeisen *et al.*, 2021 in 'Continuing education' for a list of suitable products).

6.1.6 Veterinary concerns

Immediate signs of fire ant stings are agitation, pruritus (severe itching on the skin), and erythema (superficial skin reddening). Stings usually develop into pustules in 12 to 24 hours due to local necrosis (death of tissue cells) caused by piperidine alkaloids in the fire ant venom (Figure 6.8). Affected skin may feel thickened.

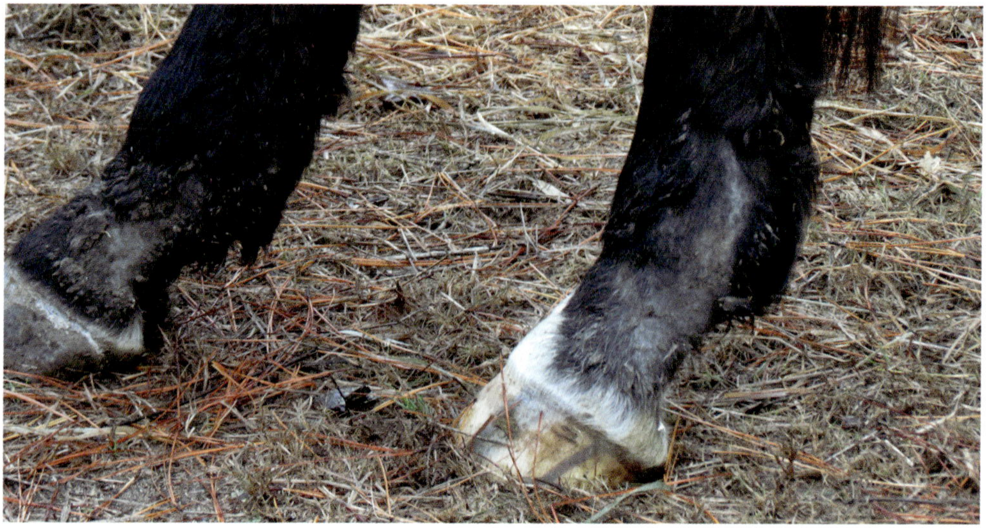

Figure 6.8. Dermatitis on leg caused by multiple fire ant bites.

While fire ant bites are rarely a veterinary concern, newborn foals can be killed if attacked by ants from a mound. Adult horses rarely have fatal interactions with ants, and anaphylactoid reactions (acute reactions as a result of exposure to a trigger) are rare without extremely high exposure to stings and high sensitivity to the fire ant venom proteins. Large numbers of stings can cause significant itching, and this may lead to additional self-harm.

If ants remain attached to the horse, they should be groomed or washed off. Contact a veterinarian if a horse has ant bites extending over a large portion of its body, or if it is experiencing extreme itching. The main goal of treatment is to reduce itching and pain using corticosteroids and nonsteroidal anti-inflammatory drugs. In some cases, antihistamines may be beneficial. Horses should be monitored, as the stings can also serve as a gateway for bacterial infections, leading to cellulitis. Good basic cleaning and wound care will help reduce this potential complication.

6.2 Africanized bees (*Apis mellifera* subspecies hybrids)

6.2.1 Pest summary (Table 6.2)

Africanized bees (Figure 6.9) are a hybrid between subspecies of the Western or European honeybee (*Apis mellifera*): the East African lowland honeybee, *Apis mellifera scutellata,* and various Western honeybee subspecies from Europe, namely the Italian honeybee, *Apis mellifera ligustica,* and the Iberian honeybee, *Apis mellifera iberiensis.* Hybridisation occurred when the East African lowland honeybee was introduced to Brazil in 1956. Hybrids then spread throughout South America and arrived in North America in 1985. The hybrids, known as Africanized bees or Africanized honeybees, have different behaviours from Western honeybee subspecies from Europe. Africanized bees like the East African lowland honeybee are much more defensive of colonies and nests than other Western honeybee subspecies. For simplicity, for the rest of this

Figure 6.9. The Africanized honeybee, *Apis mellifera* spp. *scutellata* is almost identical to Western honeybee subspecies. It is a hybrid between the East African lowland honeybee and various Western honeybee subspecies from Europe.

Table 6.2. Pest summary of Africanized honeybees, primarily *Apis mellifera scutellata* x European honeybee species hybrid.

Distribution[1]	Australia: No Canada: No Europe: No New Zealand: No **United States: Yes**
Size	Workers 12 mm (0.5 in), male drones 14 mm (0.55 in), queens 20 mm (0.8 in)
Key identification characteristics	European honeybees have a distinctive barrel-shaped body, abdomen covered in hair, golden brown with black abdominal stripes; africanized honeybees are nearly identical
Mouthparts of adults	Chewing
Feed on	Pollen and nectar
Development time[2]	16-26 days
Adult life span	5-10 weeks
Total eggs laid	1,500 per day
May be mistaken for	European honeybee subspecies, bumblebees, wasps
Impact on horses	Allergic reaction, acute renal failure, death due to complications from stings

[1] The pest distribution in the regions targeted by this publication are listed as of the date published. Areas of primary pest concern (if any) are in bold. Global distributions are listed in Section 6.2.2.
[2] Development times are approximate and can change with environmental conditions including temperature, humidity, and availability of food resources.

section, we will refer to European and Africanized honeybees. Both are Western honeybees. The former are Western honeybee subspecies from Europe and the latter are a hybrid of these and the East African lowland honeybee, which has a similar behaviour as the hybrid. When disturbed they will chase the offender for up to 400 m (437 yards), ten times as many bees from the colony will sting, and they have been known to kill humans, horses, and other animals due to complications from the greater number of stings. In popular media and films, they are sometimes called 'killer bees.' Horses are particularly susceptible because they are often confined in pastures and unable to escape an Africanized bee attack if a nest is disturbed.

6.2.2 Distribution and seasonality

While European honeybee subspecies are present on every continent except Antarctica, the Africanized honeybee hybrid is present only in northern South America and southern North America. They cannot survive harsh winters or late dry summers due to their susceptibility to food deprivation because this hybrid, like the East African lowland honeybee, does not store honey-like European honeybees. Further invasion into more northern states is expected to occur as climate change improves the environmental suitability of these locations.

6.2.3 Description

Bees have a short, narrow waist between the thorax and abdomen that gives them an hourglass shape. Africanized and European honeybees are light brown with golden yellow and brown bands. The thorax and abdomen are hairy, but the thorax appears much hairier. They have clear wings and bent antennae. The workers are approximately 12 mm (0.5 in) in length, the drones are 14 mm (0.55 in) and the queens are close to 20 mm (0.8 in). Worker bees have specialised structures called pollen baskets on their hind legs for carrying large quantities of pollen (Figure 6.10). They also have a barbed stinger on the tip of the abdomen. The barbs, or backwards-facing spines, prevent the sting from being removed. The queen looks very similar to a worker bee, except she has an enlarged abdomen and her stinger has reduced barbs. Drones have a larger head with fly-like eyes and a blunt abdomen with no stinger.

Figure 6.10. Honeybees have specialised structures on their hind legs for carrying pollen, seen here as the orange mass attached to the leg of this European honeybee (labelled A).

Africanized honeybees (Figure 6.9) look like typical European honeybees (Figure 6.10). Unlike the East African lowland honeybees, Africanized honeybees are not smaller; however, there are small differences in the wing sizes between Africanized honeybees and European honeybee subspecies that can be differentiated by entomologists or pest control professionals. The only reliable method to differentiate the hybrid from the subspecies is using molecular analysis. Mostly this hybrid will be distinguished based on behaviour, primarily its aggressive defensive activities, increased swarming behaviour, and its unusual nesting locations.

All honeybees defend their nests, but while European honeybees defend with 10 to 20 individuals, an Africanized honeybee attack, like an East African lowland honeybee attack, will include hundreds of bees. They are also more likely to attack and will follow the offender for longer distances. Swarming behaviour, which is when a large number of individuals leave the hive to form a new colony, is also different. European honeybees may swarm one to three times per year but Africanized bees swarm more than 10 times per year. Africanized honeybees are also more likely to abandon their nests at times of stress, such as food deprivation or disturbance. While European honeybees prefer to nest in large cavities, such as tree hollows and chimneys, Africanized honeybees are much less selective and will nest in smaller cavities such as buckets, old tyres, and house eaves (Figure 6.11). Africanized bees also will nest hanging from tree branches and in-ground cavities.

6.2.4 Life history

Africanized honeybees have complete metamorphosis with four life stages: egg, larva, pupa, and adult. The life cycle of Africanized honeybees is similar to European honeybees with some distinct differences. Like European honeybees, there are three castes (types) of bees in the colony: the queen, workers, and males (called drones). Unfertilised eggs become drones, and fertilised eggs become females. Female larvae that are fed pollen, nectar, and brood food become a non-reproductive adult worker. Female larvae that are destined to become queens are placed in special cells in the hive and fed a richer diet that includes 'royal jelly', a secretion produced from glands in the head of nurse worker bees. After the old queen leaves the colony by swarming, a new queen emerges from her wax cell and after a period of further development leaves the hive. She will be mated by 10 to 20 drones over 7 to 10 days in the air. Africanized honeybee colonies produce more drones, so this favours their mating over European subspecies in areas with the hybrid.

Development from egg to adult varies between the European subspecies and the hybrid and depends on the caste; European honeybee queens take 16 days, workers 21 days, and drones 24 days. Africanized honeybee queens and workers develop more quickly at 14 and 19 to 20 days, respectively. This faster development time gives the Africanized honeybees a great advantage when competing with European honeybees.

Figure 6.11. European honey bees often nest in large cavities like tree hollows (A) while Africanized honeybees may nest in smaller cavities like buckets or tyres (B), or exposed hanging from tree branches (C) or in-ground cavities.

6.2.5 Control

Africanized honeybees are generally indistinguishable from the European honeybee, which is found worldwide. Observing the bees will allow this hybrid to be distinguished based on its aggressive behaviour and unusual nesting locations. Although it should be noted that the East African lowland honeybee has similar behaviour. Management practices to eliminate as many potential nesting sites as possible or removing any bee nests if they are found can reduce the risk of horse encounters with Africanized honeybee colonies.

Cultural control

If the equine farm is located in the range of Africanized honeybees, cultural control practices like removing or discarding tyres, buckets, or other items that could be used as nest sites can help reduce the risk of colonisation and later attacks. Any protected location with a small entrance could be used as a nest site. Properties should be inspected regularly looking for potential nesting sites and nesting activity, particularly the eaves and walls of any structures (see O'Malley *et al.*, 2019 in 'Continuing education' for more information on keeping bees from nesting).

Physical and mechanical control

Exclusion

Gaps in walls that are greater than 30 mm (1.2 in) wide should be closed off with caulk or fine-mesh hardware cloth to help prevent honeybees from gaining access and forming a nest (Figure 6.12). However, this must be done before bees begin to nest. If bees are already present, they will need to be removed or killed before the hole is blocked as this could force the bees deeper into the structure.

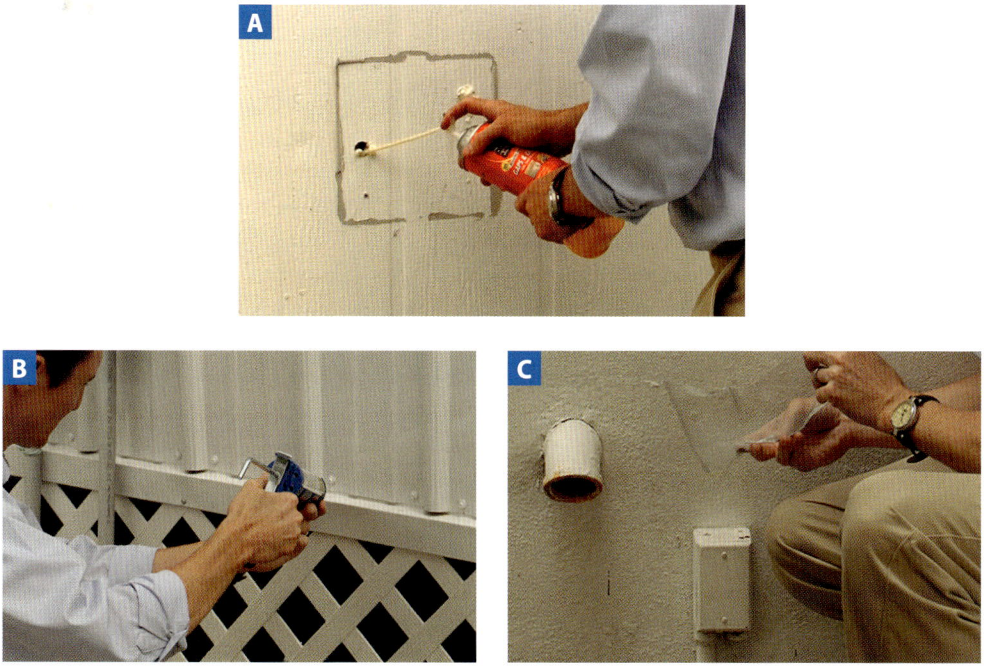

Figure 6.12. To prevent Africanized bees from building nests, close off gaps in walls with expandable foam (A), caulk (B), or fine-mesh hardware cloth (C).

Removal

Do not attempt to remove any nesting bees. If swarms of nuisance honeybees are detected nesting outside of a hive, they should be removed by a registered beekeeper or pest control operator or eradicated by a pest control operator. The pest control operator might use soap and water to immobilise and remove the swarm if safe removal is possible. Removals will likely need to be done at night (when all the worker bees are back in the nest), with full personal protective equipment, and people, horses, and other animals should be kept far away from the area, ideally inside.

Chemical control

There are no chemical controls that are recommended for controlling or repelling either European honeybees or Africanized honeybees. Licensed pest control operators can use restricted chemicals to eradicate nuisance honeybee colonies if it is not possible to remove them safely.

6.2.6 Veterinary concerns

It is rare for a horse to have more than some minor discomfort from one or two stings from a bee. Much like humans, horses will experience pain, swelling, and warmth after a bee sting. However, multiple stings can lead to severe allergic responses and will require immediate veterinary care.

Following a bee attack, horses may appear excited. Progressive clinical signs occur following this initial period and can include severe, swelling, rhabdomyolysis (muscle tissue breakdown), hypovolaemia (low blood plasma), colic, and renal (kidney) injury, which can last 3 to 4 days (Lewis and Racklyeft, 2014; Staempfli *et al.*, 1993). Envenomation from bee stings has been demonstrated to cause acute renal failure (Ribeiro *et al.*, 2020). Management may include hot and cold compresses to the affected areas, intravenous fluids, anti-inflammatory drugs, corticosteroids, anti-histamines, analgesia, and antibiotics (Fonteque *et al.*, 2018; Lewis and Racklyeft, 2014).

6.3 Eye gnats (family Chloropidae)

6.3.1 Pest summary (Table 6.3)

Eye gnats (Figure 6.13) or eye flies are small, yellow, or black, shiny flies in the family Chloropidae, which contains 160 genera and about 2,000 species worldwide. The genera *Liohippelates*, *Hippelates*, and *Siphunculina* are those with veterinary importance due to their attraction to secretions of the eyes, nose, and ears, as well as to open wounds. In this way, they irritate their hosts, but they also act as mechanical vectors of several pathogens.

6.3.2 Distribution and seasonality

Siphunculina, the tropical eye flies, are pests in the Old World, mostly Asia. *Liohippelates* and *Hippelates* are pests of North and South America. Eye gnats often form large aggregations in areas with loose sandy soil, particularly in rural areas. Eye gnats can be found throughout their range, but are particularly common in the sandy soil regions of the Southeastern USA and around irrigated fields in inland southern California. Populations tend to fluctuate with seasonal temperatures with more gnats in the summer months and fewer in the winter months.

Table 6.3. Pest summary of eye gnats, family Chloropidae.

Distribution[1]	Australia: Yes Canada: Yes Europe: Yes New Zealand: No **United States: Yes**
Size	1.4-1.6 mm (0.05-0.06 in)
Key identification characteristics	Small, shiny yellow or black, clustering behaviour around eyes, nose, mouth, and wounds
Mouthparts of adults	Sponging (non-biting)
Feed on	Body secretions around eyes, mouth, nose, and wounds
Development time[2]	2-8 weeks
Adult life span	7 days
Total eggs laid	200-400
May be mistaken for	Biting midges (Chapter 4), dung midges (Chapter 8)
Impact on horses	Vector of pathogens causing vesicular stomatitis, streptococcal infections, and respiratory viruses including equine herpesvirus 1, 2, 4, and 5, corneal ulcers, and irritation

[1] The pest distribution in the regions targeted by this publication are listed as of the date published. Areas of primary pest concern (if any) are in bold. Global distributions are listed in Section 6.3.2.
[2] Development times are approximate and can change with environmental conditions including temperature, humidity, and availability of food resources.

Figure 6.13. Eye gnat, *Liohippelates* spp.

6.3.3 Description

Eye gnats are very small, 1.4 to 1.6 mm (0.05 to 0.06 in) long (Figure 6.14), and have shiny black or grey bodies and legs that are yellow to orange-brown. Other names for these flies are grass flies, frit flies, and eye flies. As the common name eye gnat suggests, the most conspicuous feature of these gnats is the habit of the adults to hover around and visit the eyes. They may be confused with biting midges or dung midges. However, biting midges are found on the body of the horse, not congregating in the same body locations as eye gnats. Dung midges are found around piles of manure and not on horses.

6.3.4 Life history

Eye gnats have complete metamorphosis with four life stages: egg, larva, pupa, and adult. Eye gnats develop in moist, sandy soils (Figure 6.15) with substantial amounts of organic material such as manure, cut grass, hay, and crop debris. Females lay their eggs on soil that has recently been disturbed by digging, ploughing, or trampling by animals. Once the eggs hatch, larvae develop and pupate in the soil. Development time is 2 weeks to 3 months depending on soil, moisture, and temperature conditions (Burgess, 1951).

Adult male flies are not attracted to horses, but females gather around them to feed on secretions from wounds, eyes, noses, and other areas of horses and humans (Figure 6.16). They have sponging mouthparts that cannot break the skin, but females use curved spines on the hind legs to rasp the skin to increase the amounts of secretions produced, so they can be irritating.

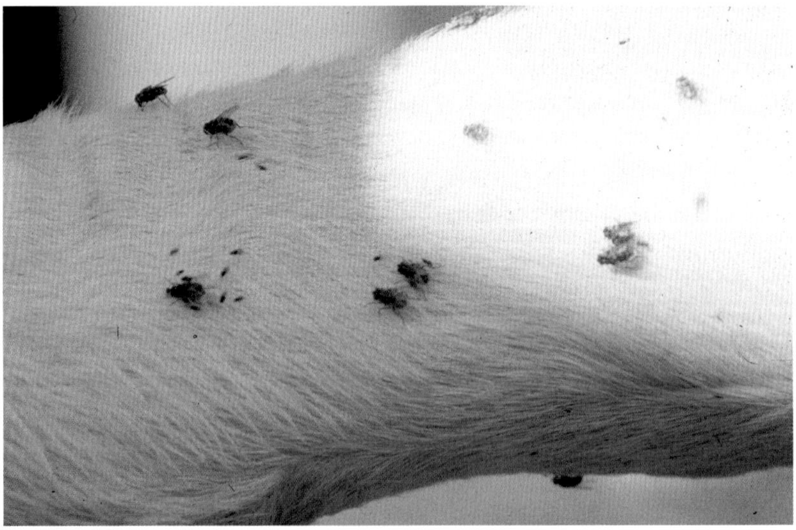

Figure 6.14. Tiny eye gnats feeding next to the much larger stable flies. The gnats are feeding on the blood droplets formed at stable fly feeding sites.

Figure 6.15. Eye gnats develop in moist and loose sandy soils, often mixed with organic matter like hay, feed, and grass clippings.

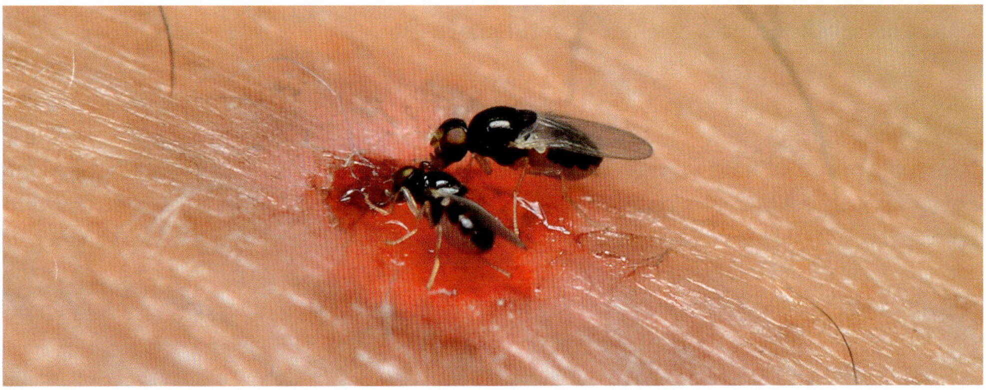

Figure 6.16. Eye gnats congregate around the eyes, nose, mouth, and wounds (pictured feeding on a human wound) and feed on secretions from these areas, often causing great annoyance.

6.3.5 Control

Due to the extreme number of eye gnats that can be produced from the soil, area-wide treatment to reduce numbers is impractical if not impossible. Cultural control should target larval development sites and include adult management.

Cultural control

The most effective way to reduce eye gnat populations is to manage development sites. However, reducing eye gnat development sites is nearly impossible in equine facilities as manure, moisture, and soil disturbance are inseparable from animals. In some cases, reducing organic matter where it can be manipulated, such as around waste piles or hay bales, may help reduce the attractiveness of these areas to eye gnats for breeding and development. Moist soils are required for development, so reducing water trough overflow, and improving drainage to prevent standing water and promote drying may help reduce egg-laying and larval survival.

At farms with manure storage areas like pits or piles, actively composting manure can significantly reduce eye gnat numbers (Hogsette, 1981) (See Augustin and Rahman, 2016 in 'Continuing education' for more information on effective composting of animal manure).

Physical and mechanical control

Flight barriers

Physical barriers, such as screened windows or netting can be a barrier to eye gnat entrance into barns, stables, run-in sheds, or other areas where horses may be housed. A 2.4 m (8 ft) fine screen fence or other mesh barriers around a paddock can reduce more than 90% of eye gnats within the paddock as this pest tends to fly close to the ground (Bethke *et al.,* 2013). This may be a bit impractical for an equine facility, but a 91 cm (36 in) high silt or erosion fence (Figure 6.17) can be used to exclude the majority of the population.

Trapping

Trapping also can be used to provide some relief from eye gnats, especially when used along with other methods of control (Figure 6.18). Multiple traps placed in lines have been shown to intercept eye gnat flight and reduce populations. To make a trap, fill a plastic container about half to three-quarters full with water and four chicken eggs or 115 g (4 oz) egg powder and mix the contents. Drill three to four holes in the top half of a two-litre (half-gallon) bottle with a 20 mm

Figure 6.17. Placing silt fencing around pastures can help with eye gnats, which fly low to the ground.

(0.75 in) drill bit and fill the bottle with the egg solution. If other size containers are used, the equivalent is an egg to water ratio of five eggs per 3.8 litres (1 gallon). Hang the traps 60 to 90 cm (2 to 3 ft) from the ground and away from access areas such as doors and windows. Traps should be checked weekly and replaced at least every 2 weeks. More frequent replacement may be necessary in warmer weather. To increase trap captures, the bottom half of the two-litre (half-gallon) bottle can be painted black.

Fly sheets, boots, and masks

As with other flies, protective clothing like fly sheets, fly boots, and fly masks can be used to reduce eye gnat annoyance and prevent injury to the eyes. These can also be used in a more targeted way over wounds to protect them from eye gnat feeding.

Chemical control

Area and soil pesticides

Eye gnat control with chemicals has been relatively ineffective due to the broad range of development sites and the difficulty with reaching developing flies beneath the soil. There are additional concerns with fly pesticide resistance and broad applications of pesticides that may significantly impact or kill non-target species including beneficial pollinators. Various pesticides, fogs, and soil treatments have been tested with very limited success and are not recommended. Applications of pesticides applied on a community level for mosquito control may reduce eye gnat numbers temporarily, but development sites quickly repopulate as the pesticide dissipates or degrades.

Figure 6.18. Homemade eye gnat traps can be made with 2-litre bottles baited with an egg and water solution.

Residual pesticides

If eye gnats are a problem in the stable, residual sprays can be used to kill adults. Residual pesticide premise sprays are applied to walls, ceilings, and other places where flies rest. These are typically sprayed and allowed to dry but affect the flies when they land on the treated surface. Residual pesticides work best when the surface they are sprayed on is clean and dry. They can last a few weeks before a reapplication is necessary. Make sure to read the label prior to purchasing a residual pesticide to ensure it is suitable for use around horses.

Topical products

Many of the same repellents that are used on horses for mosquito and other fly control can be used to provide some relief from eye gnats. Because eye gnats are active during the day, morning applications of insect repellents may offer a full day of protection, although the duration of repellency has not been evaluated with eye gnats. Applications of cream or roll-on repellents may offer protection that is longer lasting than sprays and can be applied to specific areas where gnats tend to congregate on the horse (Figure 6.19), particularly around the eyes, nose, mouth, ears, and any wounds, or wherever horses are being affected. Be careful when applying repellents that they do not get in sensitive areas like the eyes or wounds.

6.3.6 Veterinary concerns

Eye gnats are primarily a nuisance pest. However, *Hippelates plebejus* and *Hippelates nobilis*, are larger species with rasping mouthparts capable of producing a biting sensation when feeding on sensitive skin (Bigham, 1941). The persistence of their feeding activity is highly annoying to

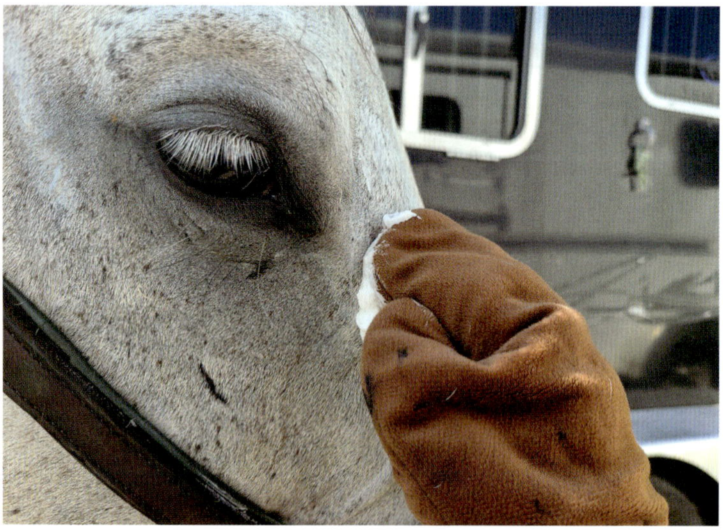

Figure 6.19. Cream repellents (shown) or roll-on repellents can be applied around areas where eye gnats congregate.

horses and their owners and is thought to lead to decreased condition and performance (Hinkle *et al.*, 2001).

Eye gnats can serve as mechanical vectors for viruses and bacteria shed in respiratory, ocular, and mammary secretions. Of particular note for horses are vesicular stomatitis, streptococcal infections, and respiratory viruses including equine herpesvirus 1, 2, 4, and 5, which have been recovered from eye gnats in the laboratory. While eye gnats have been implicated in the spread of mastitis in other species, particularly cattle, they are unlikely to be a factor in equine mastitis. In well-managed horse populations, corneal ulcers secondary to irritation caused by eye gnats are the most concerning (Figure 6.20). See Brooks (2016) in 'Continuing education' for more information on treating equine eye conditions.

6.4 Blister beetles (genus *Epicauta*)

6.4.1 Pest summary (Table 6.4)

Blister beetles are in the insect family Meloidae in the order Coleoptera (Figure 6.21). There are approximately 2,500 species worldwide. They are named due to the production of a defensive secretion, cantharidin, that causes skin blisters. Different species produce very different concentrations of the toxic compound. When these beetles are handled they demonstrate a defensive behaviour called reflex bleeding, or releasing haemolymph (insect blood equivalent) that contains the blistering compound. The largest genus, *Epicauta,* is an accidental pest of horses but the primary group of concern as these species can be found in alfalfa hay including baled hay and alfalfa fed as pellets or cubes. If these beetles are present, even if dead, they are toxic to horses if ingested causing cantharidiasis (cantharidin poisoning).

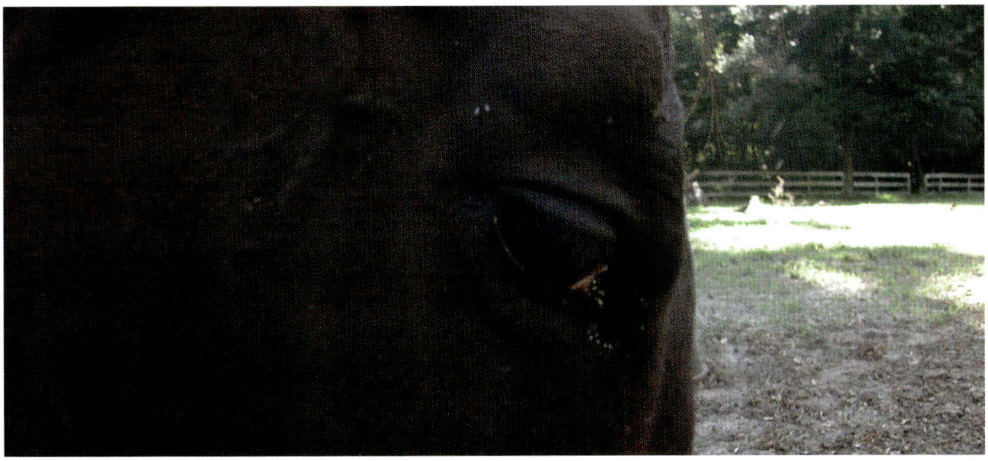

Figure 6.20. Corneal ulcers in the equine eye are a concern with eye gnats because of their persistent feeding around the eye on secretions.

Table 6.4. Pest summary of blister beetles, *Epicauta* spp.

Distribution[1]	Australia: Yes Canada: Yes Europe: Yes New Zealand: Yes **United States: Yes**
Size	3-70 mm (0.12-2.75 in)
Key identification characteristics	Long legs, elongated body, the appearance of a neck between body and head, often found in groups
Mouthparts of adults	Chewing
Feed on	Flowers and leaves
Development time[2]	9-12 months
Adult life span	3 months
Total eggs laid	50-300
May be mistaken for	Beetles in the family Carabidae (ground beetles), Oedemeridae (false blister beetles), and the Tenebrionidae subfamily Lagriinae (long-jointed beetles)
Impact on horses	Cantharidiasis (cantharidin poisoning)

[1] The pest distribution in the regions targeted by this publication are listed as of the date published. Areas of primary pest concern (if any) are in bold. Global distributions are listed in Section 6.4.2.
[2] Development times are approximate and can change with environmental conditions including temperature, humidity, and availability of food resources.

Figure 6.21. Striped blister beetle, *Epicauta vittata*, is the most common cause of cantharidin poisoning in horses.

6.4.2 Distribution and seasonality

Blister beetles are found worldwide with the greatest diversity in dry areas (semi-arid to arid). The most relevant genus, *Epicauta*, is distributed almost worldwide with species present on every continent except Antarctica and Australasia. They are a pest of crops throughout their range and can be a concern to horse health in any region where alfalfa is imported from where blister beetles are present, but most cases have been reported in the USA.

6.4.3 Description

Blister beetle adults vary in size from 3 to 70 mm (0.12 to 2.75 in). *Epicauta* spp. are 4 to 18 mm (0.16 to 0.70 in) and are most commonly associated with alfalfa hay. Blister beetles have soft bodies and long legs. Many species have aposematic (warning) colouration (i.e. bright colour markings that advertise their toxicity to predators). The head is fully exposed on a long, narrow neck.

The species of blister beetle that has been reported most often as the cause of poisoning is the striped blister beetle, *Epicauta vittata,* which is black with orange to red stripes. The margined blister beetle, *Epicauta funebris*, black blister beetle, *Epicauta pennsylvanica*, and the ashgrey blister beetle, *Epicauta fabricii,* also have led to poisoning events. These three species range in colour from grey to black (Figure 6.22).

Figure 6.22. There are several species of the *Epicauta* blister beetles that have been reported as causing poisoning to horses when they are fed infested alfalfa hay. The striped blister beetle is the most common cause (Figure 6.21), but the margined (*Epicauta funebris*) (A), black (*Epicauta pennsylvanica*) (B), and ashgrey (*Epicauta fabricii*) (C) blister beetles also have led to poisoning events.

Blister beetles are commonly confused with beetles in the family Carabidae (ground beetles), Oedemeridae (false blister beetles), and Tenebrionidae (Figure 6.23).

6.4.4 Life history

Blister beetles have complete metamorphosis with four life stages including egg, larva, pupa, and adult. Eggs are laid under stones, and when the larvae hatch they make their way to a feeding site. Blister beetle larvae are rarely seen. The smallest larvae look quite different to the later growth stages of these beetles. These small larvae have long legs and are boat-shaped. The more mature larvae have reduced legs and are grub-like. This larger stage will feed and go into a delayed development state through the winter and eventually will pupate. Immature blister beetles are not a risk to horses because they do not feed on plant leaves or flowers. Adult blister beetles do feed on flowers and leaves of plants and are often found in groups.

6.4.5 Control

Blister beetles are accidental pests of horses. These insects are only a problem when a horse ingests beetles or beetle body parts that are in the hay.

Figure 6.23. Because many blister beetles have muted colours, like grey and black, they can be confused with beetles in the families Carabidae (ground beetles) (A), Oedemeridae (false blister beetles) (B), and Tenebrionidae (C).

Cultural control

One of the preferred plants of *Epicauta* spp. blister beetle adults is alfalfa. The adults congregate on the leaves at the top of the plant where they feed on the leaves, as well as nectar and pollen in the flowers (Figure 6.24). They are most abundant in arid (dry) regions where the food that is eaten by larvae is found. They are also more common in the mid-to late-summer.

Hay management

Hay inspections can be conducted, but it is important to note that inspections of bales may not identify beetles, and dead beetles or parts of beetles may still contain active cantharidin. Instead, it is important to follow best management practices if hay is produced on-farm, or to communicate with the hay seller about where and when hay was cut, and if hay management practices were established to avoid blister beetle contamination.

In some cases, equine farm owners or managers may prefer to produce their own hay. Hay cut when blister beetle adults are active should be checked for beetles on the flowers, particularly near field edges. Baling at the same time as using a crimper to fluff and dry hay should be avoided. Hay that has been cut and left to dry (Figure 6.25) before baling is safer because the beetles will typically have time to escape. Horse owners can avoid the risk of beetle ingestion by growing their own alfalfa and using the practices described above, or by using local producers and getting to know their pest management practices (see Ward, 1997 in 'Continuing education' for information on hay production practices to reduce blister beetles), and by only buying hay that is cut outside of blister beetle season. Typically, early and late cuttings are less likely to have beetles, whereas second and third cuttings are more likely to have beetles.

Figure 6.24. Blister beetles often feed on leaves, nectar, and pollen in flowers. Beetles that feed in alfalfa are of particular concern for horses.

Figure 6.25. After cutting, leaving hay to dry before baling can allow blister beetles to escape and reduce the risk of beetles being ingested by horses later when the hay is fed.

Blister beetles can be found in any alfalfa or alfalfa product, but beetles or beetle parts in alfalfa pellets or cubes are rare. This is likely because the alfalfa sources for these products have good integrated pest management strategies for insect control and quality control practices that would minimise the risk of contamination with blister beetles. If beetles are suspected their presence can be confirmed by insect taxonomists who are often associated with educational or government institutions.

Biological control

Blister beetles have few predators because of their toxic chemicals. However, there are some birds (Figure 6.26) and other insects, like robber flies, that may feed on them. Leaving cut hay to dry will also leave beetles exposed to predation.

Chemical control

Although foliar applications of pesticides can be used to protect valuable crops from pests, they are not recommended for use in alfalfa for horse consumption. Besides the risk to horses from beetles that die in the hay, these pesticides may also negatively impact other beneficial arthropods.

6.4.6 Veterinary concerns

Although rare, blister beetles are a very serious, and often fatal, threat to horses. Cantharidiasis (cantharidin poisoning) in horses causes blistering of the entire gastrointestinal tract from mouth to anus. Additionally, the toxin causes extreme damage to the kidneys as the horse works to excrete it from the body. There is substantial variability in the amount of cantharidin in different species of blister beetles (See Boxler, 2020 in 'Continuing education'). Because of this variation, it is difficult to determine the number of beetles that make a lethal dose. A single beetle or as many as 200 have been reported as a lethal dose.

Figure 6.26. Leaving cut hay exposed while drying allows predators such as birds to eat blister beetles (pictured) before baling.

Early signs of cantharidiasis in horses will appear as colic with stretching, evidence of cramping, and reduction of appetite. However, colic alone cannot confirm cantharidin poisoning; the best diagnostic method appears to be chemical analysis of urine, ingesta from the upper gastrointestinal tract, and sometimes liver tissue. As the kidneys begin to feel the effects of the toxin, the horse will frequently posture to urinate and/or urinate very small amounts frequently. Blister beetle ingestion is a medical emergency. Horses should be taken to a hospital facility immediately (see Helmann and Edwards, 1997 and Epsy, 2016 in 'Continuing education' for more information on cantharidiasis).

6.5 Caterpillars with stinging (urticating) hairs (order Lepidoptera)

A few species of caterpillars have long hairs that can sting when they are touched. These urticating hairs can also be irritating after they have dropped off the caterpillar. Some caterpillars, like tent caterpillars (genus *Malacosma*, Section 6.5.1), and those of the bag-shelter moths or processionary caterpillars from Australia (*Ochrogaster lunifer*, Section 6.5.2) can cause serious veterinary issues. If these caterpillars (in the case of *Malacosma*) or caterpillar nests or exoskeletons (in the case of *Ochrogaster*) are ingested by a horse, the hairs from the caterpillars penetrate the lining of the gut (or gastrointestinal tract) and enable gut bacteria to be passed into the bloodstream. This can cause infection or pericarditis (swelling of the membrane that surrounds the heart), and is particularly dangerous in pregnant mares where it can lead to abortions and foetal abnormalities.

Other caterpillars with urticating hairs like those of the oak processionary moth (*Thaumetopoea processionea,* Figure 6.27) can cause skin irritation, tongue swelling, stomatitis, conjunctivitis, and respiratory distress in animals exposed to the hairs, although exposure is infrequent. These caterpillars are found throughout Europe and the Middle East. These caterpillars are recognisable by their characteristic way of travelling in procession (Figure 6.28), as do the caterpillars of

O. lunifer. Oak processionary moth caterpillars feed exclusively on oak trees and may create silken nests on the trees (Figure 6.29). These caterpillars tend to be accidental irritants to horses. However, for this and other species that may have urticating hairs, keeping horses away from host trees, caterpillars, and caterpillar nests can reduce the risk of exposure to the hairs significantly.

Figure 6.27. Some caterpillars like this oak processionary caterpillar have hairs that can cause extreme stinging sensations and skin irritation. Others can cause serious veterinary issues like abortions in horses.

Figure 6.28. The oak processionary moth caterpillars are recognisable by their characteristic way of moving in a line.

Figure 6.29. Oak processionary moth caterpillars make silken nests that can be found on oak trees.

6.5.1 Tent caterpillars (*Malacosma* spp.)

6.5.1.1 Pest summary (Table 6.5)

Tent caterpillars are the immature stages of a group of moths (Figure 6.30). This group includes 26 species in the genus *Malacosoma* (Lasiocampidae) that build silk tents in the branches of trees. They are primarily horticultural pests because they eat and defoliate trees. However, they can also be a pest of horses if horses accidentally ingest caterpillars when they are full-grown and move away from trees to fence lines, buckets, or water troughs (Figure 6.31).

6.5.1.2 Distribution and seasonality

Of the 26 species of tent caterpillars found globally, six are in North America. Two of these species are primarily associated with horse health concerns; the Eastern tent caterpillar, *Malacosoma americanum* (Figure 6.30), and the forest tent caterpillar, *Malacosoma disstria* (Figure 6.32), which are located in the USA. The Eastern tent caterpillar is probably the most well-known species due

Table 6.5. Pest summary of tent caterpillar, *Malacosoma* spp.

Distribution[1,2]	Australia: Yes **Canada: Yes** Europe: Yes New Zealand: Yes **United States: Yes**
Size	Up to 6.5 cm (2.5 in)
Key identification characteristics	Hairy, bright blue or green caterpillars with longitudinal stripes or spots; characteristic silken tents or mats on trees
Mouthparts of larvae	Chewing
Feed on	Primarily plants in the genera *Prunus*, *Malus*, and *Crataegus* and other broadleaf trees
Development time[3]	7-10 weeks
Adult life span	1-10 days
Total eggs laid	100-400
May be mistaken for	Walnut caterpillar, *Datana integerrima*, Tussock moth, *Orgyia detrita*, Fall webworm, *Hyphantria cunea*
Impact on horses	Mare reproductive loss syndrome, pericarditis, death

[1] The pest distribution in the regions targeted by this publication are listed as of the date published. Areas of primary pest concern (if any) are in bold. Global distributions are listed in Section 6.5.1.2.

[2] Processionary caterpillars like the oak processionary moth can be found in Europe and exposure can be a veterinary concern. However, exposure is infrequent.

[3] Development times are approximate and can change with environmental conditions including temperature, humidity, and availability of food resources.

Figure 6.30. Eastern tent caterpillar, *Malacosoma americanum*.

Figure 6.31. Horses can accidentally ingest tent caterpillars as they wander away from trees to find a good place to pupate.

to its association with an extraordinary number of foal deaths in Kentucky, USA between 2001 to 2002. It is present in eastern North America, from Florida to Nova Scotia, and as far west as the Rocky Mountains. The preferred host of the Eastern tent caterpillar are species of *Prunus*, *Malus*, and *Crataegus*, especially wild black cherry, *Prunus serotina*. Along with the Eastern tent caterpillar, the forest tent caterpillar has been associated with pericarditis and can be found throughout the range of hardwood forests in the USA. This species will colonise a wide variety of hardwood trees, but rarely red maple (*Acer rubrum*) or conifers.

Figure 6.32. Outbreaks of forest tent caterpillars, *Malacosoma disstria*, have been associated with pericarditis (inflammation of the membrane around the heart) in horses.

Eastern tent and forest tent caterpillars have cyclic patterns that result in outbreak years followed by years with low numbers. Certain forest conditions and weather patterns can cause a rapid increase in tent caterpillar survival, causing outbreak years. Outbreak years typically last for 2 to 5 years before the population collapses. Generally, horses are at the highest risk for encountering caterpillars in the autumn months.

6.5.1.3 Description

The cocoons spun by the larvae are made of white to yellow silk. They spin their cocoons in protected places on the ground or under eaves. The caterpillars are hairy and often brightly coloured with longitudinal stripes or spots (Figure 6.30 and 6.32). After 2 weeks as pupae, they emerge from the cocoons as adult moths. Eastern tent caterpillar moths are reddish-brown with two pale horizontal stripes. The adult forest tent caterpillar moth is buff-brown with two darker, thin horizontal lines across the wings (Figure 6.33). Tent caterpillars can be confused with other tree caterpillars that are not pests of horses including the walnut caterpillar, *Datana integerrima*, fir tussock moth, *Orgyia detrita*, and fall webworm, *Hyphantria cunea* (Figure 6.34).

6.5.1.4 Life history

Tent caterpillars have complete metamorphosis with four life stages: egg, larva, pupa (or cocoon), and adult. After mating, the female moth lays her eggs in a mass of 100 to 400 eggs around a suitable branch of a host tree and then dies (Figure 6.35). She may live for less than 24 hours. Male moths may live for up to a week. The eggs are encircled around a small branch about the diameter of a pencil or smaller. The female covers the eggs with a secretion to protect them that makes the eggs appear shiny. Although the larvae will begin to develop in the eggs, they will lay dormant until the following spring. The larvae, called caterpillars, hatch from eggs in the spring as their host plant leaves begin to unfold.

Figure 6.33. The adult moth of the eastern tent caterpillar is buff or reddish-brown with pale horizontal stripes (A). The adult forest tent caterpillar moth is buff-brown to reddish with two darker, thin horizontal lines across the wings (B).

Figure 6.34. Tent caterpillars can be confused with other tree caterpillars including the walnut caterpillar, *Datana integerrima* (A), fir tussock moth, *Orgyia detrita* (B), and the fall webworm, *Hyphantria cunea* (C).

Most tent caterpillars establish their tent in the connections between branches or trees (not at the end of limbs like webworms or other caterpillars) in the direction to catch the morning sun (Figure 6.36). Some species leave the tent to find food and return to the tent after feeding or if the temperature gets cooler. Other species establish several temporary tents throughout their development. Forest tent caterpillars are unique in that they do not create a tent, but congregate on silken mats on tree trunks or branches to moult or rest from feeding (Figure 6.37). Caterpillars feed and grow to 6.5 cm (2.5 in), often after 80% or more of their host plant has been eaten. This largest caterpillar stage presents the most risk to horses as it leaves the host tree to find a place to pupate. This behaviour may put high numbers of these caterpillars in areas where horses are fed or pastured, increasing the risk of ingestion. Total development time from egg to adult is generally 7 to 10 weeks.

6.5.1.5 Control

Populations of caterpillars are regulated, in part, by predators, parasitic wasps, and a few diseases. Natural enemies are important for keeping tent caterpillar populations successfully in check most years. Prevention and early control using cultural, physical and mechanical, and chemical options are important to reduce the risk of horses encountering tent caterpillars.

Figure 6.35. Tent caterpillar eggs are laid in shiny masses encircled around a branch.

Figure 6.36. The Eastern tent caterpillars spin the characteristic tents where the larvae return to between feeding events and during cooler weather.

Figure 6.37. Forest tent caterpillars do not construct an actual tent but form aggregations on silken pads on tree bark and foliage.

Cultural control

Property inspections

Trees around pastures should be inspected for egg masses in the winter and for tents in the spring.

Horse turnout management

Horses should not be pastured where there are caterpillar nests. If nests or caterpillars are found, they should be removed before horses are allowed back into the location where they were discovered.

Exclusion

Tent caterpillars affect horses after they are ingested, so keeping feed storage covered and checking water buckets frequently for the presence of caterpillars or shed skins can help reduce the risk of caterpillar contact with horses, particularly later in the summer or autumn when mature larvae leave the tents to pupate, which is when most horses encounter them.

Lighting management

Tent caterpillar moths, like many moth species, are attracted to lights. Turning off exterior lights may help reduce moth presence in the facility and reduce the number of egg masses laid on nearby trees.

Physical and mechanical control

Host tree management

Potential host trees of the tent caterpillars in the area should be removed in and near paddocks, and do not plant preferred host plants on the farm (plants in the genera *Prunus*, *Malus*, and *Crataegus*).

Removal

Removal of the eggs and small, young caterpillars will reduce tree damage, but the removal of the caterpillars at any stage will reduce the risk to horses caused by caterpillars wandering into pastures, waterers, or feed buckets looking for a place to pupate. Small caterpillar tents within easy reach can be removed by putting a hand in a plastic bag and grabbing the tent before turning the bag inside out around the tent, which can then be sealed and the caterpillars disposed of. Larger tents can be wound around a stick and disposed of by placing the stick and tent in a bag. Alternatively, branches holding tents can be cut and the tents sealed in a plastic bag or burned. Caterpillars may reappear on infested trees the following year even after removing all visible tents and caterpillars. It is not recommended to burn the tents on the tree as this can damage and potentially kill the host tree.

Slippery and sticky tree bands

Trees can be banded with a commercially available slippery material to impede the movement of caterpillars. This can keep caterpillars from climbing up trees. Another approach is to use sticky tape to achieve the same result. To do this, first, wrap duct tape 5 to 8 cm wide (2 to 3 in) tightly around the trunk at chest height. Then paint a band of sticky adhesive material onto the duct tape. The sticky material can be found in most hardware stores, where it is often sold for rodent traps. Make sure that the sticky material does not directly contact the bark of the tree because it could harm or kill the tree. Sticky tapes can also trap birds, bats, and other non-target species, so they should be used with caution and checked daily. Slippery and sticky bands will only protect trees from caterpillars moving up and down the trunk and not from caterpillars that hatch high in the tree.

Biological control

Entomopathogenic bacteria

Bacillus thuringiensis subspecies *kurstaki* can be effective when ingested by very young tent caterpillars. Products containing this bacteria can be purchased commercially with or without additional synthetic pesticides. Applications should be made to as much of the foliage as possible, particularly around the tent area. Caterpillar death is not immediate and may take 3 days. If caterpillars are still alive after 3 days, reapplication may be necessary as the bacteria will be killed by exposure to sunlight. The effectiveness of the bacteria will decrease as the caterpillars get bigger, and so applications to trees infested with large caterpillars are unlikely to be useful. Although the bacteria is somewhat specific to moths and butterflies and will not impact other insect groups, it can negatively affect other species of moths and butterflies, so applications should be limited to what is essential and likely to be effective.

Predators

Natural predators and parasites like parasitic wasps are important for controlling the caterpillars in outbreak years, but these are not commercially available. Insectivorous insects and birds are typically responsible for bringing numbers back down to acceptable levels. Installing birdhouses or nest boxes can increase the local population of birds that may feed on tent caterpillars.

Chemical control

Residual pesticides

Treatment of eggs with synthetic pyrethroids prevents larval emergence in the spring, but it is difficult to kill caterpillars within the tents. However, applications around the tents may help control populations. Application of residual pesticides to trees in the spring before tents appear is the most effective.

Before using any product on or around trees, check that products are labelled for use on equine farms and ensure that the chemical is compatible with the tree species. It is important to avoid unnecessary applications of pesticides to flowers or during flowering. Chemical applications during this time may cause unintentional negative effects on beneficial pollinators such as butterflies and bees.

Insecticidal soaps

Older caterpillars will move down the tree trunk to rest. If groups of caterpillars are seen, they can be sprayed directly with insecticidal soaps that can be purchased commercially. The spray must be applied directly to the caterpillars. Application to the vegetation or trunk will not be sufficient to kill the caterpillars because just crawling over these soaps will not kill the caterpillars.

Systemic pesticides

Chemicals can be injected into trunks of tall trees by certified arborists (Figure 6.38). Systemic pesticide injection is expensive and may result in significant tree injury from the injection process if it is not done with precision equipment and skill.

6.5.2. Bag-shelter moth (*Ochrogaster lunifer*)

6.5.2.1 Pest summary (Table 6.6)

Found throughout Australia, bag-shelter moth caterpillars, also called processionary caterpillars or itchy grubs, are the immature stages of a single species, *Ochrogaster lunifer*, in the family Notodotidae

Figure 6.38. Certified arborists can inject tree trunks that are susceptible to arthropod infestations, such as tent caterpillar infestations, with systemic pesticides. Image shows trunk injection of an ash tree.

Table 6.6. Pest summary of bag-shelter moth, *Ochrogaster lunifer.*

Distribution[1]	**Australia: Yes** Canada: No Europe: No New Zealand: No United States: No
Size	Adults are 4.5-6.5 cm (1.77-2.56 in). Caterpillars are 4-5 cm (1.57-1.96 in).
Key identification characteristics	Adults have brown or brownish-grey wings and a single spot on each hindwing and forewing; males may have up to four white streaks on the forewings; females have a white or golden tuft at the end of the abdomen; caterpillars are brownish, and densely covered with long pale hair
Mouthparts of larvae	Chewing
Feed on	Plants in the genera *Acacia* (wattle), *Eucalyptus* & *Corymbia* (gums), *Cassia*, *Senna*, *Acer*, *Planchonia* trees, and beefwood (*Greviellea striata*)
Development time[2]	1 year
Adult life span	A few days
Total eggs laid	150-550
May be mistaken for	Other hairy caterpillars
Impact on horses	Equine amnionitis and foetal loss

[1] The pest distribution in the regions targeted by this publication are listed as of the date published. Areas of primary pest concern (if any) are in bold. Global distributions are listed in Section 6.5.2.2.
[2] Development times are approximate and can change with environmental conditions including temperature, humidity, and availability of food resources.

(Figure 6.39). Both the immature and mature forms have hairs that can irritate the skin. Like tent caterpillars, they can be a serious veterinary health issue when horses accidentally ingest the nests, dispersing pre-pupal larva, shed exoskeletons, and pupal chambers (Perkins *et al.*, 2016).

6.5.2.2 Distribution and seasonality

This species is found throughout coastal and inland Australia. Adults may be seen in the spring, and caterpillars will be seen over the summer months.

6.5.2.3 Description

Adult moths are brown to brownish-grey with either white spots or streaks on the wings (Figure 6.40). The ground-nesting form of this species tends to have dark brown wings with a white tuft at the end of the abdomen, and are sometimes called bunny tail moths. The form that nests in the canopy has lighter brown wings with a golden tuft on the abdomen. Caterpillars are dark brown with brown heads. The body is covered in long hairs (Figure 6.39).

Figure 6.39. The caterpillar of the bag-shelter moth, *Ochrogaster lunifer*.

Figure 6.40. Adult *Ochrogaster lunifer*, the bag-shelter moth. The ground-nesting form female has darker wings with a white tuft on the abdomen (A), while the canopy nesting form female has lighter wings with a golden tuft on the abdomen (B).

6.5.2.4 Life history

This species has a one-year lifecycle and undergoes complete metamorphosis with four life stages including egg, larva, pupa, and adult. Adults emerge in October to November, mate, and females lay eggs. Egg masses can contain 150 to 550 eggs on the trunk or canopy of a host tree. Bag-shelter moth caterpillars feed on a variety of regional plants found throughout Australia, but most ubiquitously on wattle (*Acacia* spp.), and gum (*Eucalyptus* spp.) trees. Eggs hatch in 3 to 4 weeks. Caterpillars feed on the host tree from December to April. Caterpillars construct a nest made of silk, which grows as the caterpillars increase in size and the nest fills with shed skins and frass (faeces) (Figure 6.41). Nests can be located at the end of a tree branch, and sometimes they are on the trunk. In coastal areas of Australia, the nest is located on the ground at the base of the tree, while in inland areas the nest is located in the canopy. There are three primary types of nests; bag-type nests that are leathery, tear-shaped, and are primarily in the canopy, tree-hugger nests that are also leathery but adhere to the trunk or a branch and are flattened, and ground nests that are filamentous and found at the base of various *Acacia* trees. Tree-hugger nests are always found on *Corymbia tessellaris* (Moreton Bay ash) trees.

Total development time is 20 to 24 weeks followed by spending winter as larvae for 20 to 22 weeks. Caterpillars leave the nests in the autumn in a characteristic single-file procession in search of locations to pupate. Caterpillars overwinter in the ground and will pupate in the spring.

Figure 6.41. Bag-shelter moth caterpillars build nests that fill with caterpillar frass and shed skins as they develop (A+B).

6.5.2.5 Control

Management of bag-shelter moths and other processionary caterpillars is primarily focused on preventing horse encounters with the caterpillars or caterpillar nests. Horses are likely to encounter caterpillars or shed caterpillar exoskeletons near host trees, low nests on trees, or nests that have blown off trees. It is important to note that eggs, caterpillars, and nest materials can cause extreme itching or allergic reactions, including respiratory and eye conditions, in people if they are contacted so caution should be used when removing caterpillars or nests. Have a protocol in place for response to inadvertent human exposure. See Northern Territory Government, 2016 in 'Continuing education' for more information on human exposure to urticating hairs.

Cultural control

Property inspections

Inspect the trees regularly in the spring through the autumn to identify egg masses and nests (Figure 6.42). Earlier detection makes removal and prevention of additional nests easier.

Horse turnout management

Horses should not be pastured where there are caterpillar nests. Hairs from caterpillars can move up to 100 m (130 yards) from caterpillar activity (Perkins *et al.*, 2019). Ideally, if feasible, mares would be kept at least this far from any nests.

Figure 6.42. Inspect trees for white egg masses, like these ones laid at the base of this *Acacia* tree (A). The pile of frass and old exoskeletons are visible below the egg masses (A and B). These skins can be blown by the wind and accidentally ingested by horses.

Lighting management

Like tent caterpillars, bag-shelter moth adults are attracted to fluorescent and incandescent lights. To reduce the risk of adults being attracted to these lights and potentially laying eggs in the area, keep lights off after sundown in the spring when adults are active.

Physical and mechanical control

For all management of processionary caterpillars including removing egg masses, nests, and caterpillars, it is important to wear personal protective gear like gloves and eye protection. Clothing may be contaminated with microscopic hairs so it is best to wear disposable outer coveralls and change into different clothes after handling nests. For added protection, eggs and nests should not be removed during windy weather or by individuals with respiratory conditions.

Host tree management

Tree hosts including *Acacia* (wattle) trees, beefwood, and eucalypts can be removed to prevent caterpillar presence on a farm. Removal of these trees may expose old nest material, so this should be planned when pregnant mares are not in the pasture. Trees should be carefully disposed of by removal from the property.

Reducing horse exposure to caterpillars

While removing host trees is preferred, sometimes that is not feasible. Fencing trees to prevent mares from grazing near the base may reduce some risk of accidental ingestion (Figure 6.43). It is important to note that hairs can be moved by the wind, and fencing will only allow enough time to inspect trees while nests are young and should not be considered a standalone method of prevention. Keeping feed storage covered and checking water buckets or troughs frequently for the presence of caterpillars or shed skins and nests can also help reduce the risk of caterpillar contact with horses.

Removal

Low nests can be removed gently from affected trees and placed in a closed container like a bucket with a lid for transport. It is generally easier to remove egg masses and young nests. Dispose of removed material by burying. Burning may be associated with the dispersal of microscopic hairs.

Biological control

There are some natural enemies of bag-shelter moths. A tachinid fly (*Carcelimyia dispar*) is known to be a parasite of the caterpillars. It is unknown what effect these flies have on caterpillar populations, and they have not been developed into commercial biological control agents.

Figure 6.43. If host trees cannot be removed, fencing around trees may help reduce mare encounters with caterpillars and low nests.

Chemical control

There are currently no recommendations for chemical control of bag-shelter moths. Off-label use of agricultural chemicals may be illegal unless it is registered for that use or covered by a specific permit. Please check with the Environmental Protection Authority (https://www.environment.gov.au/protection) before using any chemicals to control bag-shelter moth caterpillars.

6.5.3 Veterinary concerns

Mare reproductive loss syndrome/equine amnionitis and foetal loss

Caterpillars with urticating hairs have had devastating effects on horse breeding operations in the past. The Eastern tent caterpillar and the bag-shelter moth or processionary caterpillar have been implicated in mare reproductive loss syndrome, and equine amnionitis (inflammation of the membrane around the developing foetus) and foetal loss, respectively. Pregnant mares are particularly susceptible to infections of the placenta due to migrating hairs (setae) from caterpillars (Cawdell-Smith *et al.* 2013; Todhunter *et al.*, 2014a,b; Webb *et al.*, 2004). These infections can lead to abortions called mare reproductive loss syndrome in the case of tent caterpillars in the United States, and equine amnionitis and foetal loss in the case of bag-shelter moths in Australia, resulting in the loss of the entire breeding season. There are distinct differences in these conditions and the association of caterpillars with horses that should be understood depending on the reader's region (Todhunter *et al.*, 2014a,b).

Mares are most at risk in the autumn months. No therapeutic options exist to effectively prevent these abortions. Affected horses should be carefully monitored for retained foetal membranes, and uterine lavages performed if this occurs.

It is unlikely that mares will ingest processionary caterpillars because of the urticating hairs that cause itching and other reactions. The shed exoskeletons (caterpillar cast skins), however, are very light and may be carried by the wind into fields where mares graze. Ingesting these exoskeletons or portions of nests may cause equine amnionitis and foetal loss. While tent caterpillars and bag-shelter moth caterpillars have been shown to induce mare reproductive loss syndrome, and equine amnionitis and foetal loss, respectively, other caterpillar species with similar hairs may represent some risk to pregnant mares (see Sebastian *et al.*, 2008 in 'Continuing education' for more information on mare reproductive loss syndrome).

Pericarditis

Migrating hairs from tent caterpillar ingestion have been associated with pericarditis (Bolin *et al.*, 2005; Chapuis *et al.*, 2020), an inflammation of the tissue covering the heart. In extreme cases, this can lead to death. Horses demonstrating a decline in performance in an area where tent caterpillars are present should have an ultrasound to evaluate the heart. Treatment options are available for pericarditis (Reimer, 2013), and these work best when instituted early. Pericarditis has not been associated with processionary caterpillars.

Continuing education

Fire ants: what they are and what they aren't
Vail, K., Canaday, T.A. and Parkman, P., 2020.
https://fireants.tennessee.edu/wp-content/uploads/sites/208/2020/11/About-fireant-poster_reducedsize.pdf
- This poster complete with detailed images shows the differences between fire ants and other common ant species. Images of ants, as well as ant mounds, are shown.

Fire ants
Queensland Government, 2018.
https://www.qld.gov.au/environment/plants-animals/animals/pests-diseases/fire-ants
- The first detection in Australia of red imported fire ant occurred in Brisbane in 2001 and continues to be the target of a national cost-shared eradication program. The National Red Imported Fire Ant Eradication Program is delivered by Biosecurity Queensland on behalf of the Australian Government and all state and territory governments and includes scouting and control for discovered fire ants.

Fire ants in pastures

Billeisen, T., Brandenburg, R., Waldvogel, M. and Watson, W., 2021.
https://content.ces.ncsu.edu/fire-ants-in-pastures
• Fire ants nest in sandy soils in pastures, and stings can cause severe pain and swelling in horses. Basic management of pastures and chemical control options are presented in this management guide for the United States.

Bee-proofing for Florida citizens

O'Malley, M.K., Ellis, J.D. and Neal, A.S., 2019.
https://edis.ifas.ufl.edu/publication/IN741
• Preventing Africanized bees from accessing potential nesting sites is the best way to reduce interactions between bees and horses. Remove and discard items such as tyres and buckets that could be used as nesting sites. Seal up holes and eaves where bees could enter.

Composting animal manures: a guide to the process and management of animal manure compost

Augustin, C. and Rahman, S., 2016.
https://www.ag.ndsu.edu/publications/livestock/composting-animal-manures-a-guide-to-the-process-and-management-of-animal-manure-compost
• Actively composting manure can significantly reduce eye gnat populations. However, often manure is stored in accumulation areas and passively composted, which will not have the same effectiveness. In this guide, methods of actively composting manure are presented, which can help with managing these, and other, fly pests.

Equine ophthalmology for horse owners

Brooks, D., 2016.
https://aaep.org/horsehealth/equine-ophthalmology-horse-owners
• An overall view of common equine eye problems, along with first-line treatments.

Blister beetles in alfalfa

Ward, C., 1997.
http://lubbock.tamu.edu/files/2011/10/Blister-Beetle-in-Alfalfa_6.pdf
• Cantharidiasis risk in horses can be reduced with proper management of blister beetles in hay, primarily alfalfa. In this management guide, the biology of blister beetles is reviewed, along with the identification of major species associated with cantharidiasis. Management of hay and proper cutting and drying is presented.

Dealing with blister beetles

Boxler, D., 2020.
https://beef.unl.edu/beefwatch/2020/dealing-blister-beetles
• Blister beetles produce the chemical cantharidin, which can be toxic to horses. In this article, the relative amounts of cantharidin produced by species of beetle are compared as well as some management practices for blister beetles.

Clinical features of blister beetle poisoning in equids

Helmann, R.G. and Edwards, W.C., 1997. Journal of the American Veterinary Medical Association 211: 1018-1021. https://pubmed.ncbi.nlm.nih.gov/9343547/

- There is a range of toxin amounts that must be ingested for blister beetle poisoning in horses, and it is not always fatal. Clinical signs are related to the amount of cantharidin that has been ingested. Diagnosis and treatment of blister beetle poisoning are discussed.

Blister beetle poisoning

Epsy, B., 2016.
https://aaep.org/horsehealth/blister-beetle-poisoning

- In this brief review of blister beetle poisoning, the clinical signs of poisoning are reviewed. Guidelines for veterinary care are presented.

Itchy caterpillars

Northern Territory Government 2016.
https://nt.gov.au/wellbeing/emergencies-injuries-and-accidents/bites-and-stings/itchy-caterpillars/introduction

- Australian Lepidoptera (moths and butterflies) contains several species of caterpillars that cause irritation when in contact with humans. They are armed with irritating hairs or stinging spines that serve as protection against predators. This resource provides information to identify exposure to common caterpillars with irritating hairs and provides practices to treat these exposures.

Mare reproductive loss syndrome

Sebastian, M.M., Bernar, W.V., Riddle, T.W., Latimer, C.R., Fitzgerald, T.D. and Harrison, L.R., 2008. Veterinary Pathology 45: 710-722. https://journals.sagepub.com/doi/full/10.1354/vp.45-5-710

- Mare reproductive loss syndrome is associated with tent caterpillars. This review presents the findings of the 2001-2002 outbreak of eastern tent caterpillars, pathological findings associated with this syndrome, experimental case studies, and investigations into fertility effects.

References

Bethke, J.A., Vander Mey, B. and Bates, L.M., 2013. Eye gnats. University of California Pest Notes publication 74164 http://ipm.ucanr.edu/PDF/PESTNOTES/pneyegnat.pdf.

Bigham, J.T., 1941. *Hippelates* (eye gnats) investigations in the southeastern states. Journal of Economic Entomology 34: 439-444.

Bextine, B.R. and Thorvilson, H.G., 2002. Field applications of bait-formulated *Beauveria bassiana* alginate pellets for biological control of the red imported fire ant (Hymenoptera: Formicidae). Environmental Entomology 31: 746-752. https://doi.org/10.1603/0046-225X-31.4.746

Bolin, D.C., Donahue, J.M., Vickers, M.L., Harrison, L., Sells, S., Giles, R.C., Hong, C.B., Poonacha, K.B., Roberts, J., Sebastian, M.M., Swerczek, T.W., Tramontin, R. and Williams, N., 2005. Microbiologic and pathologic findings in an epidemic of equine pericarditis. Journal of Veterinary Diagnostic Investigation 17: 38-44.

Brown, W., 2014. Natural, organic, and alternative methods for imported fire ant management. ENTO-009. Texas A&M AgriLife Extension, http://fireant.tamu.edu/files/2014/03/ENTO_009.pdf

Burgess, R.W., 1951. The life history and breeding habits of the eye gnat, *Hippelates pusio*, Loew, in the Coachella Valley, Riverside County, California. American Journal of Hygiene 53: 164-177. https://citeseerx.ist.psu.edu/viewdoc/download?doi=10.1.1.939.23&rep=rep1&type=pdf

Cawdell-Smith, A.J., Todhunter, K.H., Perkins, N.R. and Bryden, W.L., 2013. Exposure of mares to processionary caterpillars (*Ochrogaster lunifer*) in early pregnancy: an additional dimension to equine amnionitis and fetal loss. Equine Veterinary Journal 45: 755-760. https://doi.org/10.1111/evj.12044

Chapuis, R.J., Ragno, V.M., Ariza, C.A., Movasseghi, A.R., Sayi, S., Uehlinger, F.D. and Montgomery, J.B., 2020. Septic fibrinous pericarditis in 4 horses in Saskatchewan following an outbreak of forest tent caterpillars in 2017. The Canadian Veterinary Journal 61: 724-730. https://pubmed.ncbi.nlm.nih.gov/32655155/

Fonteque, J.H., Mendes, R.P., Souza, A.F., Granella, M.C.S., Schade, J., Casa, M.S., Yonezawa, L.A. and Volpato, J., 2018. Systemic toxic reaction due to multiple honeybee stings in equine: Case report. Arquivo Brasileiro de Medicina Veterinária e Zootecnia 70: 767-772.

Hinkle N.C., Scholl, P.J., Mock, D.E. and Warner, W.B., 2001. Horses. In: Geden C.J. and Hogsette J.A. (eds) Research and extension needs for integrated pest management for arthropods of veterinary importance: Proceedings of a workshop in Lincoln, Nebraska. Second Edition, pp. 261-262. Available at: https://www.ars.usda.gov/southeast-area/gainesville-fl/center-for-medical-agricultural-and-veterinary-entomology/mosquito-and-fly-research/docs/lincoln-workshop/

Hogsette, J.A., 1981. Fly control by composting manure at a south Florida equine facility. In: Patterson, R.S., Koehler, P.G., Morgan, P.B. and Harris, R.L. (eds.) Status of biological control of filth flies. USDA-SEA A 106.2: F64, pp. 105-113.

Lewis, N. and Racklyeft, D.J., 2014. Mass envenomation of a mare and foal by bees. Australian Veterinary Journal 92: 141-148.

National Park Service, 2015. Fire ants. https://www.nps.gov/fosm/learn/nature/fire-ants.htm

Perkins, L.E., Zalucki, M.P., Perkins, N.R., Cawdell-Smith, A.J., Todhutner, K.H., Bryden, W.L. and Cribb, B.W., 2016. The urticating setae of *Ochrogaster lunifer*, an Australian processionary caterpillar of veterinary importance. Medical and Veterinary Entomology 30: 241-245.

Perkins, L.W., Bribb, B.W., Pagendam, D.E. and Zalucki, M.P., 2019. Variation in morphology and airborne dispersal of the urticating apparatus of *Ochrogaster lunifer* (Lepidoptera: Notodontidae), an Australian processionary caterpillar, and implications for livestock and humans. Journal of Insect Science 19: 6.

Porter, S.D. and Calcaterra, L.A., 2013. Dispersal and competitive impacts of a third fire ant decapitating fly (*Pseudacteon obtusus*) established in North Central Florida. Biological Control 64: 66-74.

Reimer, J., 2013. Management of equine pericarditis. Equine Veterinary Education 25: 334-338. https://aaep.org/sites/default/files/issues/eve-25-7-334-337.pdf

Ribeiro, P.R., Bianchi, M.V., Henker, L.C., Gonzales, F. and Pavarini, S.P., 2020. Acute renal failure in a horse following bee sting toxicity. Ciência Rural 50(5): e20190940. http://doi.org/10.1590/0103-8478cr20190940

Todhunter, K.H., Cawdell-Smith, A.J., Bryden, W.L., Perkins, N.R. and Begg, A.P. 2014a. Processionary caterpillar setae and equine fetal loss: 1. Histopathology of experimentally exposed pregnant mares. Veterinary Pathology 51: 1117-1130.

Todhunter, K.H., Cawdell-Smith, A.J., Bryden, W.L., Perkins, N.R. and Begg, A.P., 2014b. Processionary caterpillar setae and equine fetal loss: 2. Histopathology of the fetal-placental unit from experimentally exposed mareas. Environmental Pathobiology 51: 1131-1142.

Tschinkel, W.R. and Howard, D.F., 1980. A simple non-toxic home remedy against fire ants. Journal of the Georgia Entomological Society 15:102-105.

Staempfli, H.R., Wollenberg, G., Jewell, G. and McCutcheon, L.J., 1993. Acute fatal reaction to bee stings in a mare. Equine Veterinary Education 5: 250-252.

Webb, B.A., Barney, W.E., Dahlman, D.L., DeBorde, S.N., Weer, C., Williams, N.M., Donahue, J.M. and McDowell, K.J., 2004. Eastern tent caterpillars (*Malacosoma americanum*) cause mare reproductive loss syndrome. Journal of Insect Physiology 50: 185-193.

Photo credits

Title page	Nannette Turner, Flickr, license CC BY 2.0 https://www.flickr.com/photos/brokersaunders/3821119251/in/photolist-6PEeCp-6k4sH2
Figure 6.1.	Judy Gallagher, Flickr CC BY 2.0 https://www.flickr.com/photos/52450054@N04/49149965982.
Figure 6.2.	Eli Sarnat, Bugwood.org.
Figure 6.3.	Sanford Porter, USDA, ARS.
Figure 6.4.	(A,B) United States Department of Agriculture Animal and Plant Health Inspection Service Plant Protection and Quarantine, Bugwood.org; (C) Bruce Dupree, University of Alabama Extension.
Figure 6.5.	Daniel Christensen, Wikimedia Commons, license CC BY 3.0.
Figure 6.6.	Judgefloro, Philippines, Wikimedia Commons, license CC0 1.0 Universal Public Domain Dedication https://commons.wikimedia.org/wiki/File:379Ants_of_the_Philippines_eating_cheese_curls_02.jpg.
Figure 6.7.	Texas A&M AgriLife Extension.
Figure 6.8.	Banixx™.
Figure 6.9.	Jeffrey W. Lotz, Florida Department of Agriculture and Consumer Services, Bugwood.org.
Figure 6.10.	LadyDragonflyCC, Flickr, CC BY 2.0 https://www.flickr.com/photos/19646481@N06/6983110759.
Figure 6.11.	(A) Bob Peterson, Flickr, license CC BY 2.0 https://www.flickr.com/photos/50152666@N02/21003416856; (B) Bob van der Herchen, Bee Removal & Rescue; (C) Felicien Amakpe, Cercle nature et développement CENAD NG0, Bugwood.org.
Figure 6.12.	University of Florida Honey Bee Research and Extension Laboratory.
Figure 6.13.	Lyle Buss, University of Florida.
Figure 6.14.	Frank E French, Georgia Southern University, Bugwood.org.
Figure 6.15.	Erika Machtinger, University of Florida.
Figure 6.16.	Betsy Betros.
Figure 6.17.	S.A. Johnson, University of Florida.
Figure 6.18.	Bryan Vander Mey, University of California Cooperative Extension.
Figure 6.19.	Erika Machtinger, Pennsylvania State University.
Figure 6.20.	Erika Machtinger, University of Florida.
Figure 6.21.	Judy Gallagher, Flickr CC BY 2.0 https://www.flickr.com/photos/52450054@N04/50416973936/in/album-72157652008443880.
Figure 6.22.	(A) Clemson University – United States Department of Agriculture Cooperative Extension Slide Series; (B) Joseph Berger, Bugwood.org; (C) William M. Ciesla, Forest Health Management International, Bugwood.org.

Figure 6.23. (A) Gail Hampshire, Flickr CC BY 2.0 https://www.flickr.com/photos/gails_pictures/26933714923/in/photolist-2aRYecm-2aRYejq-H33k6e-KXsbTH-H2XNBh-HRr4Mt-28br8L1; (B) Gail Hampshire, Flickr CC BY 2.0 https://www.flickr.com/photos/43272765@N04/31114774005; (C) Gail Hampshire, Flickr CC BY 2.0 https://www.flickr.com/photos/gails_pictures/28595232052/in/photolist-28mpWtN-KyS3fh-8zjuWE-dKS5vH.

Figure 6.24. LiCheng Shih, Flickr CC BY 2.0 https://www.flickr.com/photos/37180297@N08/49854617657.

Figure 6.25. Erika Machtinger, Pennsylvania State University.

Figure 6.26. Hari K Patibanda, Flickr CC BY 2.0 https://www.flickr.com/photos/34176693@N06/50251417993.

Figure 6.27. R. Altencamp, Wikimedia Commons CC BY 2.0 https://commons.wikimedia.org/wiki/File:Thaumetopoea_processionea_100613.jpg.

Figure 6.28. Dennis van de Water, Shutterstock.

Figure 6.29. INTREEGUE Photography, Shutterstock.

Figure 6.30. Judy Gallagher, Flickr CC BY 2.0 https://www.flickr.com/photos/52450054@N04/14108930101.

Figure 6.31. Daniel Potter, University of Kentucky.

Figure 6.32. NatureServe CC BY 2.0 https://www.flickr.com/photos/48711769@N08/15653644032.

Figure 6.33. (A) Andy Reago & Chrissy McClarren, Flickr https://www.flickr.com/photos/80270393@N06/18204450679 CC BY 2.0; (B) Andy Reago & Chrissy McClarren, Flickr https://www.flickr.com/photos/wildreturn/18614589722/in/photolist-ny5GGR-ny9YMt-ny4WRL-umUD17-HUVYjd-GZaqZn-tUM3xW-HUVXqE CC BY 2.0.

Figure 6.34. (A) Katja Schulz, Flickr, license CC BY 2.0 https://www.flickr.com/photos/treegrow/23162468450; (B) Gail Hampshire, Flickr, license CC BY 2.0 https://www.flickr.com/photos/43272765@N04/34501423741; (C) Judy Gallagher, Flickr, CC BY 2.0 https://www.flickr.com/photos/52450054@N04/15927451605.

Figure 6.35. Robert L. Anderson, United States Department of Agriculture Forest Service, Bugwood.org.

Figure 6.36. Steven Katovich, Bugwood.org.

Figure 6.37. Amelia Martin, Dreamstime no. 94105073.

Figure 6.38. David Cappaert, Bugwood.org.

Figure 6.39. Quarti, WikiCommons CC BY 2.0 https://commons.wikimedia.org/wiki/File:Processionary_Caterpillar_Top.JPG.

Figure 6.40. M. Uemura, The University of Queensland.

Figure 6.41. (A) Mrpbps, Flickr CC BY 2.0 https://www.flickr.com/photos/25413523@N08/19300626100; (B) N. Perkins, The University of Queensland.

Figure 6.42. L. Perkins, The University of Queensland.

Figure 6.43. Erika Machtinger, Pennsylvania State University.

Chapter 7. Other insect pests of horses

Chapter overview

This chapter covers several groups of insects that have negative impacts on horses, but in different ways. Warble flies or cattle grubs (Section 7.1), equine bot flies (Section 7.2), and screwworm flies (Section 7.3) all affect horses by developing in the skin tissue or digestive tract of horses, depending on the species. Mayflies (Section 7.4) are insects that are not generally considered to be pests but can transmit *Neorickettsia risticii*, which causes equine neorickettsiosis (also known as Potomac horse fever, PHF) if infected mayflies are ingested by horses. Sand flies (Section 7.5) are small biting flies that are common in some areas of the world and some species of sand flies can transmit pathogens causing leishmaniasis and vesicular stomatitis.

Most of these insects have very distinct features and biologies that aid in accurate identification. In many cases, cultural management practices can be used to control the insect to reduce the risk of encounters. It is important to remember that many of the health conditions associated with these insects will require veterinary treatment and supportive care.

7.1 Warble flies (genus *Hypoderma*)

7.1.1 Pest summary (Table 7.1)

Hypoderma spp. (Figure 7.1) are flies in a genus of the insect subfamily Hypodermatinae known as warble flies. Other names for these flies are heel flies and gad flies. They are also often referred to as cattle grubs during the immature stage. They are a type of bot fly in the family Oestridae. Warble flies are pests of cattle and deer but can be accidental pests of horses. The presence of the flies around a horse can irritate the animal causing them to run recklessly in the pasture and increasing the risk of injury. The immature stages, called cattle grubs, migrate through the skin and form swellings called warbles on the neck, back, and withers of horses, which can become painful and interfere with riding, blanketing, or other activities. The resulting abscesses form warbles and can be susceptible to secondary infection.

Table 7.1. Pest summary of warble flies, genus *Hypoderma*.

Distribution[1]	Australia: No Canada: Yes Europe: Yes New Zealand: No United States: Yes
Size	Adults 15 mm (0.6 in)
Key identification characteristics	Two wings, bee-like in colour, hairy body, no visible mouthparts, M wing vein is curved up to meet the top of the wing
Mouthparts of adults	Non-functioning (vestigial) mouthparts as adults
Feed on	Immature warble flies (cattle grubs) feed on pus, cells, and other secretions, adults do not feed
Development time[2]	12 months
Adult life span	3-7 days
Total eggs laid	300-600
May be mistaken for	Equine bot flies (Section 7.2), other fly bee mimics like robber flies, bees (Chapter 6)
Impact on horses	Irritation when young grubs penetrate the skin, painful warbles, secondary infection

[1] The pest distribution in the regions targeted by this publication are listed as of the date published. Areas of primary pest concern (if any) are in bold. Global distributions are listed in Section 7.1.2.

[2] Development times are approximate and can change with environmental conditions including temperature, humidity, and availability of food resources.

Figure 7.1. Warble fly, *Hypoderma* spp.

7.1.2 Distribution and seasonality

Warble flies are found on all continents in the Northern Hemisphere. The main species of concern to horse owners are the northern cattle grub (*Hypoderma bovis*), and the common cattle grub (*Hypoderma lineatum*). There have been cases of warble flies more commonly associated with deer, such as the deer warble fly (*Hypoderma diana*), infesting horses. The common cattle grub has the widest range and is found between 25° and 60°N latitude. The northern cattle grub is more limited in distribution to cooler climates. Warble flies have been eradicated from some countries, such as the UK and Denmark.

In the southern United States, the adults are active from February to March, and the grubs are present (in warbles) on animals in December. In Canada, the flies are active in summer (June to August), and the grubs are present in warbles on animals from February to May. In Europe, the adults are active in the warm spring and summer months, and the grubs are on animals from January to July. The common cattle grub is always 1 to 2 months ahead of the northern cattle grub. For example, in Europe adult common cattle grubs are active from March to May and adult northern cattle grubs from June to September.

7.1.3 Description

The adult flies are large, approximately 15 mm (0.6 in), hairy, and resemble bumblebees. They are typically brown, orange, or yellow. Warble flies can look similar to bot flies (Section 7.2), but they can be differentiated from equine bot flies using the wing veins (Figure 7.2). The M vein in equine bot flies is curved down to meet the bottom of the wing, but the same vein goes straight up to meet the top of the wing in warble flies. Warble flies can also be mistaken for flies in the family Asilidae (robber flies) or bees (Figure 7.3). Robber flies and bees are not pests or associated with horses.

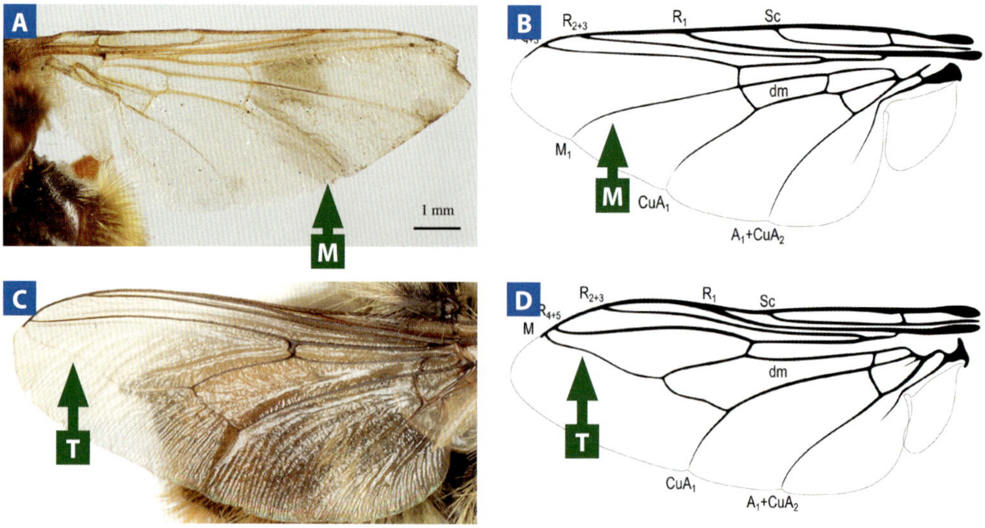

Figure 7.2. The M wing vein is curved down to meet the bottom of the wing in *Gasterophilus* (equine bot flies; A,B) (labelled M) and goes straight up to meet the top of the wing in *Hypoderma* (warble flies; C,D) (labelled T).

Figure 7.3. Warble flies may be confused with other flies that mimic bees like the robber fly *Mallophora bomboides* (A) or actual bee species like the urban digger bee, *Anthrophora urbana* (B).

Warble fly eggs are cylindrical, white, and approximately 1 mm (0.03 in) long (Figure 7.4). Common cattle grub adults lay eggs in groups of five to 25, whereas northern cattle grub adults lay eggs singly. The grubs that hatch are 1 mm (0.03 in) in length and are maggot-like; by the time they arrive at their resting sites the grubs are about 15 mm (0.6 in) in length (Figure 7.5). They darken in colour as they develop. The pupae are enclosed in a black puparium (Figure 7.5).

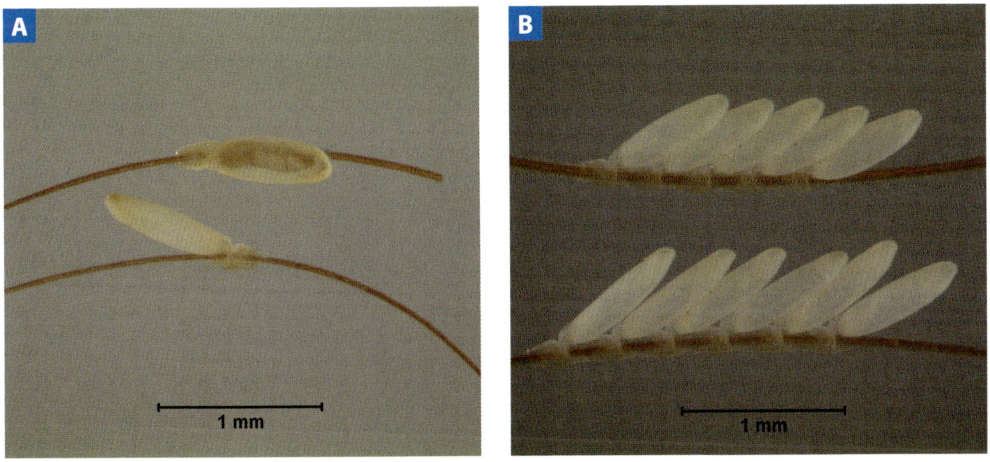

Figure 7.4. Eggs of the northern cattle grub are laid singly (A), and eggs of the common cattle grub are laid in groups (B) on the legs of large animals by warble flies, *Hypoderma* species.

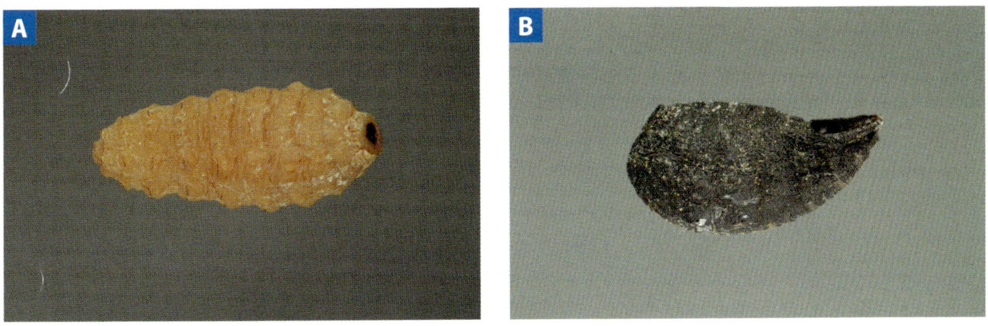

Figure 7.5. Cattle grub (*Hypoderma* spp.) larva (called a grub) (A) and an empty puparium (B).

7.1.4 Life history

Warble flies have complete metamorphosis with four life stages: egg, larva (called grubs), pupa, and adult. The adult female flies lay single or multiple eggs (depending on the fly species) on the legs of large animals on warm sunny days in late spring to early summer. In her limited life, a female may lay 300 to 600 eggs. In natural hosts like cattle and deer, the eggs hatch, and the grubs burrow into the skin using digestive enzymes. They migrate to resting sites in tissue in different areas of the body depending on the fly species. For example, the common cattle grub migrates to the spinal cord, but the northern cattle grub migrates to the oesophagus.

After a few months, the grubs migrate to the skin and use digestive enzymes to cut a breathing hole known as the warble pore when they occur in cattle. Over 3 to 4 days a warble or boil forms around the grub (Figure 7.6). Within the warble, the grub feeds on pus, dead cells, and other secretions. The grubs grow and darken in colour (Figure 7.5). After 4 to 8 weeks the grub will emerge from the breathing pore in the warble and after burrowing into the ground will pupate (or finish development into an adult fly) in the soil. Flies emerge in 1 to 3 months. The adult flies do not feed, instead they mate within 24 hours and lay eggs before dying in 3 to 7 days. Because horses are not the normal host for cattle grubs, the larvae often die within the warble without completing development.

7.1.5 Control

Warble flies exclusively develop in association with animals, so the best control methods are prevention. Monitoring for fly eggs on horse hair, changes in horse behaviour, and inflammation that may indicate warbles can help prevent or catch early-stage infestations. Some chemical treatments can also be effective.

Cultural control

Horses are most at risk in areas when they are housed near infested cattle where warble flies are known to be present. Avoid keeping cattle and horses together or in pastures that are close together (Figure 7.7). During the warble fly season, it is important to conduct regular inspections of horses for the presence of eggs glued to hairs on lower regions, unusual bumps on the horse, and monitoring for unusual behaviour that could be a response to warble fly presence, like agitation and running across pastures in both horses and cattle.

Physical and mechanical control

It is recommended to consult with a veterinarian to remove cattle grubs. Grubs can be removed through the warbles, however, younger grubs are difficult to remove because of their backwards-facing spines, and in many cases, warbles will need to be opened by a veterinarian to access the grub since cattle grubs usually do not cut a breathing hole in horses.

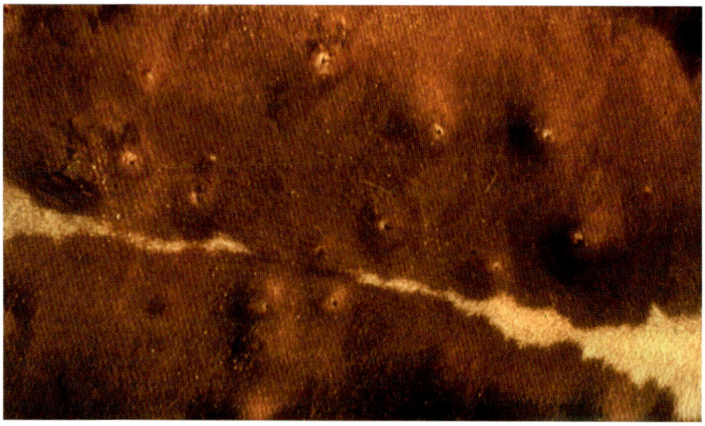

Figure 7.6. Example (from cattle) of warbles along the back of the animal.

Figure 7.7. Horses that are pastured with or near cattle are most at risk for warble fly infestation.

7.1.6 Veterinary concerns

In horses, warble flies cannot complete their life cycle, and the grubs may cause additional complications migrating through the body. In rare cases, they may infest the brain causing neurological complications.

If the grubs migrate to the back of the horse, they rarely cut a breathing hole in the skin, but they do produce a warble, which can be quite painful for the horses, especially if they are not treated, and often require removal by a veterinarian. When removing the grubs it is important to avoid rupturing the grub, as the release of the body contents can cause localized inflammation, abscesses, or anaphylactic reactions. Warbles can also serve as sites of secondary infections both while the grub is present and following removal.

Macrocyclic lactone dewormers serve as an effective treatment for horses with early-stage infections prior to warble formation. When treating horses with systemic pesticides for cattle grub control, it is important to consider the life cycle of the insect and time the treatment accordingly. The safest time to kill the grubs is from egg hatch until just before the migrating maggots reach the sites in the oesophagus, for the common cattle grub, or spinal cord, for the northern cattle grub. If the treatments are applied too late, the maggots will be killed in the resting sites, where they will prevent normal body functions and cause severe side effects. Consult with your veterinarian about timings as these will vary by region.

Note that warble flies are a notifiable disease in some countries including the UK (see 'Resources' for more information on responsible authorities).

7.2 Equine bot flies (genus *Gasterophilus*)

7.2.1 Pest summary (Table 7.2)

Species of the genus *Gasterophilus* (family Gasterophilinae) are the main bot fly pests of horses (Figure 7.8). Larvae of equine bot flies inhabit the stomach and intestine causing ulceration in horses and other equids. In general, *Gasterophilus* do not cause clinical disease, but *Gasterophilus pecorum* is the most pathogenic of the four species and can result in death if the attached larvae cause swelling in the oesophagus. It is important to note that humans can be accidental hosts of equine bot fly larvae. There are 12 species in the genus *Gasterophilus,* and most of them infest the digestive tract of horses as well as other equids like mules, donkeys, zebras, and other ungulates (hooved animals), such as reindeer and caribou.

7.2.2. Distribution and seasonality

The 12 species have a worldwide distribution. Four species of *Gasterophilus* are found infesting horses, the most common species worldwide being the horse bot fly *Gasterophilus intestinalis* and the throat bot fly *Gasterophilus nasalis*. Other species often identified in horses are the lip bot fly *Gasterophilus haemorrhoidalis*, which is present worldwide, and the dark-winged horse bot *Gasterophilus pecorum*, which is present only in the Old World.

Table 7.2. Pest summary of equine bot flies, genus *Gasterophilus*.

Distribution[1]	Australia: Yes Canada: Yes Europe: Yes New Zealand: Yes United States: Yes
Size	10-15 mm (0.39-0.59 in) in length
Key identification characteristics	Two wings, hairy body, no visible mouthparts, pointed abdomen for laying eggs, M wing vein is curved down to the bottom of the wing
Mouthparts of adults	Non-functioning (vestigial) mouthparts as adults
Feed on	Gut tissue and secretions as larvae, adults do not feed
Development time[2]	10-12 months
Adult life span	3-7 days
Total eggs laid	100-1000
May be mistaken for	Warble flies (Section 7.1), fly bee mimics like robber flies, other bees (Chapter 6)
Impact on horses	Stomach ulceration, secondary infections

[1] The pest distribution in the regions targeted by this publication are listed as of the date published. Areas of primary pest concern (if any) are in bold. Global distributions are listed in Section 7.2.2.
[2] Development times are approximate and can change with environmental conditions including temperature, humidity, and availability of food resources.

Figure 7.8. Equine bot fly, *Gasterophilus intestinalis*.

Gasterophilus flies have only one generation per year. *Gasterophilus intestinalis* and *G. nasalis* adults are most active in summer and autumn, with the greatest number of larvae found in the animals in winter. Most larvae will drop to pupate in the ground in the spring. *Gasterophilus haemorrhoidalis* adults are most active in early summer. *Gasterohilus pecorum* is most often found during the wet season.

7.2.3 Description

Adult equine bot flies are similar in appearance to warble flies. Adults are 10 to 15 mm (0.39 to 0.59 in), and they have yellow and black hairs so, like warble flies, can be confused with some robber flies and bees (Figure 7.8). They can be identified from warble flies by the wing veins. The M vein is curved down to meet the bottom of the wing in equine bot flies, but the same vein goes straight up to meet the top of the wing in warble flies (Figure 7.2).

Equine bot fly eggs are cylindrical, approximately 1 mm (0.4 in), off-white to yellow and look like flecks of paint. The larvae have hooked mouthparts and a rounded body that is covered in spines (Figure 7.9). Older larvae are 1 to 2 cm (0.4 to 0.8 in). The pupae are cylindrical, reddish-brown

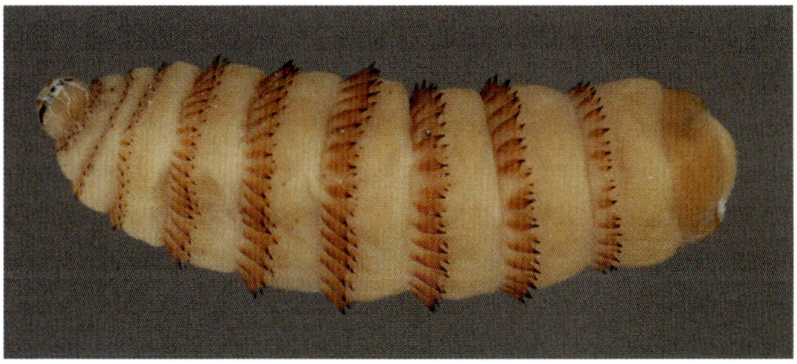

Figure 7.9. Larva of the equine bot fly, *Gasterophilus intestinalis*.

and passed out of the animal in the faeces. Equine bot fly adults do not have functional mouthparts and so do not feed or bite. The female has an elongated abdomen that curves under her body.

7.2.4 Life history

Equine bot flies have complete metamorphosis with four life stages: egg, larva (called bots), pupa, and adult. In general, equine bot flies lay their eggs directly on the host or on an intermediate arthropod vector, such as a mosquito or other fly species that come into frequent contact with horses or other ungulates. If laying eggs directly, the female will lay 100 to 400 eggs, if laying the eggs on another arthropod, the female captures the arthropod and lays 15 to 30 eggs on its abdomen. The eggs will incubate there and then drop off in response to body heat when that arthropod lands on the horse. Equine bot flies lay 100 to 1000 eggs in total, split between several hosts typically in early summer. The eggs are laid in different areas of the body depending on the species. *Gasterophilus intestinalis* lays eggs on the hairs of the legs (cannon bones and knees (Figure 7.10), *Gasterophilus nasalis* on the throat, and *Gasterophilus haemorrhoidalis* on the lips. *Gastrophilus pecorum* lays eggs on the legs of the horse but also on grasses and other forage where the horse may come into contact with them while grazing.

The horse inadvertently stimulates the hatching of the larvae from eggs on the skin through licking, rubbing, or breathing on them. Equine bot fly larvae are ingested and spend a few weeks in the soft tissue of the lips, gums, and tongue. After being swallowed, larvae attach to the lining of the stomach and intestine where they feed on gut tissue and secretions and grow. After 10 to

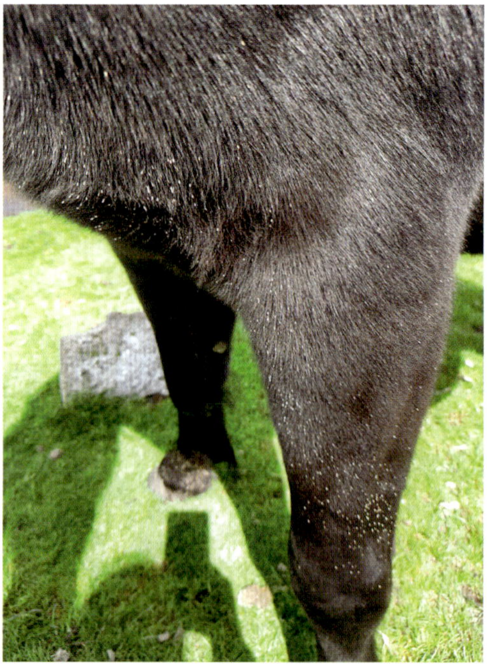

Figure 7.10. Equine bot fly eggs are off-white to yellow and look like flecks of paint.

12 months of development, they are passed out in faeces as mature larvae, which burrow into the dry manure or soil below the manure pile and pupate. This typically occurs in late winter to early spring. The adults emerge in 1 to 2 months and mate within a few days. The life cycle takes approximately 1 year.

7.2.5 Control

Although equine bot flies are rarely a serious issue for horses, it is still important to monitor for eggs on hairs and larvae in the faeces. Because bot flies do not reproduce inside the horse, they will not be identified with faecal egg counts like parasitic worms (Chapter 9). Equine bot flies exclusively develop in association with animals, so the best control methods prevent infestation. The presence of the bee-like bot flies laying eggs causes horses to panic and practice extreme avoidance behaviours that can result in injury or at least reduce time spent grazing.

Cultural control

Check the hair, particularly around the legs, throat, head, and lips, for fly eggs from late spring until the first frost when adults are active and check the faeces for larvae to prevent or catch early-stage infestations (Figure 7.11).

Remove manure from pastures and stalls frequently to prevent the accumulation of pupae. Active composting of manure will kill equine bot fly larvae. Rotate grazing in pastures to reduce equine bot fly accumulations. If other animals like sheep or cattle are present on the farm, it can be helpful to alternate these different animals among the available pastures (see Kenny, 2018 in 'Continuing education' for more information on pasture rotation).

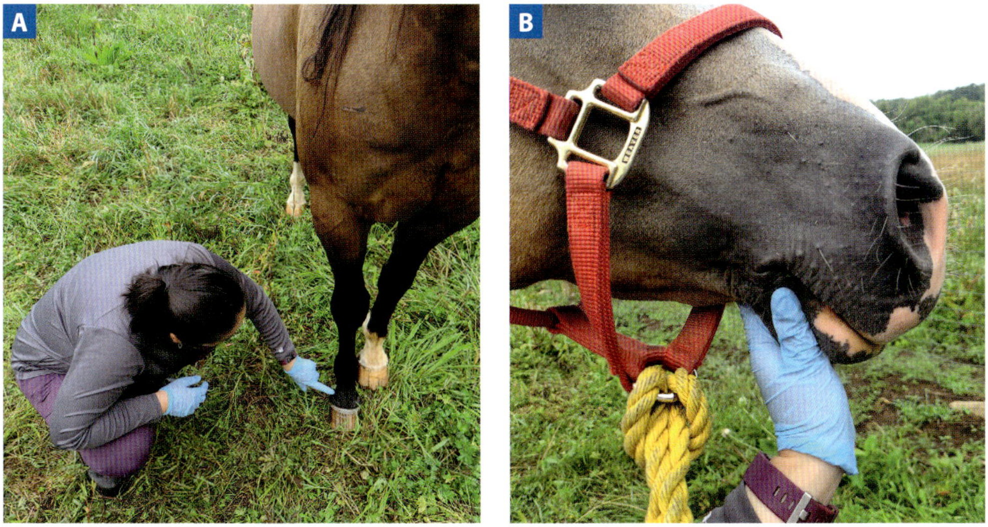

Figure 7.11. Equine bot flies lay eggs on different parts of the body, depending on the species. Check for eggs on the legs (A), jaw, throat, and lips (B).

Physical and mechanical control

Physical barriers

Fly masks with extended nose guards can be used as a deterrent if the problematic bot fly species in the area lay eggs on the face, such as the nose or lip bot fly (*Gasterophilus haemorrhoidalis*). Fly sheets and boots may also help deter bot flies from laying eggs around the shoulders and parts of the lower leg.

There is no evidence that baby oil, mineral oil, or other oils applied to the body will have any impact on preventing bot flies from laying eggs. Furthermore, there may be some serious detrimental impacts of oil application. Oils may increase skin temperature, attract dirt and debris, and potentially even cause burning if horses are in the sun.

Removal

Eggs can be removed from horses before the larvae hatch using a specially designed tool called a bot fly knife (Figure 7.12), or by clipping hair or sponging with hot water (45 °C/113 °F) to stimulate hatching and then washing off larvae. Another method is to use a sponge to apply warm water and dish detergent then rinse off the residue immediately after the mixture dries.

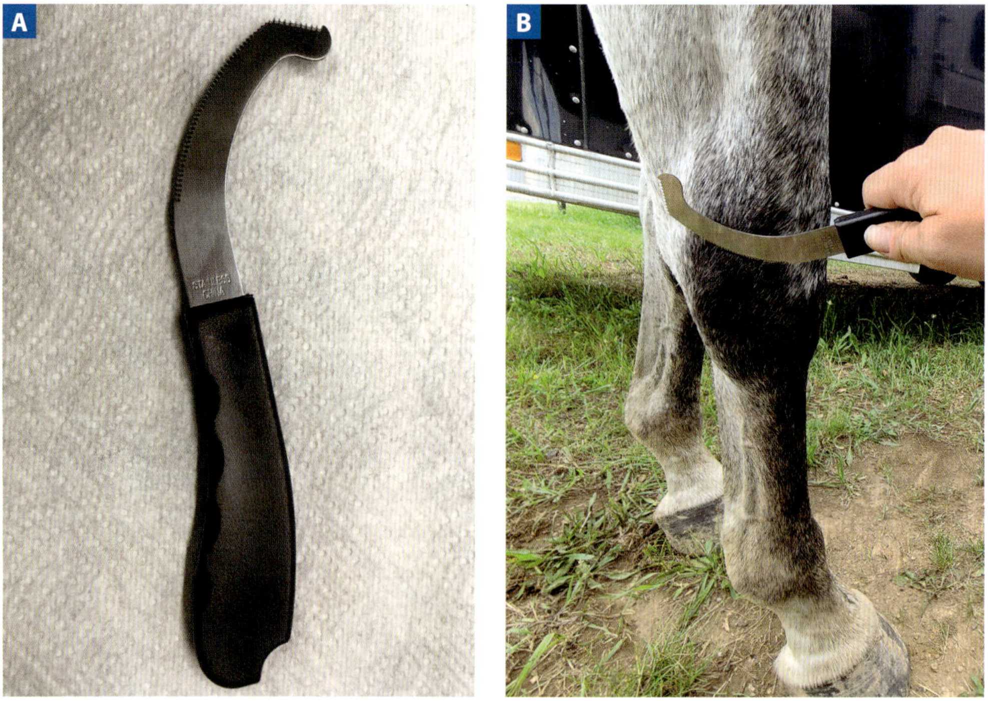

Figure 7.12. A bot fly knife (A) can be used to remove equine bot fly eggs from the hair on the legs (B) and other areas of the body.

These measures may need to be repeated daily throughout the season, and it is easy to miss a few eggs. Be careful when examining and treating horses for eggs because equine bot fly larvae can accidentally infest humans. Gloves should be worn if touching the eggs to prevent larvae from being transferred to hands, and from hands to eyes or mouth. Eggs removed from horses will dry out quickly and will not reattach to the horse if the animals roll on them in the field.

Chemical control

Repellents and other pesticides are not likely to be effective against equine bot fly eggs as they are extremely resilient. However, promoting hatching with a warm water solution that contains a pesticide may kill newly emerged larvae. Repellents do not appear to be effective against equine bot flies, mostly because the adult flies spend little time in contact with the horses.

7.2.6 Veterinary concerns

Equine bot flies are infrequently a source of severe veterinary concern, although infestations of *G. pecorum* are a serious problem in wild populations of Przewalski's horse in the desert steppes of China. Migrating larvae will rarely cause inadvertent damage to areas as they travel through. Secondary infections, such as peritonitis, can be caused during these migrations, but these are rare occurrences. Equine bot fly larvae may also cause ulceration of the stomach when they detach from the lining, leading to discomfort and colic-type symptoms. This is a rare event even with overwhelming infestations.

For the diagnosis of internal larvae, gastroscopy may be useful if a large burden is suspected, as the larvae can be seen. Treating horses with a dewormer that works on equine bot fly larvae is the best way to control infestations once the eggs have hatched. Avermectins applied orally as liquids, gels, boluses, and feed additives are the most commonly used. Treat the horse 1 month after eggs are observed in the spring and again in the autumn after the frost and the adult flies are no longer active.

7.3 Wound-invading flies: screwworms (family Calliphoridae) and other flies that develop in wounds (families Calliphoridae and Sarcophagidae)

7.3.1 Screwworms (family Calliphoridae)

7.3.1.1 Pest summary (Table 7.3)

The New World screwworms are flies in the genus *Cochliomyia* in the family Calliphoridae, which includes blue and green bottle flies and other species of what are called blow flies. There are four species in the genus and two of medical and veterinary importance. *Cochliomyia hominivorax*, the primary screwworm (Figure 7.13 and 7.14), and *Cochliomyia macellaria*, the secondary screwworm. The two species are named because the former (primary) invades the wound and the latter (secondary) will only infest an already invaded wound. The larvae of both species infest the body of animals, growing and feeding on tissues, a condition called myiasis. The primary screwworm feeds on living tissue and so is more damaging than the secondary screwworm, which feeds only on carrion or necrotic tissue of animals that are already injured or infested with primary screwworm.

The Old World screwworm, *Chrysomya bezziana* (family Calliphoridae), has a similar life history to the primary New World screwworm and so can be considered a primary screwworm. Although neither species of primary screwworms currently occurs in North America, Europe, Australia, or New Zealand, there is an ever-present risk of their introduction.

Table 7.3. Pest summary of Old and New World primary screwworms.

	Old World screwworm, *Chrysomya bezziana*	New World screwworm, *Cochliomyia homnivorax*
Distribution[1]	**Australia: No (risk of reintroduction)** Canada: No Europe: No **New Zealand: No (risk of reintroduction)** United States: No	Australia: No Canada: No Europe: No New Zealand: No **United States: No (risk of reintroduction)**
Size	8-10 mm (0.3-0.4 in)	8-10 mm (0.3-0.4 in)
Key identification characteristics	Metallic black-blue, large dark red-brown eyes, four faint lines on the thorax, abdominal segments with narrow dark bands along the posterior margins	Metallic blue, large orange eyes, three lines on the thorax
Mouthparts of adults	Sponging (non-biting)	Sponging (non-biting)
Feed on	Nectar (male and female), animal tissue (female and larvae)	Nectar (male and female), animal tissue (female and larvae)
Development time[2]	10-14 days	12-15 days
Adult life span	8-10 days	2-3 weeks
Total eggs laid	400-500	500-700
May be mistaken for	Other screwworms, metallic blow flies (Chapter 8)	Other screwworms, metallic blow flies (Chapter 8)
Impact on horses	Myiasis	Myiasis

[1] The pest distribution in the regions targeted by this publication are listed as of the date published. Areas of primary pest concern (if any) are in bold. Global distributions are listed in Section 7.3.1.2.
[2] Development times are approximate and can change with environmental conditions including temperature, humidity, and availability of food resources.

Figure 7.13. **Primary screwworm, *Cochliomyia hominivorax*, whose larvae develop in living tissue. Note the stripes on the thorax; in this species, the middle stripe (indicated with an arrow) will appear shorter than the outer two stripes. The adults are also typically more blue than blue-green.**

Figure 7.14. ***Cochliomyia hominivorax*, the primary screwworm (Fig. 7.13) and *Cochliomyia macellaria*, the secondary screwworm (A) are very similar in appearance, but the three lines on the thorax of the secondary screwworm are equal in length (indicated with an arrow) and the adult is more green-blue than blue (B). The primary screwworm is of greater horse health concern because it develops in living tissue. However, the secondary screwworm can infest dead tissue on the edge or surface of wounds.**

7.3.1.2 Distribution and seasonality

The New World primary screwworm is limited in distribution, mostly due to an eradication campaign using sterile insect technique in the USA and Mexico, but it is still present in parts of northern South America, Central America, and the Caribbean (see United States Department of Agriculture, 2017 in 'Continuing education' for more information on sterile insect technique). The secondary screwworm is more widely distributed and occurs throughout northern South America, Central America, the Caribbean, Mexico, the USA, and southern Canada. A recent incursion of primary screwworm back into the USA occurred in September 2016 in the Florida Keys with devastating impacts on the endangered key deer and five confirmed infestations in domestic animals. The outbreak was declared eradicated in March 2017 after the implementation of a multi-component response plan (see National Key Deer Refuge, US Fish and Wildlife Service, 2017 in 'Continuing education' for more information on this plan). This species was accidentally introduced into Libya in 1988 but was successfully eradicated within several years.

Old World screwworm, *Chrysomya bezziana*, is restricted to tropical areas of the Old World including Southeast Asia, tropical and subtropical Africa, some Middle Eastern countries, India, the Malay Peninsula, Indonesia, the Philippines, and Papua New Guinea. Although it has only been found once in Australia, the risk of introduction from Papua New Guinea remains high. The discovery of either species in North America, Europe, Australia, and New Zealand requires immediate reporting to respective governmental agencies so that a response can be mounted before the flies become established. In their current range, both Old and New World screwworms are most abundant during the transition from wet to dry seasons (Phillips *et al.*, 2004; Siddig *et al.*, 2005).

7.3.1.3 Description

The primary New World screwworm is a metallic blue fly about two to three times the size of a house fly, about 8 to 10 mm (0.3 to 0.4 in) (Figure 7.13 and 7.14). The adults have large orange eyes that are as wide as the body, yellow on the front of the head under the eyes, and three stripes that run down the back, with the middle stripe starting further back and appearing shorter than the outer stripes. The secondary screwworm looks similar except it is slightly smaller, about 6 to 9 mm (0.23 to 0.35 in), a more greenish-blue, and the three lines begin at the same point behind the head and the lines are all the same length (Figure 7.13). Primary and secondary screwworm adults may be mistaken for blue bottle flies (Figure 7.15) or other blow flies, which are also often metallic but lack distinct stripes on the thorax and yellow on the head under the eyes.

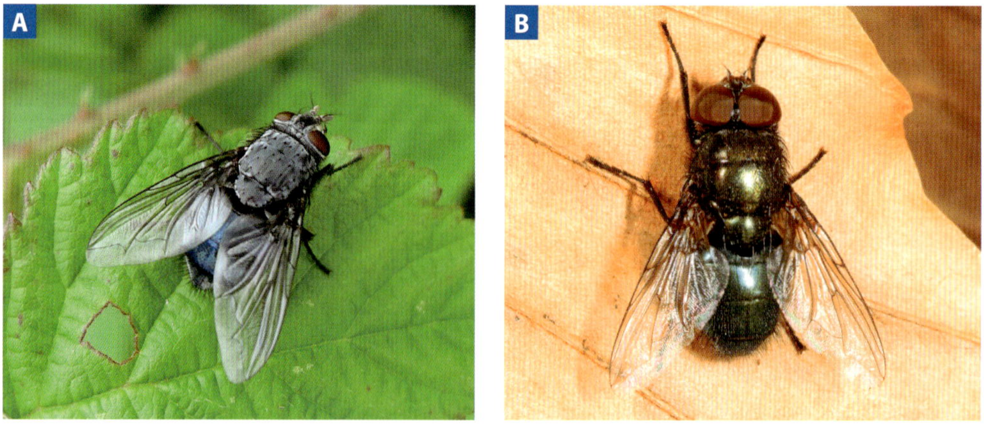

Figure 7.15. Primary and secondary screwworm adults can be mistaken for other common blow flies, as seen here, which are also metallic (discussed in Chapter 8).

Screwworm eggs are white, approximately 1 mm (0.04 in) in length. The larvae are maggot-like, legless, cream coloured, and can reach up to 17 mm (0.67 in) feeding in the wound. Mature larvae have rings of spines around the body that give the appearance of a spiral, which are the reason for the name screwworm (Figure 7.16). Distinguishing between larvae of primary and secondary screwworms is very difficult using morphological features. However, only primary screwworms feed on living tissue and will create deep lesions. If secondary screwworm larvae are present in a wound, they will be feeding on dead tissue at the edge or surface of the wound.

Old World screwworms are primary screwworms that resemble many other blow flies. They are shiny metallic blue, green, or blue-green flies that can be up to 10 mm long. The head of the fly has yellow-brown colouration below the eyes (Figure 7.17). Like the New World screwworm, the larvae have an array of spines around each segment to help hold them in place in host tissues.

Figure 7.16. Larva of *Cochliomyia hominivorax*, the primary screwworm, with the distinctive rings of spines.

Figure 7.17. Old World screwworm, *Chysomyia bezziana*, is found in many tropical Old World countries, including Papua New Guinea.

7.3.1.4 Life history

Screwworm flies have complete metamorphosis with four life stages: egg, larva, pupa, and adult. Primary and secondary screwworm flies can lay up to 350 and 200 eggs per female, respectively. During her lifetime she may lay up to 3,000 eggs. The eggs may be laid in different sites, and multiple females will lay eggs in the same site. Eggs hatch in 24 hours and then larvae feed on the carcass or tissues in the wound until maturity, which can take as little as 4 days. Once mature they migrate away from the body to pupate in soil or under leaf litter, rocks or fallen wood. The pupal stage lasts for 7 to 10 days. Once the adults emerge, they can live for up to 6 weeks. Male and female flies feed on flower nectar, and female flies additionally feed on animal tissues.

A suitable site for the primary screwworm can be any broken skin like a superficial scratch, tick bite, or wounds from castration or branding. Mucous membranes can also be infested. The umbilical cord of newborn foals during the healing process can also attract screwworms to lay eggs. Secondary screwworm (*Cochliomyia macellaria*) attacks are examples of accidental myiasis, Like most blow flies, these are common flies that normally develop in carrion. However, they also will lay eggs in wounds that have festered to the point of giving odours that attract the flies. Their larvae do not develop in healthy tissue but are limited to feeding on necrotic septic tissues. In horses, this is most common in wounds that have become infected after gelding.

7.3.1.5 Control

The primary screwworm is so damaging to its host that it is essential to prevent infestations by cleaning and caring for wounds promptly, and monitoring for and appropriately treating any wounds that have fly larvae. Regularly inspect the wounds for eggs and larvae and, if found, consult with a veterinarian. Suspected screwworm infestations in the USA, Canada, Europe, Australia, and New Zealand must be reported to appropriate officials (see 'Resources' for more information on responsible authorities).

If samples are determined to be primary screwworm, the appropriate agency will be made aware of the problem and will start investigating the extent of the issue. If the release of sterile insects is necessary (New World primary screwworms only), it will be implemented along with careful monitoring for affected animals and removal of larvae. Typically officials will then provide monitoring and control measures to eradicate the problem. Therefore, curative management is not usually done by farm owners themselves.

Cultural control

Horses imported from areas where screwworms are present should go through a quarantine procedure and be inspected for the presence of wounds and developing larvae for the first several weeks. In 2000, 17 horses were imported from South America and despite going through quarantine procedures, one infested animal escaped detection and was shipped to Florida, USA, where the infestation was later discovered and contained. This emphasises the need for diligent surveillance.

Although New World screwworm has been eradicated from the areas this book covers, it is possible that a reintroduction may occur like the 2016 reintroduction in Florida. Horses in areas where screwworm was endemic should be monitored regularly for wounds (Figure 7.18), and they should be treated as quickly as possible. This is because the area is likely to be highly suitable for screwworm development and survival. If larvae are found in wounds, a veterinarian should be consulted for removal and potential reporting to regulatory institutions.

As the fly larvae of the secondary screwworm develop in decomposing carcasses, carrion, and rotting meat, removal, burial, or composting of these development sites from the landscape will reduce the presence of flies on the property and the annoyance that comes with excess fly abundance.

Physical and mechanical control

If primary screwworm infestation is suspected, it is strongly recommended to seek veterinary assistance. Wounds will likely be sensitive and painful, necessitating veterinary involvement for the safe removal of the larvae. In addition, the veterinarian will notify the necessary officials responsible for screwworm infestation mitigation.

An attractant lure known as swormlure-4 has been shown to activate and attract New World primary screwworm. Old World screwworms are also attracted to swormlure-4, but other attractants are more effective and selective for this species (Urech *et al.*, 2012). Traps for monitoring

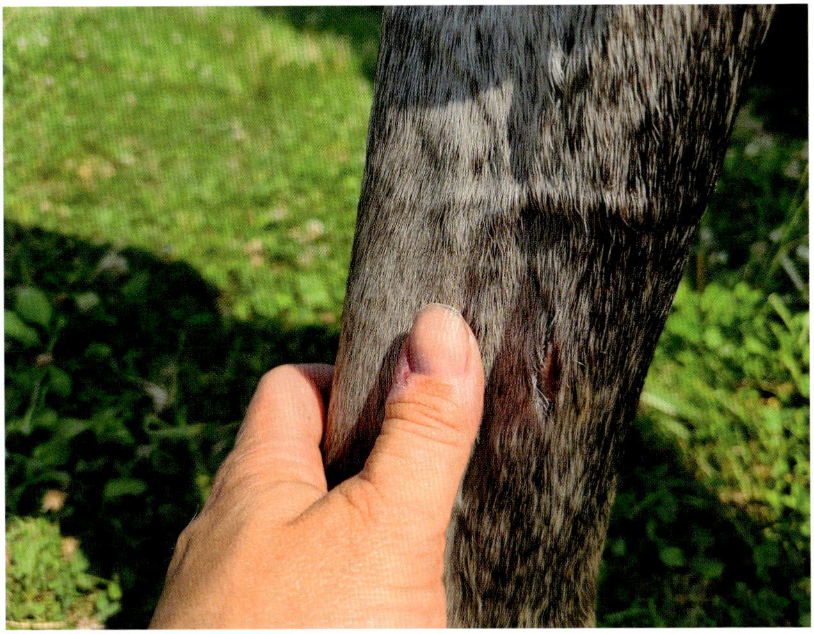

Figure 7.18. In areas where screwworms are present, animals should be inspected regularly for wounds and the presence of fly eggs and larvae.

of screwworms have been developed, although these are typically used for monitoring the success of eradication programs following the release of sterile insects by officials and they are not currently commercially available.

Sterile insect technique (SIT)

New World primary screwworm infestations have been controlled with nationally run sterile insect technique programs. If an infestation in an eradicated area is reported, for example in Florida or Texas, then it is likely that the sterile insect technique will be used by officials. The technique involves the release of large numbers of male screwworms that were previously sterilized with radiation. The sterile flies outcompete the wild fertile screwworms for female mates. The females then lay eggs, but the larvae fail to hatch. As the females only mate once, the impact of sterile insect technology on this pest is fast, and eradication has been demonstrated on multiple occasions.

7.3.1.6 Veterinary concerns

Primary screwworms, both Old and New World, are of veterinary concern due to the severity and number of wounds they cause on an animal (Figure 7.19). Untreated infestations are often so severe that they lead to the death of the animal due to overwhelming bacterial infection and toxin release from the larvae. Affected horses require immediate veterinary care to remove larvae and manage the resulting infections. Special care should also be taken to kill all larvae removed from the animal to prevent the re-establishment of screwworms in an area where they have been eradicated. Doramectin has been administered as a preventative. Doramectin should only be used under the supervision of a veterinarian as it is currently not labelled for use in horses. Primary screwworms are a reportable disease in the USA, Canada, Europe, Australia, and New Zealand. Veterinarians or agricultural authorities should be contacted immediately (see 'Resources' for more information on responsible authorities).

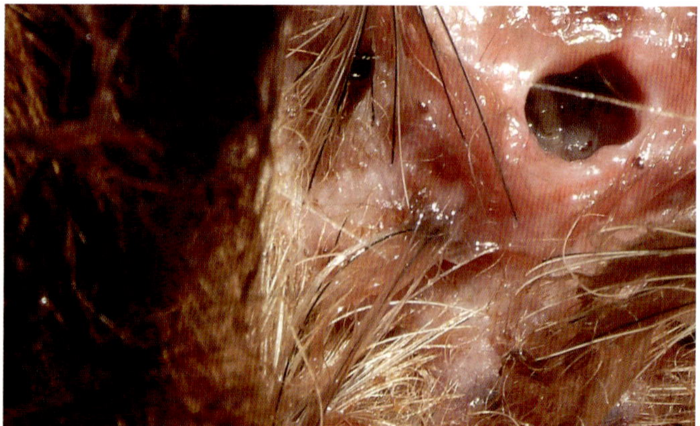

Figure 7.19. Primary screwworm larvae feed on living tissue and untreated infestations can lead to the death of the animal. Here, New World screwworm larvae can be seen developing within an abscess (upper right of photo) on the animal.

7.3.2 Other wound-invading flies (families Sarcophagidae and Calliphoridae)

In addition to primary screwworm, other fly species occur in the areas covered by this book that occasionally infest horse tissue (Hall and Wall, 1995). This includes the spotted flesh fly, *Wohlfahrtia magnifica*, the hairy maggot fly, *Chrysomya rufifacies*, and the *Lucilia* species responsible for sheep-strike.

7.3.2.1 Spotted flesh fly, *Wohlfahrtia magnifica* (family Sarcophagidae)

The spotted flesh fly (Figure 7.20) is found in southern Europe, Central Asia, the Middle East, North Africa, and China. It does not occur in North America, Australia, or New Zealand. The adult flies are about 10 mm (0.39 in) long and have alternating grey stripes on the thorax. Larvae develop in healthy host tissue and reach about 20 mm (0.79 in) in length before dropping to the ground to pupate. The entire life cycle is completed in 4 to 6 weeks.

Like the primary screwworms, this fly only develops in the tissue of living animals. Adult flies deposit batches of about 150 newly hatched larvae (not eggs) either on wounds or around the anus or genital openings of a wide range of mammals, especially dogs and sheep (Figure 7.21) (Giangaspero *et al.*, 2011). In horses, flies primarily attack the vulvar areas of mares, especially during oestrus and after foaling (Farkas and Képes, 2001), and the genitals of stallions and geldings.

7.3.2.2 Hairy maggot blow fly, *Chrysomya rufifacies* (family Calliphoridae)

The hairy maggot blow fly is native to Australia but has expanded to include warmer regions of most of the world. Adult flies are 6 to 12 mm (0.24 to 0.47 in) long and are distinctly green (Figure 7.22) but can appear blue under bright natural sunlight. Despite their common name, the larvae resemble other fly larvae except for a series of short fleshy projections along the sides of the larva;

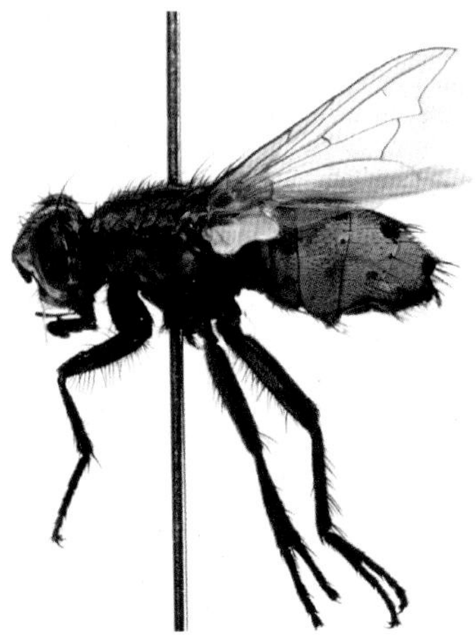

Figure 7.20. Adult spotted flesh fly, *Wohlfahrtia magnifica*.

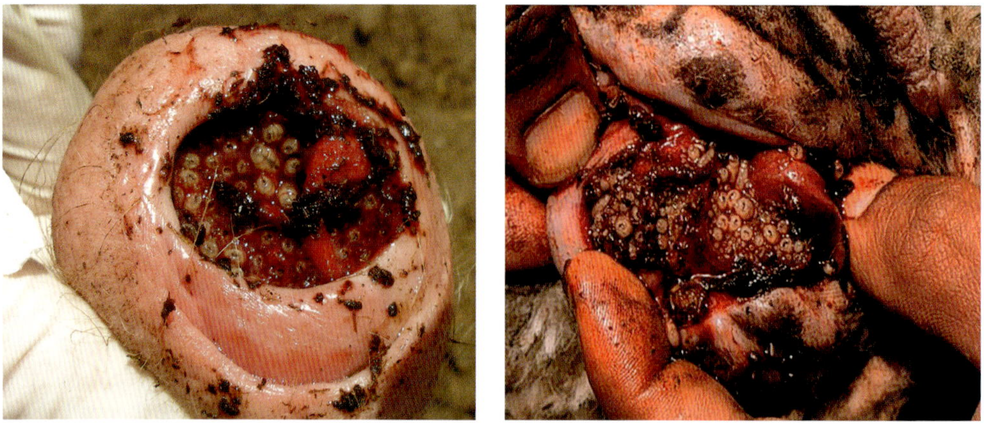

Figure 7.21. Spotted flesh fly larvae are laid on wounds or around natural orifices like the prepuce as seen here. They are particularly common in sheep.

they do not have any true hairs. The species is related to other notorious myiasis-causing (flesh invading) flies of the genus *Chrysomya* that include the Old World screwworm (*C. bezziana*), the oriental latrine fly (*C. megacephala*), and the African tropical latrine fly (*C. putoria*).

Hairy maggot blow flies are primarily decomposers of carrion that generally prefer to lay their eggs on large carcasses (Figure 7.23) (Baumgartner, 1993). An individual fly will lay about 200 eggs, which hatch within a day. Larvae remain and develop rapidly on the carcass, completing larval development in as little as 2.5 days at 29 °C. Within the carcass, the larvae also will attack and eat the larvae of other fly species. Mature larvae wander away from the carcass and pupate

Figure 7.22. Adult hairy maggot blow fly, *Chrysomya rufifacies*.

in the soil and emerge as adult flies about a week later. Adult flies can live for up to 5 weeks. This species is widely used in forensic entomology to determine the time of death in criminal cases (Flores *et al.*, 2014). This species is also a common agent of secondary myiasis when flies lay eggs in sheep that have been struck by other flies (usually *Lucilia* spp.). The flies will occasionally lay eggs in wounds of horses that have become necrotic, where they feed on the damaged tissue.

7.3.2.3 Sheep strike flies (genus *Lucilia*, family Calliphoridae)

Lucilia cuprina (Australian sheep strike fly; Figure 7.24A) and *Lucilia sericata* (green bottle fly; Figure 7.24B) are blow flies that usually develop in carrion. They are also of great importance to the sheep and goat industries because the flies will deposit eggs in the hind areas of the animals where the fleece has been soiled by urine and faeces. Larvae then feed in the fleece and on the skin surfaces, sometimes invading healthy tissue. The effects of an infestation can be devastating, sometimes leading to death because of the toxic effects of ammonia produced by the larvae.

The two flies are close relatives that are nearly indistinguishable from each other based on appearance. The Australian sheep strike fly originated in Australia but is now found in the southern USA and Mediterranean countries of the European Union. Adults are 6 to 10 mm (0.24 to 0.39 in) long. The green bottle fly originated in Europe, but now has a near-global distribution, and adults are the same size at 6 to 10 mm (0.24 to 0.39 in) long. These flies are typically among the first flies to arrive at a fresh carcass, where they deposit their eggs (Figure 7.25). Larval development typically takes about 5 days, after which mature larvae leave the carcass to pupate in the soil. The life cycle takes about 2 weeks under summer conditions.

As stated above, both species also cause severe myiasis in sheep populations wherever sheep and the flies are found. In rare cases, they will also cause myiasis in horses, usually striking the sheath in geldings and intact males.

Figure 7.23. Larvae of the hairy maggot blow fly, which are usually found developing in carrion (A+B).

Figure 7.24. Adults of the Australian sheep strike fly, *Lucilia cuprina* (A), and the green bottle fly, *Lucilia sericata* (B). Both species are common flies with a metallic green appearance.

Figure 7.25. *Lucilia* flies are among the first flies to discover and lay eggs on a fresh carcass.

7.3.2.4. Management considerations for other wound-invading flies

It is important to note that blow flies are largely beneficial insects that provide essential services as decomposers and pollinators (Chapter 8). Except for those who keep horses in areas where spotted flesh flies occur, the risk of horses contracting myiasis from these flies is very low. Good husbandry practices, especially involving sheath cleaning and wound care, are all that is needed to minimize the chance of problems.

7.3.2.5 Veterinary concerns for other wound-invading flies

Myiasis generally occurs at the tip of the penis, in fossas around the urethral process, and affecting the urethral process itself. Myiasis may also be seen in the clitoral fossa of mares, though this is less common. Rarely myiasis from these species can be seen in untended wounds on the rest of the body. Basic cleaning of the area and application of a pyrethrin ointment is often all that is needed to remove the offending maggots. Any damage to the affected tissues readily heals once cleaning has occurred. Pyrethrin ointments or wipes around the opening of the sheath will prevent reinfection from occurring. Wounds should be cleaned regularly and kept bandaged to both treat and prevent myiasis. If the wound is difficult to cover, pyrethrin ointments may be used to discourage fly activity.

7.4 Mayflies (order Ephemeroptera)

7.4.1 Pest summary (Table 7.4)

Mayflies, also called 'up-winged flies', are not true flies, but instead insects belonging to the order Ephemeroptera (Figure 7.26). There are 3,000 species in 400 genera and 42 families. The immature stages live in freshwater, and the pre-adult and adult stages are terrestrial. Typically, mayflies are not considered to be pest species. However, they are involved in the transmission of the bacteria, *Neorickettsia risticii*, which causes equine neorickettsiosis in horses. This condition was previously called Potomac horse fever. Equine neorickettsiosis has a complex life cycle that includes trematodes (flukes), snails, and aquatic insects. Horses become infected by accidentally eating mayflies infected with the bacteria. Other insects including caddisflies, damselflies, dragonflies, and stoneflies also can become infected with the bacteria and lead to infection in horses. However, mayflies are the insects typically associated with outbreaks of equine neorickettsiosis because they often emerge in mass numbers, and so they are difficult for horses to avoid (Figure 7.27).

7.4.2 Distribution and seasonality

Mayflies are present in freshwater bodies throughout the world except for Antarctica. Throughout much of their range, they are not considered to be a pest and are an important part of the aquatic community, where they provide a food source for fish. Equine neorickettsiosis, caused by the bacteria *Neorickettsia risticii*, was first identified in areas surrounding the Potomac River, northwest of Washington D.C. However, like mayflies, the range of this pathogen is global, and it has been described in the USA, Canada, South America (Venezuela, Uruguay, and Brazil), India, Europe (Italy, France, and the Netherlands), and Australia.

7.4.3 Description

Adult mayflies have long abdomens and long wings that close behind the mayfly like a butterfly but do not fold up over the abdomen. The membranous triangular wings are held upright at rest (Figure 7.28). They have an enlarged thorax due to the muscles that power the wing movements.

Table 7.4. Pest summary of mayflies, order Ephemeroptera.

Distribution[1]	Australia: Yes Canada: Yes Europe: Yes New Zealand: Yes United States: Yes
Size	2.5-32.0 mm (0.1-1.3 in)
Key identification characteristics	Long abdomen, long membranous wings, slender body
Mouthparts of adults	Non-functioning mouthparts
Feed on	Adults do not feed
Development time[2]	> 1 year
Adult life span	Several days
Total eggs laid	400 to 3,000
May be mistaken for	Mosquitoes (Chapter 4), crane flies (Chapter 8)
Impact on horses	Equine neorickettsiosis (Potomac horse fever PHF), caused by the bacteria *Neorickettsia risticii*

[1] The pest distribution in the regions targeted by this publication are listed as of the date published. Areas of primary pest concern (if any) are in bold. Global distributions are listed in Section 7.4.2.

[2] Development times are approximate and can change with environmental conditions including temperature, humidity, and availability of food resources.

Figure 7.26. Mayfly (order Ephemeroptera) adult.

Figure 7.27. When mayflies emerge in mass numbers it can be difficult for horses living near mayfly development sites to avoid accidental consumption.

They have short antennae and large eyes. They have a long cylindrical abdomen with two or three tail-like appendages (cerci) at the tip.

Immature mayflies (nymphs) have a long, cylindrical body that is slightly flattened (Figure 7.29). The head, with its strong chewing mouthparts, points forwards or down. The thorax has three segments with a leg on each, and the abdomen has ten segments with up to seven pairs of gills. At the tip of the abdomen, there are two to three fine projections. Most of the nymphal stages are aquatic, but the final stage, a pre-adult stage, is terrestrial. The pre-adult stage has wings but is sexually immature and not as colourful as the adults.

7.4.4 Life history

Mayflies have incomplete metamorphosis with three life stages: egg, nymph or naiad, and adult. Female mayflies lay 400 to 3,000 eggs on the surface of the water, and the eggs sink to the bottom. As mayflies develop in fresh, flowing water, the female typically lays her eggs upstream of good development sites. Eggs can hatch in a few days up to a year. Mayflies can overwinter in the egg stage. The nymphs hatch from the eggs and go through 10 to 50 moults over several months to a year depending on species and the temperature of the water. The nymphs may be herbivores primarily feeding on algae, detritivores feeding on detritus, or predators feeding on the larvae of smaller insects. The pre-adult stages often emerge simultaneously within a mayfly population and are a preferred food of many fish. The adults do not feed and lack functional mouthparts. Adults live a day to a few days and will mate, and females will lay eggs during that time.

Figure 7.28. Adult mayflies hold their wings above the body and have filamentous projections (cerci) at the tip of the abdomen.

Figure 7.29. Mayfly nymphs are aquatic, have somewhat flattened bodies, and several filamentous projections (cerci) behind the abdomen.

7.4.5 Control

Mayflies are accidental pests of horses, and they generally develop in ecologically sensitive wetland habitats. Control is mostly limited to the prevention of ingestion by reducing attraction to equine facilities and covering food and water containers when not in use. Horses that graze near a freshwater stream or river in the spring or summer, have access to irrigated pasture, or are stabled close enough to a freshwater source that insects could be blown into the grazing or feeding area by the wind are at increased risk of ingesting infected insects.

Cultural control

If the property contains any flowing freshwater, maintaining the vegetation as a barrier could help to keep mayflies away from horses (Figure 7.30). Keeping horses away from the water could also be beneficial, particularly in mayfly season.

During mayfly emergence periods, turning off lights outside the stable or around the property will help reduce the attraction of adult mayflies from neighbouring water bodies at dawn and dusk (Figure 7.31). Preventing mayfly access to areas where horses might ingest mayflies can reduce the risk of pathogen transmission. Keep food covered to prevent dead mayflies from being ingested. Clean out water buckets and troughs regularly, at least daily during mass mayfly emergence, which is usually very short-lived (a few days or less), and clean feed areas to prevent contamination of food and water with dead mayflies. Consider excluding horses from pastures near water or where mayflies have been identified during the emergence period.

Figure 7.30. Vegetation barriers along water bodies that neighbour horse paddocks can help to reduce contact between horses and mayflies.

Figure 7.31. Mayflies, as well as many other flying insects, are attracted to lights left on at night. Turning off lights around equine facilities will reduce the attraction of flying pest insects to properties.

7.4.6 Veterinary concerns

The bacteria that causes equine neorickettsiosis (Potomac horse fever), *Neorickettsia risticii*, has a complicated path to infecting horses. The vector of the bacteria is a flatworm (a trematode, which are also known as flukes), which has a life cycle that includes freshwater snails and vertebrates that eat insects. Immature worms hatch and find a host, which can include mayfly nymphs, infecting the immature mayflies with the bacteria. After infecting an insect host like a mayfly, the worms typically infect fish, amphibians, birds, and bats when the mayfly is eaten. In these animals, the worms reproduce and eggs are deposited in the water when the animals defecate. However, pre-adult and adult mayflies infected with the worms and bacteria may be accidentally ingested by horses after dying in food buckets, water buckets, water troughs, or on pasture grasses after emerging from the water. The bacteria then infects the horse, potentially causing equine neorickettsiosis.

In a study examining infection of dead mayflies around a horse show grounds in Minnesota where an outbreak of six infected horses occurred, one in ten mayfly samples collected from the show facility were positive for the bacteria (Wilson *et al.*, 2006). Other potential routes of infection for horses are drinking water contaminated by the worms and being directly infected by the worms when standing in water.

After ingestion of an infected arthropod or worm, the incubation period is 1 to 3 weeks. The bacteria is released into the gastrointestinal tract and reproduces in the cells. It will eventually reproduce in the blood resulting in an infection that causes colitis, preventing the absorption of water and electrolytes. In severe cases, this can cause death. It is estimated that equine neorickettsiosis can be fatal in 5 to 30% of horses (Pusterla and Madigan, 2013).

Infected horses displaying fevers, diarrhoea, anorexia, depression, or sore feet should be seen by a veterinarian immediately. Twice daily oxytetracycline for 5 days and increased hydration can be effective if started early. In severe cases, a blood transfusion may be necessary. Response to treatment is usually seen within 12 to 24 hours (Bertin *et al.*, 2013). Horses with laminitis, a painful condition of the feet, as a result of infection should be kept in a stall with supportive bedding, icing, and pain relief medication.

Check with a veterinarian regarding the time of year when symptoms are most frequently observed. A vaccine is available and has been shown to reduce the severity of symptoms. Timing should be based on the local emergence of mayflies for best efficacy. Because the vaccine is not fully protective, all control measures possible should be instituted.

7.5 Sand flies (family Phlebotominae)

7.5.1 Pest summary (Table 7.5)

Sand flies in the family Psychodidae, subfamily Phlebotominae, are small, hairy, blood-feeding flies. There are several other flies that are sometimes referred to as sandflies, such as *Culicoides* biting midges (Ceratopogonidae, see Chapter 4) or certain species of horse fly (Tabanidae, see Chapter 4).

In the Phlebotominae, there are several genera and species of concern that transmit pathogens, and the bites can be a nuisance to horses, humans, and other animals. Flies in the genus *Lutzomyia* (Figure 7.32) transmit vesicular stomatitis in the Americas and the parasite that causes leishmaniasis in Central and South America; flies in the genus *Phlebotomus* (Figure 7.33) transmit the parasites that cause leishmaniasis in Europe to horses, humans, and other animals (See European Centre for Disease Prevention and Control, 2020 in 'Continuing education' for more information).

Table 7.5. Pest summary of sand flies, family Phlebotominae.

Distribution[1]	Australia: Yes Canada: Yes Europe: Yes New Zealand: No United States: Yes
Size	3 mm (0.12 in)
Key identification characteristics	Hairy small flies, long legs and antennae
Mouthpart of adults	Piercing, sucking
Feed on	Blood (females), sugar (males and females)
Development time[2]	50 days
Adult life span	<2 weeks
Total eggs laid	30-70 eggs per batch
May be mistaken for	*Culicoides* biting midges (Chapter 4)
Impact on horses	Leishmaniasis, vesicular stomatitis

[1] The pest distribution in the regions targeted by this publication are listed as of the date published. Areas of primary pest concern (if any) are in bold. Global distributions are listed in Section 7.5.2.

[2] Development times are approximate and can change with environmental conditions including temperature, humidity, and availability of food resources.

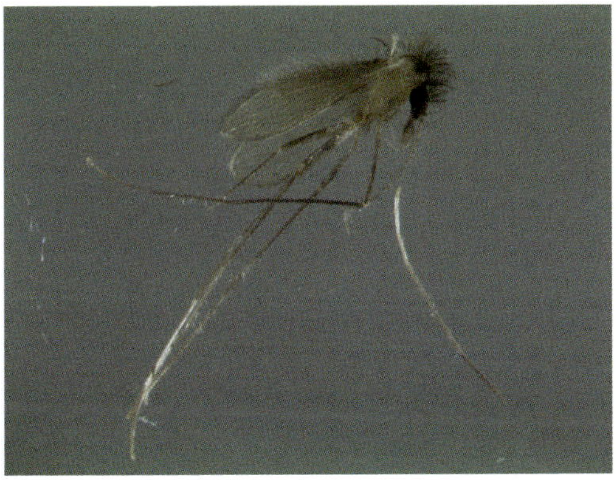

Figure 7.32. Sand fly, *Lutzomyia shannoni*. Species in the *Lutzomyia* genus transmit the parasites that cause leishmaniasis and vesicular stomatitis in Central and South America.

7.5.2 Distribution and seasonality

Phlebotomine sand flies are found within the 50°N and 40°S parallels, but they are not found in New Zealand or the Pacific Islands. This insect group occurs throughout the tropics and subtropics, as well as temperate zones. Sand flies are widespread throughout the Mediterranean area. Increasing temperatures due to climate change is predicted to extend the range of these insects.

7.5.3 Description

Sand flies are small, approximately 3 mm (0.12 in), and hairy. They have long legs and long beaded antennae. *Lutzomyia* vary in colour by species from almost white to nearly black (Figure 7.32). *Phlebotomus* are yellow with obvious black eyes (Figure 7.33). Both genera have a humped thorax and oval hairy wings that are held erect. Eggs are approximately 0.3 mm (0.01 in), dark oval, and laid in batches. Larvae have dark heads and lighter segmented bristly bodies. On the tip of the abdomen, the larvae have two long hairs (Figure 7.34).

7.5.4 Life history

Sand flies have complete metamorphosis with four life stages: egg, larva, pupa, and adult. After taking a blood meal, females lay batches of 30 to 70 eggs in moist (but not aquatic) habitats with a supply of organic material. Some species develop in the dung of burrowing mammals. Larvae hatch from the eggs in 20 days, and then the larvae develop through four instars over 18 days feeding on organic material. In more temperate climates, sand flies spend the winter as larvae. Once fully developed, they will move to a drier location to pupate. Adults emerge after 7 to 12 days. Both sexes feed on carbohydrates, but females also feed on blood, which gives them the necessary nutrients for egg production. The adults are nocturnal, and they shelter during the day

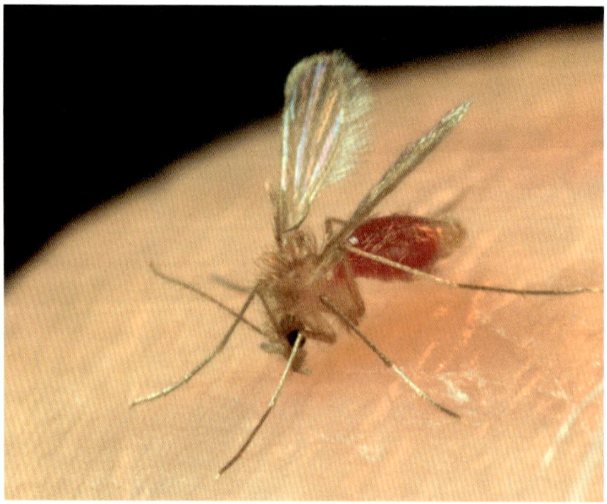

Figure 7.33. Sand flies in the genus *Phlebotomus* transmit the parasites that cause leishmaniasis in the Old World. These insects have a humped thorax and oval, hairy wings.

Figure 7.34. Four larval stages (from left) and pupa (right) of *Phlebotomus papatasi*; note the presence of long hairs at the tip of the abdomen on the larvae.

in dark warm locations such as among leaf litter, in animal burrows (Figure 7.35), and cracks and crevices. When flying, they take short flights as they are weak fliers and often appear to be hopping.

7.5.5 Control

Sand fly control is difficult due to the development sites being difficult to locate and hard to treat. Reduction of sand fly numbers and their natural rodent hosts around the facility is the best option (see Alexander and Maroli, 2003 in 'Continuing education' for a comprehensive review of control options).

Figure 7.35. Some species of sand flies develop in burrows of animals such as rabbits and rodents.

Cultural control

Sanitation can remove organic waste close to horses that may attract the rodent hosts of sand flies as well as sand flies themselves. Sealing cracks in walls, sewers, and basements may help exclude sand flies and rodents from living close to horses.

Eliminating sand fly resting and development sites from close to where horses are pastured or stabled can help to reduce numbers on horses, as they are not strong fliers and rarely move more than a few hundred metres from their development site to feed. For example, removing organic waste, unwanted vegetation, wood piles, and termite mounds may disrupt the development sites of sand fly larvae. Exclusion and removal of wildlife from the property, particularly rodents and other small mammals, can help reduce the number of sand flies. After animals are removed, fill mammal burrows. Wild animal trapping should comply with all regional regulations, and wild animals should never be handled by anyone other than a professional wildlife trapper or biologist.

Physical and mechanical control

Seal dark, cool, moist areas that could be adult sand fly resting sites, such as sewage drains and holes and cracks in walls. Seal cracks in cool, dark, and humid places, such as animal shelters and stables. Sand flies are smaller than mosquitoes and can likely get through conventional mosquito netting, so very fine mesh, such as no-see-um or biting midge netting, is required to exclude sand flies from areas (100 holes/cm^2 or 625 holes/in^2).

Very little is known about the effectiveness of traps for managing sand flies. In one study, a variety of trap features were evaluated against *Phlebotomus papatasi* (Müller *et al.*, 2015). These researchers concluded that the most effective traps used a combination of carbon dioxide and ultraviolet light. They also suggested that placing numerous traps could be useful for reducing sand fly populations.

Biological control

The bacteria *Bacillus sphaericus*, which is closely related to the mosquito control agent *Bacillus thuringiensis israelensis*, has been used for sand fly management. When sugar solutions containing *Bacillus sphaericus* were applied to vegetation in sand fly habitats in an attractive toxic sugar bait system, adult flies took the bacteria back to their larval development sites resulting in reduced larval and adult populations (Robert *et al.*, 1997). However, there are no commercially available biological control agents for sand flies.

Chemical control

Treated fly sheets, boots, and masks

Fly sheets, boots, and masks with permethrin bound to the fabric fibres can be purchased to repel flies. The repellent action of these products generally lasts 20 to 40 washes, depending on the product. The duration of effectiveness may change with exposure to sunlight and weather patterns, so observe horses periodically to make sure it is still providing the protection desired. Fly sheets purchased without pesticide treatment can also be sent to a company that specializes in binding permethrin to fabric for treatment. It is important to recognize that do-it-yourself treatment of fabrics with pesticides that do not list fabric treatment for horses on the label may be a violation of the law. In addition, horses are sensitive to chemical pesticides and off-label treatment may cause skin reactions.

Residual pesticides

Adult sand flies can be controlled by spraying residual pesticides, mainly pyrethroids, in run-in sheds, stables, and other buildings (Figure 7.36). Residual pesticides will work best on clean surfaces and will last a few weeks. Space sprays can also reduce numbers temporarily. Bifenthrin, permethrin, cypermethrin, and lambda-cyhalothrin are some of the compounds used to control sand flies (Bowles *et al.*, 2015). Applications of pesticides to kill immature sand flies are not recommended because development sites are randomly distributed and inconspicuously throughout the landscape so can be challenging to identify, and the scale of pesticide application would be costly, time-consuming, and potentially environmentally damaging.

Sugar baits and host treatment

Novel ways to apply pesticides have been investigated with sand flies. Just as with the biological control agent, *Bacillus sphaericus*, pesticides can be added to sugar baits. Attractive toxic sugar baits containing boric acid and spinosad applied to resting sites have successfully reduced sand fly populations in some arid areas where flowers are scarce (Müller and Schlein, 2011) (see

Figure 7.36. Residual pesticides that are sprayed on clean areas that sand flies may rest may help reduce the number of sand flies in an area.

Malamud-Roam, 2014 in 'Continuing education' for more information on toxic sugar baits). Feeding known hosts of sand flies with feed-through pesticides that will remain active in the faeces is being studied as a way to get pesticides into the hard-to-locate larval development sites. In a field study, frequent applications of a rodent bait containing 0.005% fipronil reduced fleas and sand flies in rodent burrows (Poché *et al.*, 2018). There was an 86% reduction in sand flies leaving treated rodent burrows and none of those present were blood-fed (Tsurim *et al.*, 2021). As yet, this technology is not commercially available and fipronil is not labelled for this use. Feeding wild rodents pesticides that are not labelled for this purpose may not only be a violation of the law, but also a violation of wildlife protection regulations.

7.5.6 Veterinary concerns

Sand flies serve as the carrier for the parasites causing leishmaniasis. Serological data indicate that exposure of horses to *Leishmania* parasites differs by region (Mhadhbi and Sassi, 2020). Infected horses will have crusts and nodules, which open to weeping sores in areas where sand flies commonly feed (Figure 7.37). The ears, scrotum or udder, neck, and legs are most commonly affected. Lesions on horses typically regress spontaneously so the efficacy of treatments can be difficult to assess. See Reuss (2013) in 'Continuing education' for more information on leishmaniasis in horses.

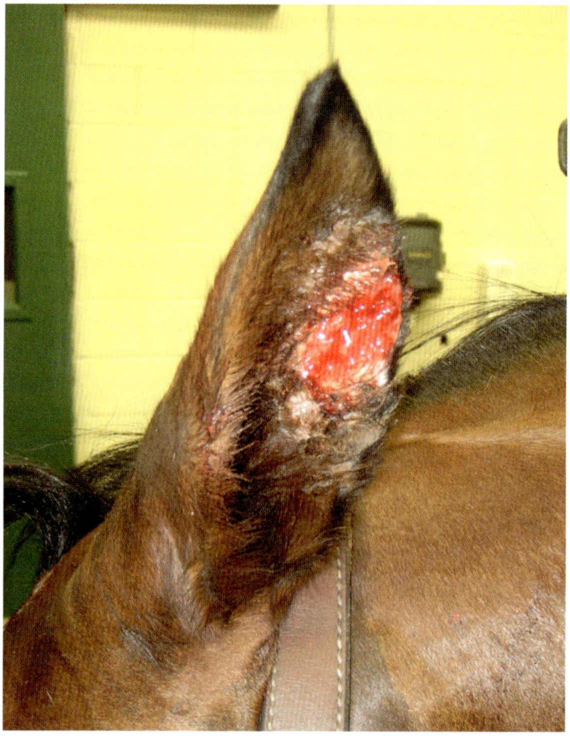

Figure 7.37. Sand flies are the carrier of parasites that cause the open sores associated with leishmaniasis.

Continuing education

Screwworm – key deer information

National Key Deer Refuge, US Fish and Wildlife Service, 2017.
https://www.fws.gov/refuge/National_Key_Deer_Refuge/News_Releases.html
- Information on the 2016 infestation of the New World screwworm and subsequent eradication from southern Florida is provided in both reports and news releases.

Common green bottle fly, sheep blow fly

Anderson, M. and Kaufman, P.E., 2011.
UF/IFAS Featured Creatures, publication number EENY-502, https://entnemdept.ufl.edu/creatures/livestock/flies/lucilia_sericata.htm.
- Information on the biology of green bottle flies.

Hairy maggot blow fly

Byrd, J., 1998.
UF/IFAS Featured Creatures, publication number: EENY-23, https://entnemdept.ufl.edu/creatures/livestock/hairy_maggot_blow_fly.htm.
- Fact sheet on the biology of hairy maggot blow flies

Old World screw-worm fly: a diagnostic manual, 3rd ed.

Animal Health Australia, Canberra, 2017.
https://www.animalhealthaustralia.com.au/wp-content/uploads/SWF-Manual_Full_digital.pdf
- An excellent handbook on Old World screwworm and the risk of its introduction to Australia

How to make rotational grazing work on your horse farm

Kenny, L., 2018.
https://extension.psu.edu/how-to-make-rotational-grazing-work-on-your-horse-farm
- Rotational grazing or pasture rotation is a management practice to improve forage and keep soils healthy. Rotation reduces overgrazing and can help prevent some pests. This guide details how to conduct rotational grazing on an equine facility.

New World screwworm ready reference guide – sterile insect response

United States Department of Agriculture, 2017.
https://www.aphis.usda.gov/animal_health/emergency_management/downloads/nws_rrg_sterileinsectresponse.pdf
- This guide details the sterile insect technique and how it has been used to eradicate the New World screwworm from the United States.

Phlebotomine sand flies – factsheet for experts

European Centre for Disease Prevention and Control, 2020.
https://www.ecdc.europa.eu/en/disease-vectors/facts/phlebotomine-sand-flies
- This resource includes detailed descriptions of distribution, identification, the transmission of pathogens, and public health control measures in Europe.

Control of phlebotomine sandflies

Alexander, B. and Maroli, M., 2003. Medical and Veterinary Entomology 17: 1-18.
https://doi.org/10.1046/j.1365-2915.2003.00420.x

- Phlebotomine sand flies transmit many pathogens causing disease. The current advantages and disadvantages of different chemical compounds for sand fly control and other control methods are discussed.

Attractive toxic sugar baits for vector control

Malamud-Roam, K., 2014.
https://ir4.rutgers.edu/PublicHealth/PHP%20articles/Toxic%20Sugar%20Baits_Layout%201.pdf

- This information sheet describes toxic sugar baits for mosquito and sand fly control.

Review of equine cutaneous leishmaniasis: not just a foreign animal disease

Reuss, S.M., 2013.
https://aaep.org/sites/default/files/issues/MedicineReuss.pdf

- Leishmaniasis has been reported in horses around the world. Leishmaniasis can be reported in horses in the United States but generally after international travel. However, cases of leishmaniasis have been reported in horses without a record of travel. This review article discusses the clinical manifestations of the disease in horses.

Resources

Resources are listed in alphabetical order and may not be inclusive of all organizations, companies, materials, or products available.

CFIA (Canadian Food Inspection Agency) handle the reporting of notifiable diseases in Canada. https://inspection.canada.ca/animal-health/terrestrial-animals/diseases/reportable/eng/1303768471142/1303768544412

DEFRA (Department for Environment, Food, and Rural Affairs) handle the reporting of notifiable diseases in the UK. https://www.gov.uk/government/collections/notifiable-diseases-in-animals

Insect Shield® Insect Repellent Fabric Treatment. https://www.insectshield.com/

OIE (World Organization for Animal Health) national agencies are responsible for reporting notifiable diseases to the OIE. https://www.oie.int/en/who-we-are/mission/. Responsible agencies for each member country are listed here: https://www.oie.int/en/who-we-are/members/

USDA-APHIS (United States Department of Agriculture-Animal and Plant Health Inspection Service) handle the reporting of notifiable diseases in the USA. https://www.aphis.usda.gov/aphis/ourfocus/animalhealth/nvap/NVAP-Reference-Guide/Animal-Health-Emergency-Management/Notifiable-Diseases-and-Conditions

References

Baumgartner, D.L., 1993. Review of *Chrysomya rufifacies* (Diptera: Calliphoridae). Journal of Medical Entomology 30: 338-352.

Bertin, F.R., Reising, A., Slovis, N.M., Constable, P.D. and Taylor, S.D., 2013. Clinical and clinicopathological factors associated with survival in 44 horses with equine neorickettsiosis (Potomac horse fever). Journal of Veterinary Internal Medicine 27: 1528-1534. https://doi.org/10.1111/jvim.12209

Bowles, D.E., Brich, S.C., Linthicum, K.L., Johnson, R.N., Linton, Y. and White, G.B., 2015. Sand flies (Diptera: Psychodidae: Phlebotomine) significance, surveillance, and control in contingency operations. Armed forces Pest Management Board Technical Guide No. 49. Available at: https://www.acq.osd.mil/eie/afpmb/docs/techguides/tg49.pdf.

Farkas, R. and Képes, G.Y., 2001. Traumatic myiasis of horses caused by *Wohlfahrtia magnifica*. Acta Veterinaria Hungarica 49: 311-318.

Flores, M., Longnecker, M. and Tomberlin, J.K., 2014. Effects of temperature and tissue type on *Chrysomya rufifacies* (Diptera: Calliphoridae) (Macquart) development. Forensic Science International 245: 24-29.

Giangaspero, A., Traversa, D., Trentini, R., Scala, A. and Otranto, D., 2011. Traumatic myiasis by *Wohlfahrtia magnifica* in Italy. Veterinary Parasitology 175: 109-112.

Hall, M. and Wall, R., 1995. Myiasis of humans and domestic animals. Advances in Parasitology 35: 257-334.

Mhadhbi, M. and Sassi, M., 2020. Infection of the equine population by *Leishmania* parasites. Equine Veterinary Journal 52: 28-33. https://doi.org/10.1111/evj.13178

Müller, G.C. and Schlein, Y., 2011. Different methods of using attractive sugar baits (ATSB) for the control of *Phlebotomus papatasi*. Journal of Vector Ecology 36: S64-S70.

Müller, G.C., Hogsette, J.A., Kline, D.L., Beier, J.C., Revay, E.E. and Xue, R.D., 2015. Response of the sand fly *Phlebotomus papatasi* to visual, physical and chemical attraction features in the field. Acta Tropica 141: 32-36.

Phillips, P.L., Welch, J.B. and Kramer, M., 2004. Seasonal and spatial distributions of adult screwworms (Diptera: Calliphoridae) in the Panama Canal area, Republic of Panama. Journal of Medical Entomology 41: 121-129.

Poché, D.M., Torres-Poché, Z., Yeszhanov, A., Poché, R.M., Belyaev, A., Dvořák, V., Sayakova, Z., Polyakova, L. and Aimakhanov, B., 2018. Field evaluation of a 0.005% fipronil bait, orally administered to *Rhombomys opimus*, for control of fleas (Siphonaptera: Pulicidae) and phlebotomine sand flies (Diptera: Psychodidae) in the Central Asian Republic of Kazakhstan. PLoS Neglected Tropical Diseases 12: e0006630. https://doi.org/10.1371/journal.pntd.0006630

Pusterla, N. and Madigan, J., 2013. *Neorickettsia risticii*. In: Sellon, D.C. and Long M. (eds.) Equine infectious diseases E-book, 2nd ed., pp. 347-351.

Robert, L.L., Perich, M.J., Schlein, Y., Jacobson, R.L., Wirtz, R.A., Lawyer, P.G. and Githure, J.I., 1997. Phlebotomine sand fly control using bait-fed adults to carry the larvicide *Bacillus sphaericus* to the larval habitat. Journal of the American Mosquito Control Association 13: 140-144. PMID: 9249650.

Siddig, A., Al Jowary, S., Al Izzi, M., Hopkins, J., Hall, M.J.R. and Slingenbergh, J., 2005. Seasonality of Old World screwworm myiasis in the Mesopotamia valley in Iraq. Medical and Veterinary Entomology 19: 140-150.

Tsurim, I., Wasserberg, G., Natan, G.B. and Abramsky, Z., 2021. Systemic control of cutaneous leishmaniasis sand-fly vectors: fipronil-treated rodent bait is effective in reducing *Phlebotomus papatasi* (Diptera:

Psychodidae) female emergence rate from rodent burrows. Journal of Medical Entomology 58: 974-978. https://doi.org/10.1093/jme/tjaa201

Urech, R., Green, P.E., Brown, G.W., Spradbery, J.P., Tozer, R.S., Mayer, D.G. and Kan, Y.T., 2012. Field assessment of synthetic attractants and traps for the Old World screw-worm fly, *Chrysomya bezziana*. Veterinary Parasitology, 187: 486-490.

Wilson, J., Pusterla, N., Bengfort, J.M. and Arney, L., 2006. Incrimination of mayflies as a vector of Potomac horse fever in an outbreak in Minnesota. Proceedings of the American Association of Equine Practitioners 52: 324-328.

Photo credits

Title page	Amanda, Flickr, CC BY 2.0 https://www.flickr.com/photos/82538355@N00/1635266084.
Figure 7.1.	Gail Hampshire, Flickr, CC-BY-2.0 https://www.flickr.com/photos/43272765@N04/22355940790.
Figure 7.2.	(A) Dong Zhang and Xinyu Li, Beijing Forestry University; (B,D) Emma Weeks (drawings by Adam Tofilski); (C) British Natural History Museum, www.nhm.ac.uk.
Figure 7.3.	(A) Bob Peterson, Flickr, license CC BY 2.0 https://www.flickr.com/photos/pondapple/7343371940; (B) Melissa McMasters, Flickr, license CC BY 2.0 https://www.flickr.com/photos/132545975@N04/50334569961.
Figure 7.4.	Lyle Buss, University of Florida.
Figure 7.5.	Lyle Buss, University of Florida.
Figure 7.6.	J. Weintraub, Ag. Canada.
Figure 7.7.	Isabelle Blanchemain, Flickr CC BY 2.0 https://www.flickr.com/photos/87805257@N00/18083654360.
Figure 7.8.	Lyle Buss, University of Florida.
Figure 7.9.	Lyle Buss, University of Florida.
Figure 7.10.	Emma Weeks.
Figure 7.11.	Erika Machtinger, Pennsylvania State University.
Figure 7.12.	(A) Chris Geden, USDA, ARS; (B) Erika Machtinger, Pennsylvania State University.
Figure 7.13.	Judy Gallagher, Flickr, license CC BY 2.0 https://www.flickr.com/photos/52450054@N04/33364645563/in/photolist-SergRb-DXbVHx-SQjwHH-QXAriW.
Figure 7.14.	(A) Judy Gallagher, Flickr, license CC BY 2.0 https://www.flickr.com/photos/52450054@N04/24909739517; (B) Judy Gallagher, Flickr, license CC BY 2.0 https://www.flickr.com/photos/52450054@N04/49397666902/in/photolist-2iywaHX-2hwBD4R-2iRaH7D-2ig6Xo7-2ijjHti-2dnhMQK; (C) Judy Gallagher, Flickr, license CC BY 2.0 https://www.flickr.com/photos/52450054@N04/49434106051.
Figure 7.15.	(A) Gail Hampshire, Flickr, license CC BY 2.0 https://www.flickr.com/photos/gails_pictures/7683243424/; (B) Judy Gallagher, Flickr, CC BY 2.0 https://www.flickr.com/photos/52450054@N04/31778926706.
Figure 7.16.	CSIRO, Australia. https://creativecommons.org/licenses/by/3.0/deed.en
Figure 7.17.	Pest and Diseases Images Library, bugwood.org.
Figure 7.18.	Erika Machtinger, Pennsylvania State University.
Figure 7.19.	Astridlorena, Wikimedia Commons, license CC BY 3.0 https://commons.wikimedia.org/w/index.php?curid=12601772.

Figure 7.20. London Natural History Museum, London, UK.

Figure 7.21. Photos: (A) Alexander Hall;, (B) Martin Hall, London Natural History Museum, license CC BY, https://veterinaryforensicentomology.myspecies.info/node/23.

Figure 7.22. James Niland, Flickr, license CC BY 2.0, https://www.flickr.com/photos/bareego/5351006391/in/photostream/.

Figure 7.23. (A) Martin J.R. Hall, Natural History Museum, London, UK; (B) Natural History Museum, London, UK.

Figure 7.24. (A) Graham Wise, license CC BY 2.0, https://commons.wikimedia.org/wiki/File:Australian_sheep_blowfly.jpg; (B) Calibas, public domain, https://commons.wikimedia.org/wiki/File:Green_bottle_fly.jpg.

Figure 7.25. Susan Ellis, Bugwood.org, license CC BY 2.0, https://www.invasive.org/browse/detail.cfm?&imgnum=5369361.

Figure 7.26. Johan J. Ingles-Le Nobel, Flickr, license CC BY-ND 2.0 https://www.flickr.com/photos/43147325@N08/26994737770.

Figure 7.27. Thomas & Dianne Jones, Flickr, license CC BY 2.0 https://www.flickr.com/photos/96332550@N00/4705285744.

Figure 7.28. Joseph Berger, Bugwood.org, license CC BY 3.0, https://www.invasive.org/browse/detail.cfm?imgnum=5387684.

Figure 7.29. Dave Huth, Flickr, license CC BY 2.0, https://www.flickr.com/photos/39798370@N00/5789175336.

Figure 7.30. United States Department of Agriculture, National Agroforestry Center, Flickr, license CC BY 2.0 https://www.flickr.com/photos/139938511@N02/26802235240.

Figure 7.31. Woodleywonderworks, Flickr, license CC BY 2.0 https://www.flickr.com/photos/73645804@N00/7408790886.

Figure 7.32. Lyle Buss, University of Florida.

Figure 7.33. EdRowton Photography.

Figure 7.34. Ray Wilson, UK.

Figure 7.35. Mary Gillham Archive Project, Flickr, license CC BY 2.0 https://www.flickr.com/photos/marygillhamarchiveproject/37899113514/in/photolist-ZK1S4C.

Figure 7.36. Erika Machtinger, Pennsylvania State University.

Figure 7.37. Sarah M Reuss, VMD, DACVIM.

Chapter 8. Beneficial and harmless arthropods

Chapter overview

This chapter will highlight common arthropods that may be observed on equine farms that have a beneficial role in equine pest management and/or the ecological community. Their roles include decomposition of manure, such as by dung beetles (Section 8.1), and pest control, like parasitoids and predators of filth flies (Sections 8.2, 8.3, and 8.4). Other insects are discussed that are commonly seen on farms that have more subtle yet vitally important beneficial effects as decomposers like soldier flies (Section 8.5), and dung midges and scavenger flies (Section 8.6). Pollinators like blow flies (Section 8.9), and insects that are crucial parts of the broader food web like crane flies (Section 8.7) and non-biting midges (Section 8.8) also are covered in this chapter. The roles these insects play are important enough that preservation of these insects through careful management of pests is recommended. See SARE Outreach (2005) in 'Continuing education' for more information on protecting beneficial insects on a farm.

8.1 Dung beetles (family Scarabaeidae)

8.1.1 Summary

Dung beetles (Figure 8.1), also known as scarab beetles, are in the family Scarabaeidae and feed on faeces (called dung or manure interchangeably). There are three main subfamilies of dung eating beetles, the Scarabaeinae, or the true dung beetles of which there are more than 5,000 species, the Aphodiinae, or small dung beetles, and the Geotrupinae, the earth-boring scarab beetles. Dung beetles are beneficial because they increase dung decomposition by feeding on, and often burying, dung in pastures. Reductions in dung at the surface reduces the presence of, and issues caused by, gut parasites and insect pests whose larvae feed on dung, notably house flies, stable flies, and in cattle dung, face flies and horn flies. Tunnelling and burying also increase pasture quality by soil aeration and improved nutrient cycling and uptake by plants. Additionally, they disperse and bury seeds present in animal dung assisting in seed recruitment. Dung beetles are reported to save the USA millions of dollars each year by burying above-ground livestock waste (Losey and Vaughan, 2006).

8.1.2 Distribution and seasonality

Dung beetles are found on all continents except Antarctica. These beetles can be found where animal faeces are present, including grassland, farmland and pastures, and forests. Dung beetle species that thrive in pasture settings have been introduced to many areas where they did not exist previously, including Australia and North and South America (Doube *et al.*, 2014). Dung beetles are active throughout the year, but are most active spring through autumn. Their activity is influenced by rainfall, with more egg laying occurring during periods of rainy weather when moisture levels in dung are highest.

Figure 8.1. Dung beetles are essential for the breakdown of animal manure.

8.1.3 Description

Dung beetles vary in size from a few millimetres to several centimetres. The Scarabaeinae are larger, 6 to 30 mm (0.2 to 1.2 in), whereas the Aphodiinae are small beetles at 4 to 6 mm (0.16 to 0.20 in). Geotrupinae are medium-sized beetles, typically 14 to 20 mm (0.15 to 0.20 in).

In general, dung beetles are roundish, dark coloured, sometimes shiny, and can have a metallic sheen. The legs are strong for rolling and burrowing (Figure 8.2). The wings are transparent or white and only visible during flight. When on the ground the wings are folded under the hard outer shell of the beetle called the elytra. Males of some species have a horn on the top of the head or on the thorax. Dung beetle larvae, called grubs, complete their development within dung. They are cream coloured with brown legs and heads. The larvae can also be recognised as beetles because of their characteristic C-shaped body (Figure 8.3).

Figure 8.2. Dung beetle adults may be dull or shiny but always have strong legs for digging.

Figure 8.3. Dung beetle larvae have a characteristic C-shape like many other beetle grubs.

8.1.4 Life history

Dung beetle adults are attracted to fresh dung. There are three types of dung beetles named because of their dung handling behaviour (Hanski and Cambefort, 2014). 'Rollers' collect and roll dung into a ball that is then buried and used as a food source or breeding chamber. 'Tunnelers' bury the dung where they find it and 'dwellers' feed on the dung where it landed. While all are beneficial for nutrient recycling, rollers and tunnelers are particularly beneficial due to the removal of the dung from the surface and soil aeration.

Immature and adult dung beetles feed on dung, but some species may also feed on mushrooms, carrion, and decomposing leaves and fruit. The male and female beetles may work together to bury the dung before mating. The female then lays her eggs in the dung. Roller females typically lay one egg per ball of dung. Depending on the species, one or both the parents of rollers and tunnellers may stay with the larvae that develop through three stages while feeding on the dung over a 2- to 4-month period. The parent providing care cleans the larvae and removes their faeces. The larvae will pupate in the dung. When they emerge as adults they leave the dung in search of fresh manure. The lifespan of a dung beetle is 3 to 5 years.

8.1.5 Management considerations

Because dung beetles feed on manure as adults and larvae, it is important to preserve populations to promote their contributions to the biological control of manure-developing flies and parasitic worms. Manure management practices to reduce other pests like filth flies should be balanced with dung beetle conservation. Dragging, harrowing, and other manure manipulation should continue to be conducted in the spring through late summer, but reducing these activities when manure-developing pests are not as numerous can promote dung beetle activity.

Avoid treating manure both directly with pesticides and indirectly through long-lasting deworming, antiparasitic, or systemic pesticide treatments, such as ivermectin. See Sands and Wall (2018) in 'Continuing education' for more information on the effect of parasiticides on dung beetles. Several active ingredients in dewormers can have negative impacts on dung beetles, which further supports the targeted use of these products (see Chapter 10 and Cutress, 2021 in 'Continuing education' for suggestions on protecting dung beetles on a farm). Insect growth regulators (IGRs) targeting filth flies that use cyromazine and diflubenzuron are not known to harm dung beetles.

8.2 Fly predators in manure and bedding

8.2.1 Summary

Filth fly (Chapter 3) eggs and newly hatched larvae are attacked by many predators that reduce fly numbers in a largely unseen way. Although there are many species of predators that feed on filth fly eggs, the two most prominent species are the beetle *Carcinops pumilio* (Figure 8.4) and the mite *Macrocheles muscaedomesticae* (Figure 8.5). These two species are often present at the same time and have complementary characteristics for reducing fly numbers (Geden, 1990). *Carcinops* beetles, sometimes called hister beetles, are available from at least one commercial source in the USA (see Table 8.1 in 'Resources' for insectaries selling biological control agents). There are no commercial sources of the *Macrocheles* mites. An interesting feature of this mite's life history is that the mites move from one location to another by attaching themselves to flies and hitchhiking their way to new surroundings, a behaviour known as phoresy (Figure 8.6).

8.2.2 Distribution and seasonality

Both *Carcinops* beetles and *Macrocheles* mites are globally distributed and are commonly found in most places where filth fly eggs and larvae are present. *Carcinops* beetles have a strong affinity for bird droppings and are common in poultry operations, but can be found in other animal manures as well. *Carcinops* are most abundant in late spring through summer (Kaufman *et al.*, 2002). For *Macrocheles* mites, phoresy is more common in spring and summer than autumn and winter (Carmona Rodrigueiro and Pires do Prado, 2004), but mites can be found in manure whenever fly larvae are active.

Figure 8.4. ***Carcinops pumilio*** **adults (pictured) and larvae feed on the eggs of filth flies.**

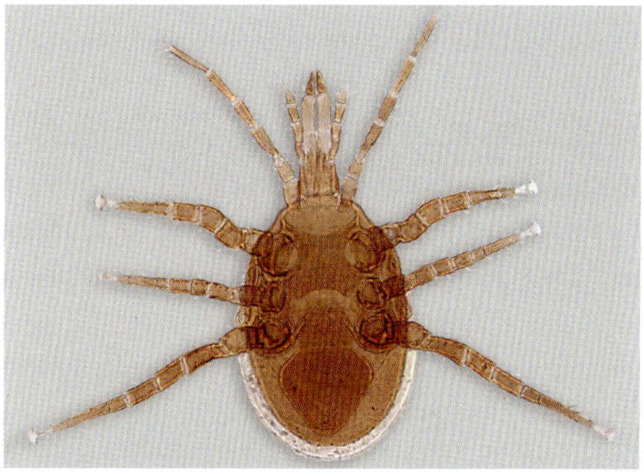

Figure 8.5. ***Macrocheles muscaedomesticae*** **adults (female pictured) feed on the eggs of filth flies.**

Figure 8.6. ***Macrocheles*** **mites attach themselves to adult flies (pictured) when the local environment stops providing adequate prey availability (fly eggs). They detach from the fly when it visits a new location where fly eggs are in greater abundance.**

8.2.3 Description

Carcinops adult beetles are small, black, oval shaped-beetles that are about 1.5 to 3.0 mm (0.6 to 1.2 in) in length (Figure 8.4). Adults and larvae are both predators on fly eggs and young fly larvae.

Macrocheles mite adults are small (about 1 mm or 0.4 in), orange-brown mites. They hold their front pair of legs in a forward manner that falsely suggests the presence of long antennae (Figure 8.5). These mites are only present in manure and will not be found on horses like parasitic mites (Chapter 5).

8.2.4 Life history

Carcinops beetles have a life cycle with complete metamorphosis and four stages: egg, larva, pupa, and adult stages. Development of *Carcinops* beetles from egg to adult takes 22 to 30 days under temperate summer weather conditions. Adult beetles are exceptionally long-lived, with lifespans that can reach 3 years. Beetle adults consume 15 to 80 fly eggs per day (Geden *et al.*, 1988) and lay two to 10 eggs per day of their own. Beetles will fly from an area if prey levels become too low to support their populations. Beetle larvae (Figure 8.7) also eat fly immatures.

Macrocheles mites have a life cycle with incomplete metamorphosis: egg, larva, nymph, and adult. In contrast to *Carcinops* beetles, *Macrocheles* mites can develop from egg to adult in as little as 2 to 4 days. Adult mites consume five to 20 fly eggs per day and deposit up to 10 eggs of their own per day (Geden and Axtell, 1988). The adults live for 2 to 3 weeks.

Carcinops beetles and *Macrocheles* mites have complementary characteristics that make them more effective at fly control where they are found together. *Carcinops* beetles eat many immature filth flies and are long-lived, but the beetle's long development time makes it hard for their populations to keep up when fly numbers are rising. In contrast, the very short development time of *Macrocheles* mites can mean that these predators can increase their numbers quickly in

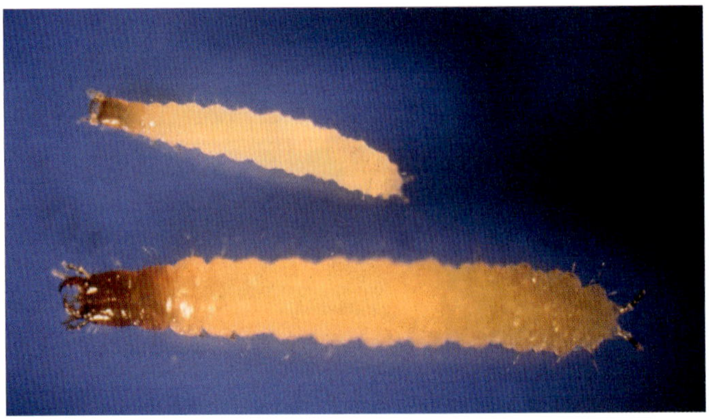

Figure 8.7. Both adult *Carcinops* beetles (Figure 8.4) and their larvae (shown here) are avid predators of fly eggs and young larvae.

response to fly numbers. Their habit of hitchhiking on flies also allows them to deploy rapidly to new fly development sites.

Although *Carcinops* beetles and *Macrocheles* mites are the most common predators of filth flies, there are many other species of mites and beetles, especially rove beetles (family Staphylinidae), that occur naturally and commonly prey on fly eggs in animal manure (Figure 8.8). Rove beetles are generalist predators, meaning they will eat many different small arthropods. These beetles are characteristically elongated with very short hardened forewings that leave most of the upper abdomen exposed and may be confused for earwigs.

8.2.5 Management considerations

Beetle and mite populations can be conserved by avoiding the direct application of pesticides to manure and other fly larval development sites. *Carcinops* beetles and *Macrocheles* mites are most likely to be found in semi-permanent manure areas like manure pits or piles. Although *Carcinops* beetles can be purchased in the USA there is no information on the effectiveness of beetle releases on equine farms. Manure should be kept dry to promote beetle and mite populations, as flies can thrive in manure that is too wet for predators to navigate.

8.3 Filth fly parasitoid wasps (family Pteromalidae)

8.3.1 Summary

Some small parasitoid wasps in the hymenopteran family Pteromalidae can kill developing filth fly pupae. Some of these species are highly beneficial for manure-developing fly pest management and should be protected (Machtinger *et al.*, 2015). These wasps are tiny and will not sting or harm a horse or human, or any other non-insect animal. Although they have a stinger, the stinger can only be used against fly pupae. Some species of parasitoid wasps in the genera *Muscidifurax* and *Spalangia* (Figure 8.9) are the most common species attacking filth flies such as the house and

Figure 8.8. Rove beetles (family Staphylinidae) are also common fly predators in animal manure. Rove beetles are characterised by having short, hardened forewings that leave most of the upper abdomen exposed.

stable fly in the United States. The adult wasps lay their eggs in the pupae of these fly species, which then are killed and eaten by the developing wasp larvae. Although mostly unnoticed, these wasps are common wherever filth flies occur, although at low numbers. Natural populations can be supplemented by purchasing wasps from commercial insectaries (described in Chapter 3) (see Table 8.1 in 'Resources' for a list of insectaries selling biological control agents).

Species of these parasitoid genera are commercially available for adding to the natural populations on the farm for pest management. The commercial vendors sometimes market them as fly predators or similar language to mitigate negative associations with the words 'wasp' and 'parasitic'. *Spalangia* species are the most common naturally on equine farms in the USA and have a preference for horse manure. *Nasonia* species are also sold for control of filth flies from commercial sources, but there is no evidence that this species is effective against filth flies found on equine farms (Machtinger *et al.*, 2015).

8.3.2 Distribution and seasonality

Pteromalid parasitoid wasps have a worldwide distribution and, with few exceptions, are all parasitic on insects and arachnids. The genera with strong livestock associations have a broad geographic range. For example, *Muscidifurax* is present in the Americas (North, Central, and South), Europe, Africa, Australasia, and Asia. *Spalangia* species are commonly found in equine facilities whereas *Muscidifurax* species are more likely to be associated with other livestock, including sheep, dairy, beef, and poultry facilities, although these preferences may be regional. Parasitoid populations increase in response to increasing fly populations so are most common in late spring through autumn.

Figure 8.9. Parasitic wasps in the genera *Spalangia* (left) and *Muscidifurax* (right) are common and important natural enemies of house flies and stable flies and can be purchased from commercial insectaries.

8.3.3 Description

Filth fly parasitoids are tiny wasps, about 1 to 3 mm (0.04 to 1.20 in) in length. These parasitoid wasps can be black, brown, or bronze. The adult is the only stage that can be seen, although they are not often observed. Immature stages occur inside the fly pupae. The eggs are opaque, oval, and 0.4 mm (0.02 in) in length. The larvae are pale, segmented, and resemble a beetle grub, and range in size from 0.5 to 2.5 mm (0.02 to 0.10 in) in length as they develop. The pupa looks like a white adult with its wings folded up and lacks a protective shell (puparium) as it is safe inside the fly pupae.

8.3.4 Life history

Filth fly parasitoids have a life cycle with complete metamorphosis with egg, larva, pupa and adult stages. An adult parasitoid locates a fly by searching in areas where fly larvae and pupae can be found developing such as in manure and waste bedding. A female adult parasitoid wasp drills through the outer shell of the fly pupa with her stinger, stings the fly pupa to kill it, and lays her eggs on the fly pupa within the puparium (Figure 8.10). When the wasp larvae hatch 2 to 3 days later they attach to the fly larvae with their mouthparts and feed. They will then pupate and remain as pupae for 8 to 12 days before emerging from the fly puparia as adults. Both sexes are ready to mate upon emergence from the fly. Total development time is 2 to 4 weeks for *Muscidifurax* and 3 to 5 weeks for *Spalangia*.

8.3.5 Management considerations

Parasitoids are best conserved by avoiding the application of larvicides to manure. Parasitoids are sensitive to the pesticides used in space sprays, but occasional use of such sprays is safe as most parasitoids are protected from spray droplets because of their location. Automated sprayers with frequent spray releases should be avoided because they increase the risk of killing exposed

Figure 8.10. Female *Muscidifurax raptor* drilling into a fly pupa to sting and laying an egg.

parasitoids. When applying residual pesticides to stable premises, avoid letting the spray path pass over manure, the floor, and the bottom of the walls. Fly baits and treatment of the animal with topical pesticides will not harm parasitoids. Parasitoids can be purchased from commercial insectaries to supplement naturally occurring populations (Geden and Hogsette, 2006, Table 8.1 in 'Resources'). As with the fly predators discussed above, manure should be kept dry to promote parasitoid effectiveness. See Machtinger *et al.* (2015) in 'Continuing education' for more information on using parasitoids as biological control agents for filth flies.

8.4 Black dump flies (genus *Hydrotaea*, a.k.a. *Ophyra*)

8.4.1 Summary

Black dump flies, *Hydrotaea aenescens*, are in the family Muscidae, which also includes the pest species house fly, stable fly, face fly, and horn fly. Black dump flies were known for many years as belonging to the genus *Ophyra*, and are referred to by that name in most of the older literature on their use as biological control agents. There have been efforts to develop black dump flies into biological control agents to get these flies to dominate over or eliminate house flies (Turner and Carter, 1990). Black dump flies are also available (as pupae) from commercial insectaries (see Table 8.1 in 'Resources' for a list of insectaries selling biological control agents). The objective of these approaches is to tip the fly balance on the farms away from house flies and in favour of black dump flies. Although black dump flies are not considered pests like house flies, they are still capable of carrying human and other animal pathogens and can be a nuisance in large numbers.

Figure 8.11. Black dump fly adults are shiny black metallic flies whose larvae will kill larvae of pest species such as house flies and stable flies.

8.4.2 Distribution and seasonality

The flies are found throughout North and South America and most of Europe. They are found frequently in concentrated animal facilities like poultry and swine farms, but they can be found in other animal facilities as well. Population increases of this species will correspond to when house flies (Chapter 3) are common.

8.4.3 Description

Adults are shiny black flies, about 6 mm (0.24 in) long, and resemble house flies and blow flies in their general appearance and by having sponging mouthparts (Figure 8.11). Their black colour distinguishes them from house flies, but they can be confused with the black blow fly (*Phormia regina*). Black dump flies differ in appearance from black blow flies by their smaller size, fewer hairs, darker colouration, and possession of orange-yellow palps associated with the mouthparts. In addition, the fourth wing vein in black dump flies is straight, whereas this vein is strongly angled in *P. regina* (Figure 8.12). Black dump fly larvae resemble house fly larvae (Figures 8.13). Adults and larvae can be distinguished from other common flies using the pictorial keys provided by the Centers for Disease Control and Prevention (see CDC in 'Continuing education').

8.4.4 Life history

Black dump flies have a life cycle with complete metamorphosis with egg, larva, pupa, and adult stages. Adult flies spend most of their time on or near the manure surface, with females flies laying about 200 eggs over their lifetime (Hogsette and Jacobs, 2003). The larvae develop in the manure and pupate in or near the manure. Development time from egg deposition to adult fly requires 2 to 5 weeks depending on the quality of the material they are developing in and temperature.

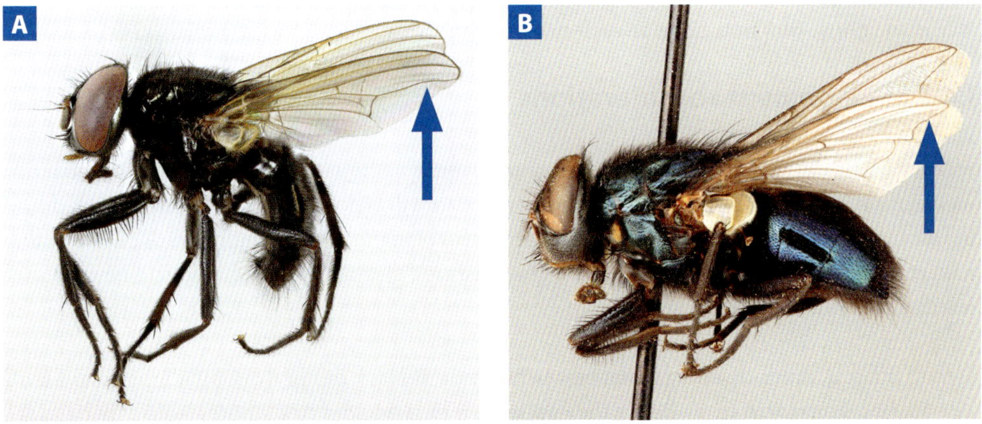

Figure 8.12. Black dump flies (A) can be confused with black blow flies (B) but the former are darker, less hairy and have a straight fourth wing vein rather than an angled vein (arrow).

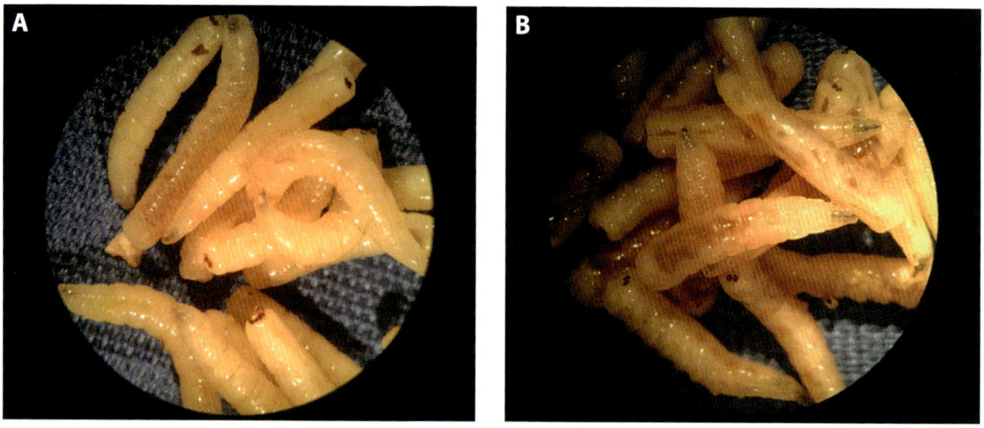

Figure 8.13. Larvae of black dump flies (A) are similar in appearance to those of house flies (B) and stable flies (not pictured).

Black dump fly biology is similar to that of the house fly except for two important differences. First, black dump flies tend to stay in and near their larval development habitat and are not generally attracted to animals, including humans, or human activity. Second, larvae of black dump flies are facultative predators of house fly larvae. In other words, they can complete development in manure alone but will attack and kill house fly larvae when they are present (Farkas and Jantnyik, 1990). Each larva can kill 10 to 20 house fly larvae per day (Geden *et al.*, 1988).

Because black dump fly larvae kill house fly larvae and the adult flies generally stay away from most human and other animal activity, several attempts have been made to encourage black dump flies to become dominant over house flies, especially on poultry farms. Results of these efforts have been mixed. It is worth noting that although black dump flies are less attracted to human activity than house flies, they are capable of carrying the same human and other animal pathogens that have made house flies notorious as vectors of pathogens causing diseases (Szalanski *et al.*, 2004).

8.4.5 Management considerations

Because black dump flies develop in the same habitat as house flies and stable flies, manure management practices to reduce filth fly presence also will reduce black dump flies. Black dump flies are most likely to establish in semi-permanent areas of manure, like manure pits or piles. Routine management in fields and stables (e.g. dragging pastures, cleaning stalls) should still be a priority to reduce filth fly numbers.

As with other manure-dwelling beneficial arthropods, if black dump fly populations are present, they are best conserved by avoiding the application of larvicides to manure. House fly baits are compatible with black dump flies. Natural populations can be supplemented by buying black dump fly pupae from insectaries in the USA and Europe, but there is no information on the effectiveness of these releases on equine facilities and there are concerns about potential pathogen transmission by these flies.

8.5 Soldier flies (*Hermetia illucens*)

8.5.1 Summary

Soldier flies are a family of 2,700 species in 380 genera that are present throughout the world, known as Stratiomyidae. Their name comes from the appearance of the antennae, which are straight and forward-projecting as though they are a soldier 'at attention'. The most well-known species is the black soldier fly, *Hermetia illucens*. They are wasp-like in appearance (Figure 8.14) and their larvae can be found in manure. However, they do not sting or act as a livestock pest. Instead, they have a beneficial role in the decomposition of organic material and can reduce the waste accrued in livestock facilities (Gold *et al.*, 2018). This process as well as the production of insects as protein for animal feed have been commercialised (Wang and Shelomi, 2017). The activity of the larvae churning through manure changes the physical properties of the substrate in a way that aids in manure breakdown and composting.

8.5.2 Distribution and seasonality

Black soldier flies (*H. illucens*) have a global distribution following their spread from the tropical areas of North, Central, and South America. Soldier flies as a family are present throughout the world in damp locations with decaying organic matter. Adults only fly in the warmer months in late spring to early autumn and they overwinter as larvae.

8.5.3 Description

Black soldier flies are medium-sized flies, approximately 16 mm (0.6 in) in length. They have an elongated black body with a metallic sheen that can be blue or green (Figure 8.14). The abdomen has two transparent windows that give the appearance of a wasp-like waist. The legs have white tips and the antennae are long, about twice the length of the head. They have two membranous

Figure 8.14. Adult black soldier flies can sometimes be mistaken for wasps but they lack a stinger and they have two rather than four wings.

wings, which they fold and lay down on the abdomen when resting. They can be differentiated from wasps by the lack of a stinger and the presence of two rather than four wings.

Black soldier fly eggs are oval, off white, and approximately 1 mm (0.04 in). The larvae are dull and off-white with a small head with chewing mouthparts. Larvae grow from 1 mm (0.04 in) to 27 mm (1.00 in) in length and 6 mm (0.2 in) in width. They are dark brown with obvious segmentation that look like ridges (Figure 8.15). The pupae look similar to the larvae as the exoskeleton hardens and darkens.

8.5.4 Life history

The black soldier fly life cycle has complete metamorphosis including egg, larva, pupa, and adult. After being mated by a male, a female black soldier fly lays an egg mass of approximately 500 eggs near a suitable larval development site. The larvae hatch from the eggs in 4 days and feed voraciously on organic matter as they develop for about 14 days (Figure 8.16). Before pupation, the larvae move to a dry sheltered area. The pupal period lasts about 2 weeks. Adults do not feed and rely on the energy stores from the larval stage until they can mate approximately 2 days after emerging.

8.5.5 Management considerations

As black soldier fly larvae are in manure, to preserve populations to promote the biological control of manure-developing flies and parasitic worms, avoid treating manure both directly with pesticides and indirectly through long-lasting systemic dewormers and antiparasitic treatments.

Figure 8.15. Soldier fly pupae are dark brown with obvious segmentation.

Figure 8.16. Black soldier fly larvae are important decomposers and have been studied as a source of protein in food for humans and other animals. In some areas, they are reared in large numbers for this purpose.

8.6 Dung midges and black scavenger flies (families Scatopsidae and Sepsidae)

8.6.1 Summary

There are two families of tiny flies that are often found on horse manure, sometimes in very large numbers. Dung midges are in the family Scatopsidae. There are around 325 described species, but likely there are many more species that have never been described. Black scavenger flies are in the family Sepsidae, which includes about 400 species. They are both decomposers of organic material including manure. Although they are often mistaken for other manure-developing flies, they are harmless to horses, humans, and other animals. Information about these flies is limited due to their harmless nature, so descriptions below are based on what is available for a small number of species. It is worth remembering that these small flies are fully grown for their species and do not grow up to become larger pest flies.

8.6.2 Distribution and seasonality

Both black scavenger flies and dung midges are distributed worldwide. Seasonality varies by species and they can have one, two, or three peaks in activity depending on the species (Kozánek and Roller, 1997; Marshall, 2012; Randall *et al.*, 1981). However, numbers are typically higher in the warmer summer months.

8.6.3 Description

Adult dung midges are small dark-coloured flies that are typically 1 to 3 mm (0.04 to 0.12 in) long. While they look a little like black fly adults, they do not have the characteristic humped back appearance of black flies (Chapter 4). They have clear membranous wings that are typically longer than the abdomen and short thick antennae (Figure 8.17). They are often spotted on flower heads of Umbelliferae (i.e. celery, carrot, or parsley family) or Compositae (i.e. aster, daisy, composite, or sunflower family) (Figure 8.18). Black scavenger flies are somewhat larger, 2 to 7 mm (0.08 to 0.28 in) long, usually shiny black flies with a narrow waist that gives them an ant-like appearance (Figure 8.17), although this can be difficult to distinguish looking at them from above and because of their small size. Both types of flies are often seen in large numbers hopping and walking over the surface of fresh manure (Figure 8.19). Black scavenger flies often flick their wings energetically above the body while walking.

In one of the better-studied dung midge species (*Scatopse fuscipes*), the eggs are laid in an egg mass that is produced in a long strand but is folded into approximately six rows of a few hundred eggs. Within the rows the eggs are 0.3 mm (0.012 in) in length and point upwards. The larvae have well-developed antennae and they appear slightly flattened.

Figure 8.17. Dung midges are in the family Scatopsidae (A). Black scavenger flies, which are in the Sepsidae (B), resemble ants, and often wave their wings over the body when walking.

Figure 8.18. Dung midges will often visit flowers, like the flies visiting this daisy.

Figure 8.19. Black scavenger flies (shown here) and dung midges are often found in large numbers on the surface of freshly dropped manure.

8.6.4 Life history

Dung midges and black scavenger flies have a life cycle with complete metamorphosis with four life stages: egg, larva, pupa, and adult. Eggs are laid in a mass by the female in a suitable habitat. The larvae feed on decaying organic matter and so are typically found on wet decomposing material such as manure, rotting wood, mushrooms, and soil. The larvae hatch after approximately 3 days and the larval stage lasts approximately 2 weeks with four instars. The pupal stage lasts approximately 4 days, with females emerging first. When dung midge adults emerge from the pupae they are ready to mate and after the female lays her egg mass she dies. The adult life span of dung midges is less than 2 days, but black scavenger flies live longer. Black scavenger flies are most commonly found on fresh manure of various types, including horse manure, and develop from egg to adult in 8 to 14 days depending on fly species, temperature, and type of manure (Laux *et al.*, 2019). Manure serves not only as a site for egg-laying and larval development in these flies but as a gathering place for mating rituals. Black scavenger flies have mostly been studied because of their elaborate mating structures, territorial behaviour, and courtship rituals (Puniamoorthy *et al.*, 2009).

8.6.5 Management considerations

Because these small flies develop in manure as larvae, populations can be conserved by avoiding treating manure both directly with pesticides and indirectly through long-lasting systemic dewormers and antiparasitic treatments. As stated above, these harmless small flies do not grow up to become larger pest species. Their presence on manure is harmless to humans and other animals.

8.7 Crane flies (family Tipulidae)

8.7.1 Summary

The Tipulidae or crane flies (otherwise known as mosquito hawks, mosquito killers, or daddy long-legs) are frequently mistaken for large mosquitoes (Figure 8.20). However, they do not bite or sting or cause any harm to human beings, horses, or other animals. Although they do not have direct benefit on equine farms, both larvae and adults are important prey items for many insectivorous animals including other insects, spiders, fish, birds, amphibians, and mammals (Pritchard, 1983).

8.7.2 Distribution and seasonality

The Tipulidae is a very large family of insects that includes at least 15,000 known species. They are found on every continent except Antarctica. The seasonality is dependent upon the species but there are usually larger numbers of adults following a rainy season for larval development.

Figure 8.20. Crane flies are often mistaken for large mosquitoes.

8.7.3 Description

Adult crane flies have a long thin body, wings and legs; the legs can easily be shed. They are medium to large flies of 7 to 35 mm (0.3 to 1.4 in) and typically yellow, brown, or grey (Figures 20 and 21). They are much bigger than mosquitoes in general and they do not have the long pointy proboscis on the head that is present in mosquitoes for feeding on blood (Figure 8.21). The abdomen of female crane flies ends in a point and the males have a blunt abdomen. This is due to the presence of an ovipositor at the end of the abdomen in females, a tube-like organ used for laying eggs. These flies do not have the elongated filaments on the abdomen associated with mayflies (Chapter 7).

Crane fly eggs are typically black. Crane fly larvae have a leathery appearance and are known as leatherjackets (Figure 8.22). The larvae are long, up to 50 mm (almost 2 in), and cylindrical.

Figure 8.21. Although crane flies are sometimes mistaken for large mosquitoes, they have much longer legs and mouthparts that cannot pierce the skin (A+B).

Figure 8.22. Crane fly larva, with typical leathery appearance.

8.7.4 Life history

Tipulidae have a life cycle with complete metamorphosis including egg, larva, pupa and adult. Adult females lay eggs on a variety of substrates including moist soil, dry soil, algae, or on the water surface depending on the species. Larvae are aquatic or semi-aquatic in freshwater (Figure 8.22). They mostly feed on decaying vegetation and plant roots. Some species feed on the aquatic larvae of other insect species. They overwinter as larvae or pupae in moist soil, vegetation or under water. Before pupation the larvae typically move to a drier area. Adults are mostly non-feeding and live for 10 to 15 days.

8.7.5 Management considerations

While there is no need to be concerned about crane fly adults or larvae in or around the stables they can be undesirable in and around the home. This group is highly attracted to lights, so turning off outside lights at night can help to reduce the numbers that are attracted to buildings and then go indoors.

8.8 Non-biting midges (family Chironomidae)

8.8.1 Summary

Although the biting midges of the Ceratopogonidae (Chapter 4) are known for their relentless biting and causing sweet itch in horses, non-biting midges of the family Chironomidae (chironomid midges) are relatively harmless to humans, horses, and other animals. Midges in this family can look like biting midges or mosquitoes but lack the necessary mouthparts for piercing skin and feeding on blood (Figure 8.23). They may be considered a pest when they emerge in large numbers, affecting visibility through windscreens, causing allergies, and damaging paint with their frass (droppings) (Figure 8.24). However, their benefits as decomposers and as part of the food chain far outweigh any inconvenience to humans. Larvae consume organic debris in aquatic habitats. They are a vital food source as larvae and adults for insects, fish, amphibians, birds, and mammals (particularly bats) (Armitage *et al.*, 2012).

Figure 8.23. Female midge resting on a window. Midges are attracted to lights, including those associated with homes.

Figure 8.24. The fronts of cars can become covered with dead midges during a major emergence event.

8.8.2 Distribution and seasonality

The Chironomidae family has a global distribution with species present throughout the world. There is even a wingless species (*Belgica antarctica*) that is the largest purely terrestrial animal in Antarctica, and the only insect that can survive year-round on that continent. The seasonality is dependent upon the species but there are usually larger numbers of adults in the late spring to summer following a rainy season for improved larval survival during development. They overwinter as mature larvae and then pupate and emerge as adults in spring.

8.8.3 Description

Adult chironomid midges are 1 to 10 mm (0.04 to 0.40 in) in length with thin legs and narrow wings. The males usually have feathery antennae (Figure 8.25). They are distinguished from biting midges (Ceratopogonidae, Chapter 4) by their larger size, longer thinner appearance, lack of biting mouthparts, and the feathery antennae of males in this family. Chironomid larvae are cylindrical and have small prolegs (stubby false legs) on the first segment after the head and the last segment. Many chironomid larvae appear red and are called bloodworms as a result (Figure 8.26). The colour comes from a haemoglobin-like substance that allows the larvae to breathe in the low oxygen content of their environment.

8.8.4 Life history

Females typically lay their eggs in a gelatinous egg mass of up to 1000 eggs on the surface of the water. Eggs sink to the bottom and when the larvae hatch they burrow into the sediment. The larvae feed on organic matter in the water and sediment. Larval chironomids are aquatic or semi-aquatic and can be found in a variety of habitats from tidal zones to bromeliads to moist soil to

Figure 8.25. Male chironomid midge with feathery antennae. Despite their passing resemblance to biting flies, these midges do not have mouthparts capable of piercing skin.

Figure 8.26. Larvae of some chironomids are called bloodworms because they have a haemoglobin-like substance that gives them a red colour.

sewage. On equine farms, they can be found in water mixed with soil and organic debris or even in water troughs. Due to their ability to survive in extreme conditions, species that can survive in the low oxygen environments of polluted waters are used as indicator species for low-quality water. The larvae develop over 2 to 7 weeks and pupate in their burrows. After 3 days the pupae swim to the surface and the adults emerge. The adults mate in swarms as they emerge at the surface. Adults of some species feed and some do not. When they do feed, they typically eat high-sugar items (such as nectar and honeydew), pollen, and fresh fly droppings.

8.8.5 Management considerations

While there is no need to be concerned about chironomid midge adults or larvae in and around stables, they can be undesirable particularly in large numbers. These midges are highly attracted to lights, so turning off outside lights at night can help to reduce numbers that are attracted to buildings and then go indoors.

Production of large numbers of chironomids usually occurs from organic-rich water systems. Check neighbouring water bodies for nutrient content. Reduce the use of fertiliser around water bodies and wet areas. This will reduce the production of midges and also allow other species to inhabit the area that will predate on the midge larvae.

Chemical treatments of water bodies for midge control should be avoided if possible due to the likely impact on non-target insects. In addition, bodies of water are often locally or regionally regulated. See Bay (2003) and Waldvogel *et al.* (2019) in 'Continuing education' for more information about sustainable chironomid management.

8.9 Blow flies (family Calliphoridae)

8.9.1 Summary

The family Calliphoridae, or as they are more commonly known blow flies, carrion flies, bluebottles, and greenbottles, are a group whose larvae feed on dead animals (carrion) and manure. They can generally be distinguished by a metallic sheen of blue, green, or black (Figure 8.27). There are around 1,200 species worldwide. Other than a few select groups, such as screwworms, that have been discussed earlier (Chapter 7), they do not cause any issues for horse owners except in cases of accidental myiasis (flesh invasion) (Chapter 7).

Figure 8.27. Blow flies are important decomposers that usually can be recognised by their shiny metallic colour.

Blow flies are critical as natural agents for decomposition of carrion and, in some cases, manure. The larvae of some species have been used to treat septic wounds because of the presence of antibiotics in their saliva (Yakovlkev *et al.*, 2019). Some species are important pollinators of flowers with strong rotting meat odour (see SARE Outreach, 2010 in 'Continuing education' for more information about blow fly pollinators). There is also a large and growing field of forensic entomology, in which the presence, species, and age of blow fly larvae on a cadaver can be used to estimate the time of death (Byrd and Tomberlin, 2019). Finally, blow flies can be mass-produced and used in livestock feed (Prado e Castro and Ameixa, 2021). A handful of species lay eggs and develop in animal wounds and body orifices, as discussed in Chapter 7. The vast majority of blow flies have no interest in horses or other animals unless they have an unattended open sore or potentially thrush.

8.9.2 Distribution and seasonality

Representative species of the Calliphoridae family are present throughout the world and some species, such as the blue bottle fly, *Calliphora vomitoria*, and the common green bottle fly, *Lucilia sericata*, are cosmopolitan and almost global in distribution. They are typically more prevalent in the warmer months. There is some seasonality to blow flies so different species may be observed throughout the year.

8.9.3 Description

Adult blow flies are typically shiny with metallic blue, black, or green bodies (Figure 8.28). They are slightly larger than a house fly at 8 to 10 mm (0.3 to 0.4 in). Like house flies they have sponging mouthparts. They have short three-segmented antennae. They can be distinguished from other large flies by using the identification keys in Scott and Littig (1962) and French *et al.* (2016).

Blow fly eggs look like grains of rice, they are yellow to white and approximately 1.5 mm (0.06 in) in length (Figure 8.29). The larvae are smooth cream-coloured maggots and 9 to 22 mm (0.35 to 0.87 in) in length. The pupae form from the last larval skin that becomes hardened and darker in colour. Larvae and pupae can be identified by the posterior spiracles (breathing holes), but it is easier to allow them to emerge as adults before attempting identification.

8.9.4 Life history

Blow flies have a life cycle with complete metamorphosis: egg, larva, pupa, and adult. After mating the female needs a protein meal to develop and lay eggs. The female will typically scavenge from carrion to obtain this meal and then lay her eggs in the carcass of the dead animal (Figure 8.29). Blow fly females will lay approximately 150 to 200 eggs per batch and up to 2,000 eggs in their lifetime. After 8 to 24 hours the larvae hatch, they feed and develop for 3 to 4 days depending upon the temperature. The larvae use chemical and mechanical means to digest the proteins in their food. They produce enzymes in their waste and they have grinding mouth parts known as mouth hooks. When fully developed the larva will leave the carrion and burrow into the surrounding sediment (like soil) to pupate. The pupation period lasts 1 to 2 weeks depending on temperature before the adult emerges. Adults are known to feed on nectar as a source of carbohydrates for their food-seeking flights. They are particularly attracted to flowers with odours that resemble rotting meat and act as a pollinator for these species.

Figure 8.28. Common blow flies are the blue bottle fly, *California vomitoria* (A), the green bottle fly, *Lucilia sericata* (B), and the black blow fly, *Phormia regina* (C).

Figure 8.29. Blow flies are the first insects to visit a fresh carcass or piece of raw meat, and quickly lay masses of yellow to white eggs (top right).

8.9.5 Management considerations

The great majority of blow flies are harmless or serve as beneficial pollinators and decomposers. Important exceptions are the sheep-strike flies and screwworms because they prefer infesting living hosts. Check that blow flies on the property are not screwworms, and inspect animals regularly for wounds. See Chapter 7 for more information on screwworm distribution and management.

Many blow flies can develop in a wide variety of decomposing organic waste. If a large number of blow flies suddenly appear at the facility it might be worth trying to find the source. Larval blow flies typically feed on carrion, which is limited, and so the situation will self-resolve, but can also be present in poorly managed compost piles, particularly those with grass clippings, vegetable waste, or garbage. Avoid mixing manure and household waste in compost piles. Another source could be a burst sewage pipe or overflowing drain. Garbage bins can also draw large numbers of blow flies.

Some jug traps that are marketed for house flies use odour baits that will collect blow flies as well (Geden *et al.*, 2009). Because many of these species can detect odours that they find attractive from a long-range, in some settings these traps may be primarily filled with blow flies and not the target house flies. This is especially true once the traps have collected enough flies that the dead decaying flies contribute to the bait's attractiveness to blow flies. It is important to clean and replace the contents of these jug traps frequently (weekly or more often in warm weather) to avoid trapping blow flies, which are beneficial decomposer and pollinator insects in the environment (Figure 8.30).

Traps that are intentionally designed for blow flies usually use very malodourous baits such as rotting meat or chemical blends that simulate its smell. Most of the information on blow fly traps has been done to collect either primary screwworm or the *Lucilia* flies responsible for sheep-strike in the UK and Australia.

Figure 8.30. Blow flies can be attracted to dead decaying flies in jug traps. Several different metallic-coloured blow flies can be seen in this jug trap. It is important to replace the traps or change the bait often to limit how many beneficial blow flies are trapped.

8.10 Yellow/golden dung flies (*Scathophaga stercoraria*)

8.10.1 Summary

Yellow dung flies belong to the family Scathophagidae, which is related to the family that includes house flies, stable flies, and horn flies (Muscidae). There are about 500 species in the family, but the yellow dung fly is by far the best-known. Yellow dung flies are medium-sized flies whose males have a distinctly yellow hairy appearance that are commonly found visiting fresh dung of several large mammals, including horses (Figure 8.31). Adult flies feed on nectar and prey on smaller insects that visit manure. Larvae develop in manure and aid in its decomposition. They are strong fliers with courtship rituals that include territorial behaviour by the males. These charismatic flies are harmless to horses and other animals.

8.10.2 Distribution and seasonality

Yellow dung flies, also called golden dung flies, are found in Europe and in Northern latitudes in North America, but do not occur in Australia or New Zealand. There is typically a spring emergence of flies that have overwintered as larvae and pupae. They are common in most places from March to November, sometimes declining in mid-summer when temperatures are highest. These flies are sensitive to heat, with mortality increasing at temperatures over 25 °C (77 °F) (Blanckenhorn *et al.*, 2014; Gibbons, 1987).

8.10.3 Description

Adult flies are 7 to 13 mm (0.27 to 0.51 in) long, with male flies usually substantially larger than the females. Males have a distinctively hairy yellow appearance that is more muted in females (Figure 8.32). The flies are most often noticed around manure pats, where males compete fiercely to mate with females when they come to lay eggs. Up to 400 flies at a time are sometimes seen on manure pats, with many of them forming mating pairs. For a review of the biology of these flies, see Blanckenhorn *et al.* (2010).

Figure 8.31. Male yellow dung fly.

Figure 8.32. Mating pair of yellow dung flies. Note the larger size and brighter appearance of the male.

8.10.4 Life history

Male yellow dung flies are predators that catch smaller insects around dung, including house flies and horn flies (Figure 8.33). They also visit flowers for nectar and to prey on small insects that also visit flower heads. When not actively hunting, males spend their time on dung waiting for females to approach for mating. Females are also predators but mostly hunt in vegetation; they only come to dung to mate and lay eggs. The flies mate on the dung surface, with copulation taking up to 50 minutes, The mated females then lay 30 to 90 eggs each on the dung before moving off to resume feeding in nearby vegetation. The flies only visit very fresh dung, which loses its attractiveness to them within 2 hours of dropping. Larvae develop in the manure and pupate in the soil or encrusted parts of the manure. Development from egg to adult can be completed in 17 days or as long as 80 days depending on temperature. Adult flies can live for up to a month. When larvae detect shorter day lengths and temperatures of around 10 °C, they develop into diapausing (hibernating) pupae that can survive the cold winter months ahead.

Figure 8.33. Yellow dung flies move from one dung pat to another, always choosing manure that is less than 2 hours old.

8.10.5 Management considerations

Yellow dung flies are harmless insects that also provide beneficial services by preying on pest flies as adults and aiding in manure breakdown as larvae. As with other manure-dwelling beneficial arthropods, populations can be conserved by minimising treating manure both directly with pesticides and indirectly through long-lasting systemic dewormers and antiparasitic treatments.

Continuing education

Strategies to enhance beneficials

SARE Outreach, 2005.
https://www.sare.org/publications/manage-insects-on-your-farm/principles-of-ecologically-based-pest-management/strategies-to-enhance-beneficials/
- A guide to protecting and increasing populations of beneficial insects on a farm.

Sustained parasiticide use in cattle farming affects dung beetles assemblages

Sands, B. and Wall, R., 2018. Agriculture, Ecosystems & Environment 265: 226-235.
https://doi.org/10.1016/j.agee.2018.06.012
- A manuscript describing the effect of macrocyclic lactones and pyrethroids on cattle farms on dung beetle diversity and abundance.

Dung, drugs, and disease: the interaction between dung beetles and farming

Cutress, D., 2021.
https://businesswales.gov.wales/farmingconnect/news-and-events/technical-articles/dung-drugs-and-disease-interaction-between-dung-beetles-and-farming
- An article describing the biology and behaviour of dung beetles as well as the effect they have on agriculture and the effect that agriculture can have on them and suggestions on how to protect them on the farm.

Use of pupal parasitoids as biological control agents of filth flies on equine facilities

Machtinger, E.T., Geden, C.J., Kaufman, P.E. and House A.M., 2015. Journal of Integrated Pest Management 6: 16. https://doi.org/10.1093/jipm/pmv015
- The use of pupal parasitoids on equine farms is becoming more popular on equine farms for fly control. However, there is limited information on how to use parasitoids effectively. In this review, the biology of available parasitoids and recommendations for use are provided.

Diptera: pictorial key to principal families of public health importance

Centers for Disease Control and Prevention.
https://www.cdc.gov/nceh/ehs/docs/pictorial_keys/flies.pdf
- An illustrated key to adult and immature Diptera of medical and veterinary importance. Note that black dump flies in these keys are referred to as *Ophyra* species.

Chironomid midges

Bay, E.C., 2003.
https://s3.wp.wsu.edu/uploads/sites/408/2015/02/PLS-45-Chironomid-Midges.pdf

• An article detailing the biology, life history and importance of non-biting midges. Management suggestions are provided.

Biology and control of non-biting aquatic midges
Waldvogel, M., Apperson, C. and Bambara, S., 2019.
https://content.ces.ncsu.edu/biology-and-control-of-non-biting-aquatic-midges
• An article with several suggestions about the management of non-biting midges.

Managing alternative pollinators: the blue bottle fly
SARE Outreach, 2010.

Table 8.1. Companies that produce and sell biological control agents for flies. This list may not be comprehensive.

Country/company	Location	Web address
United States		
Arbico Organics	Oro Valley, AZ	https://www.arbico-organics.com/
Beneficial Insectary	Redding, CA	https://www.insectary.com/
Kunafin	Quemado, TX	https://www.kunafin.com/
IPM Laboratories, Inc	Cayuga, NY	https://www.ipmlabs.com/
Rincon-Vitova	Ventura, CA	https://rinconvitova.com/
Spalding Labs	Reno, NV	https://spalding-labs.com/
Europe		
Koppert	Global	https://www.koppert.com/
Biobest	Westerlo, Belgium	https://www.biobestgroup.com/
Australia		
Bugs for bugs	Toowoomba, Queensland	https://bugsforbugs.com.au/

https://www.sare.org/publications/managing-alternative-pollinators/chapter-eight-other-managed-pollinators/the-blue-bottle-fly/
• An article describing the use of blue bottle flies as pollinators of vegetable crops.

Resources

Resources are listed in alphabetical order and may not be inclusive of all companies, materials, or products available.

References

Armitage, P.D., Pinder, L.C. and Cranston, P.S. (eds.), 2012. The Chironomidae: biology and ecology of non-biting midges. Springer Science & Business Media, Berlin, Germany, 447 pp.

Blanckenhorn, W.U., Pemberton, A.J., Bussière, L.F., Roembke, J. and Floate, K.D., 2010. A review of the natural history and laboratory culture methods for the yellow dung fly, *Scathophaga stercoraria*. Journal of Insect Science 10: 11. https://doi.org/10.1673/031.010.1101

Blanckenhorn, W.U., Gautier, R., Nick, M., Puniamoorthy, N. and Schäfer, M.A., 2014. Stage-and sex-specific heat tolerance in the yellow dung fly *Scathophaga stercoraria*. Journal of Thermal Biology 46: 1-9.

Byrd, J.H. and Tomberlin, J.K. (eds.), 2019. Forensic entomology: the utility of arthropods in legal investigations. CRC press, Taylor and Francis, Boca Raton, FL, USA, 620 pp.

Carmona Rodrigueiro, T.S. and Pired do Prado A., 2004. *Macrocheles muscaedomesticae* (Acari, Macrochelidae) and a species of *Uroseius* (Acari, Polyaspididae) phoretic on *Musca domestica* (Diptera, Muscidae): effects on dispersal and colonization of poultry manure. Iheringia Série Zoologia 94: 181-185. https://doi.org/10.1590/S0073-47212004000200011

Doube, B.M., Macqueen, A., Ridsdill-Smith, T.J. and Weir, T.A., 2014. Native and introduced dung beetles in Australia. In: Hanski, I. and Cambefort, Y. (eds.) Dung beetle ecology. Princeton University Press, Princeton, NJ, USA, pp. 255-278.

Farkas, R. and Jantnyik, T., 1990. Laboratory studies on *Hydrotaea aenescens* as predator of house fly larvae (Diptera: Muscidae). Parasitologia Hungarica 23: 103-108.

French, D., Craig, T., Hogsette, J., Pelzel-McCluskey, A., Mittel, L., Morgan, K. and Vaala, W., 2016. External parasite and vector control guidelines. American Association of Equine Practitioners, Lexington, KY, USA, 54 pp. Available at: https://aaep.org/sites/default/files/2021-02/AAEP_ExternalParasites_071316_Final.pdf.

Geden, C.J., 1990. The role of coleopteran and acarine predators in house fly population regulation in poultry production facilities. In: Rutz, D.A. and Patterson, R.S. (eds.) Biocontrol of arthropods affecting livestock and poultry. Westview Press, Boulder, CO, pp. 177-200.

Geden, C.J. and Axtell, R.C., 1988. Predation by *Carcinops pumilio* (Coleoptera: Histeridae) and *Macrocheles muscaedomesticae* (Acarina: Macrochelidae) on the house fly (Diptera: Muscidae): functional response, and effects of temperature and availability of alternative prey. Environmental Entomology 17: 739-744.

Geden, C.J. and Hogsette, J.A., 2006. Suppression of house flies (Diptera: Muscidae) in Florida poultry houses by sustained releases of *Muscidifurax raptorellus* and *Spalangia cameroni* (Hymenoptera: Pteromalidae). Environmental Entomology 35: 75-82.

Geden, C.J., Stinner, R.E. and Axtell, R.C., 1988. Predation by predators of the house fly in poultry manure: effects of predator density, feeding history, interspecific interference, and field conditions. Environmental Entomology 17: 320-329.

Geden, C.J., Szumlas, D.E. and Walker, T.W., 2009. Evaluation of commercial and field-expedient baited traps for house flies, *Musca domestica* L. (Diptera: Muscidae). Journal of Vector Ecology 34: 99-103.

Gibbons, D.S., 1987. The causes of seasonal changes in numbers of the yellow dung fly, *Scathophaga stercoraria* (Diptera: Scathophagidae). Ecological Entomology 12: 173-185.

Gold, M., Tomberlin, J.K., Diener, S., Zurbrügg, C. and Mathys, A., 2018. Decomposition of biowaste macronutrients, microbes, and chemicals in black soldier fly larval treatment: a review. Waste Management 82: 302-318.

Hanski, I. and Cambefort, Y. (eds.), 2014. Dung beetle ecology. Princeton University Press, Princeton, NJ, USA.

Hogsette, J.A. and Jacobs, R.D., 2003. The black dump fly: a larval predator of house flies. Publications from USDA-ARS/UNL Faculty. 994. Available at: https://digitalcommons.unl.edu/usdaarsfacpub/994.

Kaufman, P.E., Rutz, D.A. and Waldron, J.K., 2002. Seasonal variation in *Carcinops pumilio* (Coleoptera: Histeridae) dispersal and potential for suppression of dispersal behaviour. Journal of Medical Entomology 39: 106-111. https://doi.org/10.1603/0022-2585-39.1.106

Kozánek, M. and Roller, L., 1997. A study of scatopsid (Diptera: Scatopsidae) communities in Slovakia. Biologia, Bratislava 52: 637-646.

Laux, A., Wegmann, A., Roy, J., Gourgoulianni, N., Blanckenhorn, W.U. and Rohner, P.T., 2019. The role of larval substrate specialization and female oviposition in mediating species diversity of closely-related sepsid flies (Diptera: Sepsidae). European Journal of Entomology 116: 75-84.

Losey, J.E. and Vaughan, M., 2006. The economic value of ecological services provided by insects. BioScience 56: 311-323.

Machtinger, E.T. and Geden, C.J., 2018. Biological control of livestock pests: parasitoids. In: Garros, C., Bouyer, J., Takken, W. and Smallegange, R.C. (eds.) Prevention and control of pests and vector-borne diseases in the livestock industry. Ecology and Control of Vector-borne Diseases Volume 5. Wageningen Academic Publishers, Wageningen, the Netherlands, pp. 299-335.

Machtinger, E.T., Geden, C.J., Kaufman, P.E. and House A.M., 2015. Use of pupal parasitoids as biological control agents of filth flies on equine facilities. Journal of Integrated Pest Management 6: 16. http://doi.org/10.1093/jipm/pmv015

Marshall, S.E., 2012. Flies. Firefly Books, Buffalo, NY, USA, 616 pp.

Prado e Castro, C. and Ameixa, O.M.C.C., 2021. Blow flies (Diptera: Calliphoridae) promising candidates as animal feed ingredients. Journal of Insects as Food and Feed 7: 1065-1076. https://doi.org/10.3920/JIFF2021.0020

Pritchard, G., 1983. Biology of Tipulidae. Annual Review of Entomology 28: 1-22.

Puniamoorthy, N., Ismail, M.R.B., Tan, D.S.H. and Meier, R., 2009. From kissing to belly stridulation: comparative analysis reveals surprising diversity, rapid evolution, and much homoplasy in the mating behaviour of 27 species of sepsid flies (Diptera: Sepsidae). Journal of Evolutionary Biology 22: 2146-2156.

Randall, M., Coulson, J.C. and Butterfield, J., 1981. The distribution and biology of Sepsidae (Diptera) in upland regions of northern England. Ecological Entomology 6: 183-190.

Scott, H.G. and Littig, K.S., 1962. Flies of public health importance and their control. US Department of Health, Education, and Welfare, Public Health Service, Center for Disease Control, Publication no. 772. Available at: https://stacks.cdc.gov/view/cdc/7611.

Szalanski, A.L., Owens, C.B., Mckay, T. and Steelman, C.D., 2004. Detection of *Campylobacter* and *Escherichia coli* O157:H7 from filth flies by polymerase chain reaction. Medical and Veterinary Entomology 18: 241-246.

Turner, E.C. and Carter, L., 1990. Mass rearing and introduction of *Ophyra aenescens* (Wiedemann) (Diptera: Muscidae) in high-rise caged layer houses to reduce house fly populations. Journal of Agricultural Entomology 7: 247-257.

Wang, Y.S. and Shelomi, M., 2017. Review of black soldier fly (*Hermetia illucens*) as animal feed and human food. Foods 6: 91. https://doi.org/10.3390/foods6100091

Yakovlev, A.Y., Kruglikova, A.A. and Chernysh, S.I., 2019. Calliphoridae flies in medical biotechnology. Entomological Review 99: 292-301.

Photo credits

Title page Michael Mayer, Flickr, license CC BY 2.0 https://www.flickr.com/photos/12602671@N04/11309891283.

Figure 8.1. Mikhail Gnatkovskiy, Shutterstock no. 1089655040.
Figure 8.2. CSIRO, CC BY 3.0.
Figure 8.3. Jan Mastnik, Sweden, Shutterstock no. 112552385.
Figure 8.4. Erika Machtinger, Pennsylvania State University.
Figure 8.5. Pavel Klimov, Bee Mite ID (idtools.org/id/mites/beemites.
Figure 8.6. Albert Lleal/Minden Pictures.
Figure 8.7. Phillip E. Kaufman, Texas A & M.
Figure 8.8. (A) Mike Talbot, Flickr, license CC BY 2.0 https://www.flickr.com/photos/25258702@N04/3793645406; (B) Jeans Photos, Flickr, license CC-BY-2.0 https://www.flickr.com/photos/63479603@N00/9977095904.
Figure 8.9. Chris Geden, USDA, ARS.
Figure 8.10. United States Department of Agriculture, Office of Communications, image k9933.
Figure 8.11. Chris Geden, USDA, ARS, Florida, USA.
Figure 8.12. (A) Janet Graham, Flickr, license CC BY 2.0 https://www.flickr.com/photos/130093583@N04/25158969162; (B) Ken Perry, Flickr, license CCO 1.0 https://en.wikipedia.org/wiki/Phormia_regina#/media/File:Side_view_of_Phormia_regina.jpg.
Figure 8.13. Chris Geden, USDA, ARS.
Figure 8.14. officialhmp, Shutterstock no. 1788085229.
Figure 8.15. Max D Solomko, Shutterstock no. 661155715.
Figure 8.16. J.A. Cammack.
Figure 8.17. (A) Chris Moody, New Zealand, Shutterstock no. 414076231; (B) © entomart, public domain.
Figure 8.18. Kalle Rahkola/Wikimedia Commons.
Figure 8.19. Erika Machtinger, Pennsylvania State University.
Figure 8.20. Judy Gallagher, Flickr, license CC BY 2.0 https://www.flickr.com/photos/52450054@N04/25171292269.
Figure 8.21. (A) Sue Rickhuss, Pixabay; (B) Phillip McErlean, Ireland, Flickr, license CC BY-ND 2.0 https://www.flickr.com/photos/64320477@N05/35805270586.
Figure 8.22. Katja Schulz, 'Treegrow', Washington, D.C., Flickr, license CC-BY-2.0 https://www.flickr.com/photos/treegrow/39514918175.
Figure 8.23. Martin Cooper, Flickr, license CC BY 2.0 https://www.flickr.com/photos/92899351@N08/50311483517.
Figure 8.24. Michael Righi, Flickr, license CC BY 2.0 https://www.flickr.com/photos/53969873@N00/40549520.
Figure 8.25. Khlungcenter, Thailand, Shutterstock no. 442461613.
Figure 8.26. Oleksanrd Savchuk, Ukraine, Shutterstock no. 370842377.
Figure 8.27. Jan Schoepe, Shutterstock no. 716106118.
Figure 8.28. (A) Cherdchai Chaivimol, Shutterstock no. 1139444822; (B) Dean Morley, Flickr, license CC-BY-ND-2.0 https://www.flickr.com/photos/33465428@N02/3807627595; (C) Cherdchai Chaivimol, Shutterstock no. 1539929105.
Figure 8.29. Ostariyanov, Russia, Shutterstock no. 1117805069.
Figure 8.30. Erika Machtinger, Pennsylvania State University.
Figure 8.31. Janet Graham, Flickr CC BY 2.0.
Figure 8.32. Orangeaurochs, Flickr CC BY2.0.
Figure 8.33. Colors United, Shutterstock.

Chapter 9. Parasitic worms of horses

Chapter overview

This chapter covers the life history and management of the most common parasitic worms in horses excluding *Habronema* and *Thelazia* nematodes which are covered in Chapter 3. . Detailed information on biology, management, and veterinary concerns of parasitic worms of horses with respect to each country covered by this book are listed in 'Continuing education'. There are two main phyla of worms that affect horses. The nematodes (phylum Nematoda) include large and small strongyles, ascarids, and pinworms. Horses also are parasitised by tapeworms (phylum Platyhelminthes). Management of all groups is similar and is mainly focused on manure management and chemical treatment.

9.1 Large strongyles and small strongyles

9.1.1 Parasite summary

Parasitic worms of the genus *Strongyles* are nematodes from the family Strongylidae. Strongyles are the most common internal parasitic worm species for horses. Common species of large strongyles in horses are *Strongylus vulgaris*, *Strongylus endentatus*, and *Strongylus equinus*. Small strongyles, collectively referred to as cyathostomes, are the most common and pathogenic parasites of horses (Corning, 2009). There are at least 40 different species of small strongyles. Identifying species is a challenge, and most horses will have multiple species at the same time (Bredtmann *et al.*, 2017). All strongyles have similar life cycles.

9.1.2 Distribution and seasonality

Strongylosis has been reported from all parts of the world. Season has no impact on the prevalence of strongyle infections but shedding intensity of strongyle eggs is affected by season and significantly higher egg excretion is recorded in spring and summer (Nielsen, 2012; Nielsen *et al.*, 2007; Poynter, 1954). Hot dry spring weather (pre-wet season) is the most unfavourable for larval development, migration and survival (Hutchinson *et al.*, 1989).

9.1.3 Description

Adult worms can be identified either when passed in faeces, or during a necropsy. Encysted small strongyles can be visualised in the wall of the large colon during necropsy. Adults will be seen within the large colon contents.

The most common way to visualise adults is following administration of an anthelmintic. Horses with large worm burdens will pass adults in the faeces over several days, allowing for easy identification. Small strongyles appear as 1.0 to 1.5 cm (0.4 to 0.6 in) red, narrow worms, often in very large numbers (Figure 9.1). Large strongyles are rarely seen in well-managed horse populations. When seen they appear very similar to small strongyles but are larger in size, 2.5 to 5.0 cm (1 to 2 in).

9.1.4 Life history

Large strongyles: *S. vulgaris*, *S. edentatus*, and *S. equinus*

The *Strongyles vulgaris* life cycle is as follows: the first stage to third stage larvae (L1 to L3) live in soil, grass, and faeces. L3 is the infective form and it is ingested by horses from water film on pasture grasses. Once inside the horse, the fourth stage larvae (L4) migrate from the intestine through blood vessel walls, particularly the cranial mesenteric artery, where they cause abscesses that rupture to free the adult stage. The eggs are passed out with faeces, and larvae develop to the infective L3 stage outside the animal. Development from egg to L3 is completed in 1 to 6 weeks, depending on temperature and other conditions. Generally, *S. vulgaris* lives in the caecum,

Figure 9.1. Nematodes (small strongyles shown here) can be seen easily in manure after administering a dewormer.

sometimes the large colon. The life cycles of *Strongyles edentatus* and *S. equinus* are the same except the L4 migrate through the liver instead of through blood vessels. The larvae develop to adults in 6 to 11 months.

Small strongyles

The life cycle is similar for all species in that the L1 to L3 live in vegetation and faeces. The L3 migrate to water film on vegetation where they are ingested by the horse. Inside the horse the larvae either mature to the adult stage or to L4. L4 may encyst in the large colon wall to wait for better conditions. Adults live in the large colon, where they release eggs to be passed into the faeces. Once passed from the horse, L1 hatch from the eggs to start the cycle over. The larvae mature to adults in 6 to 10 weeks after ingestion by the horse. Within the horse, they feed on blood and other tissues.

9.1.5 Control

Cultural control

Good pasture management is key to the cultural control of parasitic worms. Removing manure from pastures offers the best reduction in pasture parasitic worm load. However, for large acreages, this can be time consuming and may not be possible.

Manure removed from stalls should not be spread directly on fields where horses graze as this can spread parasitic worm larvae. Good composting techniques have been shown to dramatically lower *Strongyles* counts in manure after as little as 4 weeks. Eight weeks of composting is generally the minimum recommendation. Compost must reach a temperature of 55 to 65 °C (130 to 150 °F) for 5 to 7 days to kill parasitic worm eggs. Turning compost every few weeks, or using an aerated static pile composting system will readily create these conditions. Aerated static pile systems use forced air that is driven through the compost in a way that maintains the desired temperature conditions without needing to turn the piles.

Regular faecal egg counts are essential for the management of parasitic worms (Figure 9.2). Faecal egg counts should be performed a minimum of once yearly. Timing is best determined by a veterinarian as it varies with the location, farm, and age of the horse. See The Brooke Hospital for Animals (2016) in 'Continuing education' for more information about faecal egg counts. Horses with high faecal egg counts (more than 350 eggs per gram of manure) are a low percentage of the population. Current estimates are around 5 to 7%. However, these horses are responsible for 90% of the parasitic worms deposited on the pasture. Using faecal egg counts, horses with high numbers of parasitic worms can be identified and targeted. Faecal egg counts should be performed on incoming horses prior to turnout on pasture. This allows horses with high egg counts to be appropriately treated before they can contaminate the pasture.

Physical and mechanical control

Dragging fields (Figure 9.3) will break up manure clumps causing exposure to sunlight and desiccation of parasitic worm eggs and larvae. Dragging also will break up piles of vegetation, which provide shelter for parasitic worms. Ideally, dragging should be conducted in pastures that are being rested. However, careful monitoring of environmental conditions is important. Dragging should be done when temperatures are greater than 29 °C (85 °F), and rain is not forecast for at least 12 hours. This can be challenging in certain geographic areas.

Twice weekly removal of manure from pastures can be as successful as a deworming program with anthelmintics. This also allows horses to graze fields more uniformly instead of avoiding faecal deposition areas and 'island grazing'.

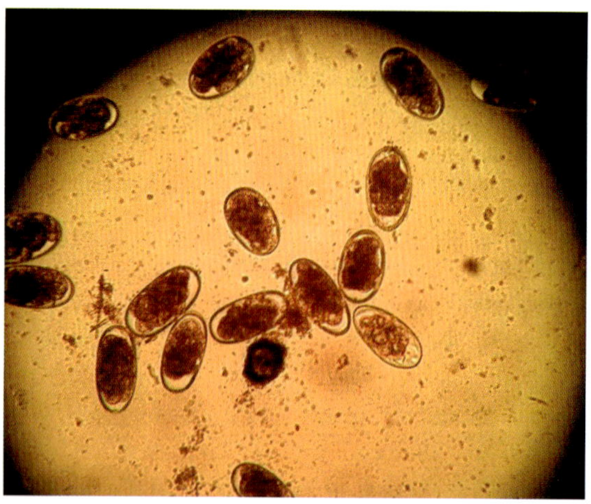

Figure 9.2. ***Strongyles* eggs under a microscope.**

Figure 9.3. Using a drag on pastures will break up manure clumps and expose parasitic worm eggs to sunlight if weather conditions are sunny and hot putting them at risk of desiccation.

Biological control

There has been some work with the fungal species *Duddingtonia flagrans* as a feed-through parasitic worm control supplement (Hernández *et al.*, 2016). The fungal spore is fed to grazing animals who deposit it on the field through the manure. This fungal species then attacks parasitic worm larvae in the environment. Extensive research has been done with sheep and other ruminants as well as horses (Braga *et al.*, 2020; Healey *et al.*, 2018a,b). A commercial product (BioWorma®, see 'Resources') is currently available in the USA, Australia, and New Zealand.

Chemical control

Macrocyclic lactones (ivermectin and similar drugs) are the cornerstone of chemical control for the *Strongyles* species. However, judicious use of these products is warranted as recent cases of resistance in Ireland and the USA have been reported (Nielsen *et al.*, 2020). It is recommended that faecal egg counts be used to determine which horses require product and when it should be administered (Papini *et al.*, 2020). The selected horses are then treated using weight estimates to be sure an appropriate dose is given (Figure 9.4). Dewormers should be administered after the mouth has been rinsed, and the horse should be monitored to be sure all of the product is swallowed. Worldwide, small strongyles have been shown to have significant resistance to all other currently available drug classes.

9.1.6 Veterinary concerns

Large strongyles were once a dreaded parasite for veterinarians. These parasitic worms cause clots in the major blood vessels supplying the small intestine causing life-threatening necrosis. With the advent of modern dewormers, these parasitic worms have largely been managed. With their long life cycle, large strongyles are managed easily with dewormers administered as infrequently as once yearly.

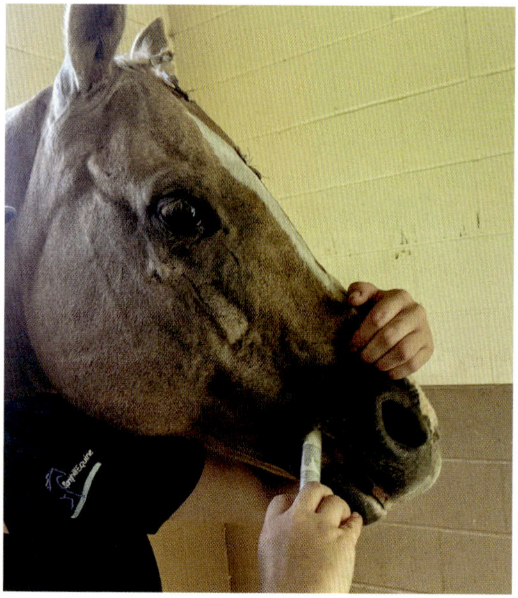

Figure 9.4. Administering a deworming drug.

Horses with low numbers of small strongyles are at little risk for health issues. Problems occur when large numbers of small strongyles have encysted into a dormant state in the large colon wall, and are then released. This release causes small defects in the mucosa of the gastrointestinal tract, causing profound inflammation that leads to a reduction of absorption of nutrients while allowing bacteria to enter the bloodstream. Careful management through targeted deworming using faecal egg counts to guide decisions will avoid this scenario.

9.2 Ascarids (genus *Parascaris*)

9.2.1 Parasite summary

Parascaris is a genus of nematodes in the family Ascarididae (Figure 9.5). It contains two species that parasitise horses, *Parascaris equorum* and *Parascaris univalens*. Parascarids live in the small intestine and are primarily reported in young horses. Most horses develop a natural immunity to *Parascaris* around 8 to 18 months of age (Clayton and Duncan, 1979). The prevalence of infection in horses younger than 1 year is 31 to 61%; in those older than 1 year, it is 25%.

9.2.2 Distribution and seasonality

Parascaris has a worldwide distribution. Foals are infected in the spring shortly after foaling. Impactions are seen in the autumn when parasites have had time to grow enough in size and number to cause blockages. These worms are present throughout the year and do not have a distinctive seasonality.

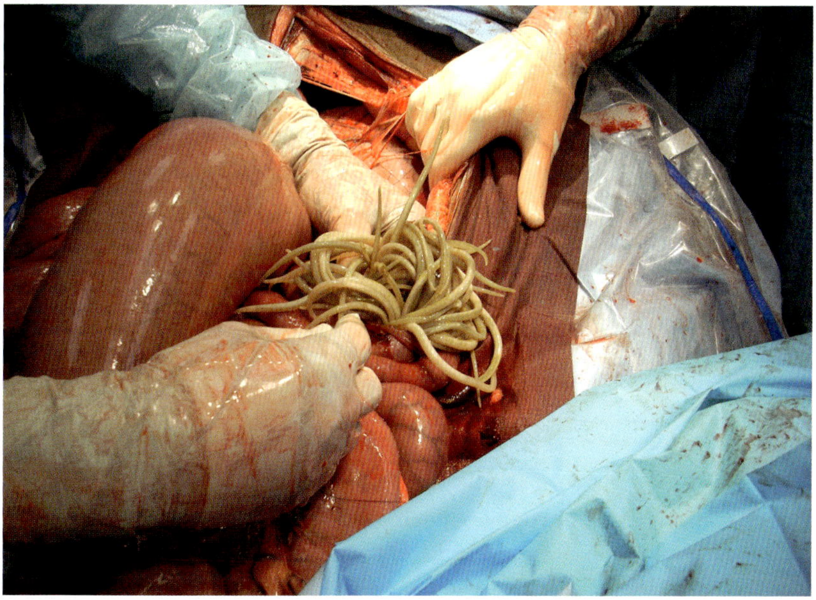

Figure 9.5. Ascarids seen at surgery after causing a blockage of the small intestine.

9.2.3 Description

Parascaris can be readily seen when passed in manure, most often following the administration of a dewormer. Adults can be quite long, 25 to 50 cm (9.5 to 19.0 in), and have a large, about 8 cm (3 in), off-white to yellowish thick body with tapered ends. Eggs can be seen during faecal egg evaluations. Eggs are brown, about 100 μm wide, almost spherical, and have a thick shell (Figure 9.6). For this species, in particular, faecal egg counts are not a reliable indicator of potential worm burden for the horse.

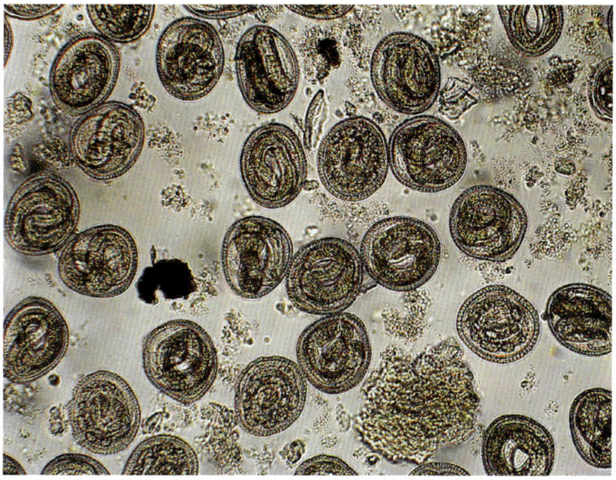

Figure 9.6. In contrast to strongyle and tapeworm eggs, ascarid eggs are nearly round and have a thick shell.

9.2.4 Life history

The eggs are laid by female worms in the intestine and then they are shed during defecation. In the soil, the larvae within the eggs develop from L1 to L2 within a 2 week period. The eggs containing L2 are present in the environment, where they can survive for 5 to 10 years. When the L2-containing eggs are consumed by horses, the L2 hatch after which they migrate through the body, primarily the liver, and lungs for approximately 30 days. Once in the lungs the larvae are coughed up and swallowed, before returning to the small intestine to mature to the L4 then to the adult stage. The life cycle takes approximately 3 months.

9.2.5 Control

Cultural control

Because this is primary a pest of young animals, pregnant mares should be managed to limit infestation. Mares should be in clean paddocks and manure removed frequently from paddocks. Stalls can be sanitised prior to foaling, and udders cleaned to remove any eggs that may be present because of laying down on infested soil.

Manure removal from pasture areas is key to cultural control. The eggs containing L1 to L2 can continue to persist in the environment even following manure removal. *Parascaris* can be minimised by mowing the grass short, removing manure, dragging the pastures, then allowing the pasture to remain fallow for 8 to 10 weeks.

Physical and mechanical control

Dragging fields to break up manure piles will help with the management of *Parascaris* numbers. However, *Parascaris* is much less susceptible to environmental factors than members of the *Strongylus* genus due to a thick covering on the eggs.

Biological control

See comments above in the strongyles section about the use of fungi for nematode management.

Chemical control

Parascarids are the most difficult to control of the important internal parasites of the horse. Widespread resistance to macrocyclic lactones already exists, and this species readily develops resistance to the other commonly used deworming products (reviewed in Leathwick *et al.*, 2017). Benzimidazoles and pyrantels are effective in most circumstances but their use should be closely monitored and directed by a veterinarian to avoid resistance for as long as possible (Nielsen, 2016).

9.2.6 Veterinary concerns

Parascaris species can cause severe colic in young horses due to large numbers of parasites in the gastrointestinal tract. Young horses should be carefully assessed by a veterinarian prior to deworming. A massive die-off will cause the parasitic worms to create impactions, which will require surgical removal (Tatz *et al.*, 2012). These cases are often managed with prophylactic mineral oil administration via nasogastric tube to facilitate the passage of dead worms. Some horses also will have anaphylactic-type reactions to large die-offs, and must be managed symptomatically. Fortunately, most horses develop immunity to *Parascaris* by 2 years of age (Clayton, 1986).

9.3 Pinworms (*Oxyuris equi*)

9.3.1 Pest summary

Oxyuris equi is the equine pinworm that is present in the small and dorsal colon. Pinworms do not cause significant disease or illness but can cause secondary damage to property and animals due to itching. There does appear to be individual susceptibility differences. Within a herd of horses, a small number will be affected, while the majority will resist infection.

9.3.2 Distribution and seasonality

Pinworms have a worldwide distribution. They have a weak seasonal pattern but are less common in the spring than in the other seasons

9.3.3 Description

Pinworms rarely can be seen in freshly passed faeces from infested horses. Adults are about 8 to 9 cm (3.0 to 3.5 in) long white worms with two-thirds of the parasite being uniformly round, one-third tapering rapidly to a point with variable transparency. Infested horses will exhibit characteristic tail rubbing behaviour due to the presence of eggs on the perianal area.

Eggs are deposited around the anus by the female parasitic worm where they appear as an off-white to pale yellow crust (Figure 9.7). The eggs are translucent and sticky and are commonly found on surfaces where horses are rubbing tails due to the itching caused by the infestation. Eggs are oval-shaped, 80-90 μm long, and can be distinguished from strongyles eggs by the presence of a cap (operculum) at one end of the egg (Figure 9.8).

9.3.4 Life history

Female adult pinworms protrude from the anus (Figure 9.9) and lay eggs on the anus and perianal area. During each laying event, a female lays from 8,000 to 60,000 eggs. The heat from the host's body causes the eggs to develop from L1 to L2 to an infectious L3 stage within the egg in 5 days. The infective eggs are then shed into the environment where they are ingested by the same and other horses. The L3 stage hatches and enters the mucosa of the large intestine and migrates to the large colon where the L4 then develops into adults. The life cycle takes 5 months (Reinemeyer and Nielsen, 2014).

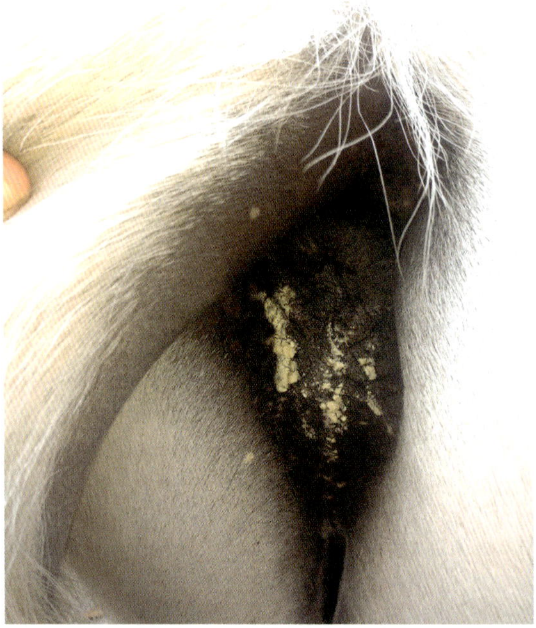

Figure 9.7. ***Oxyuris equi*** **eggs are deposited around the anus by adult female worms.**

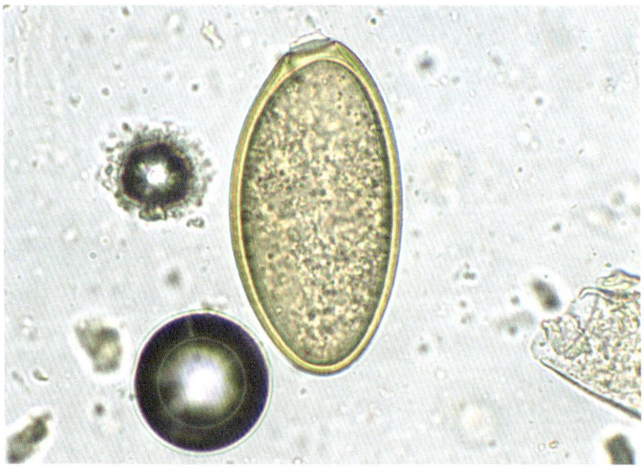

Figure 9.8. Egg of ***Oxyuris equi*****, which has a distinctive cap (operculum) at one end (top in this image).**

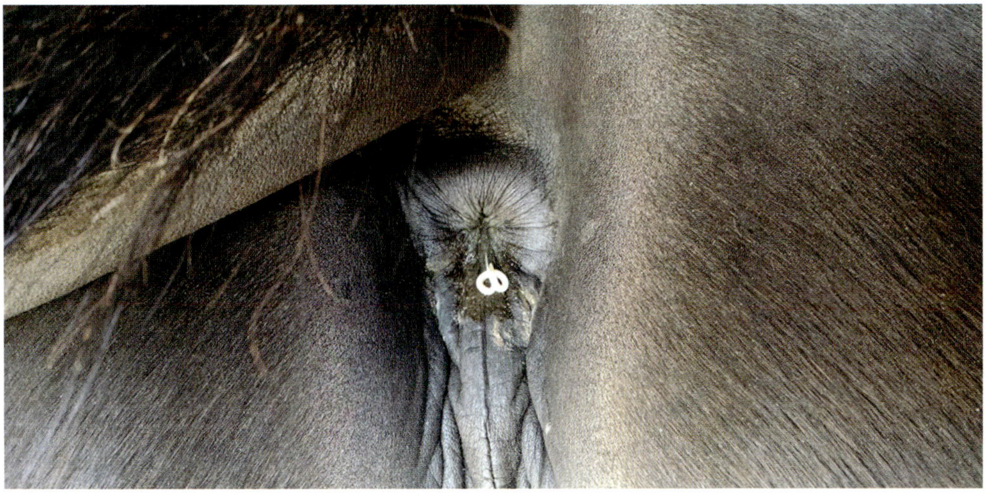

Figure 9.9. Adult pinworm outside anus of a horse.

9.3.5 Control

Cultural control

Surfaces can be washed with soap and water to minimise exposure to *Oxyuris* eggs in the environment. It is also recommended that equipment not be shared by horses known to have active *Oxyuris* infestations. Grooming is an essential component of management to check horses daily and recognise infestation and remove eggs and material. Washing the perineum area with soap and water will remove and destroy the eggs.

Chemical control

Oxyuris are very sensitive to pyrantels but may have some level of resistance to macrocyclic lactones (Wolf *et al.*, 2014). Deworming with these products when infection is identified will effectively eliminate it from that horse. However, due to environmental contamination reinfection will recur quickly.

9.3.6 Veterinary concerns

Oxyuris do not cause any direct damage to the gastrointestinal tract, or the animal they infect. Veterinary concerns lie with damage to the skin secondary to self-mutilation due to itching (Figure 9.10). Horses may lose condition when heavily infested. Infected horses scratch consistently, fail to thrive, and may bear self-inflicted wounds. In severe cases, there may be anorexia and anxiety.

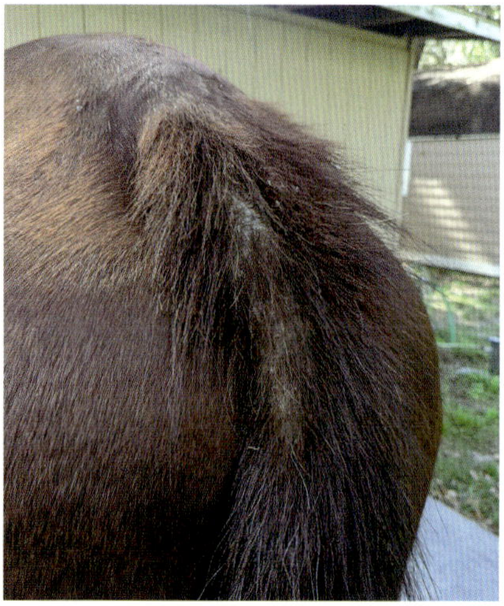

Figure 9.10. Damage caused by rubbing behaviour in response to equine pinworms.

9.4 Tapeworms (genera *Anoplocephala* and *Paranoplocephala*)

9.4.1 Parasite summary

Anoplocephala (Figure 9.11) and *Paranoplocephala* are genera of tapeworms in the family Anoplocephalidae. There are three species that infest equines, *Anoplocephala perfoliata*, *Anoplocephala magna*, and *Paranoplocephala mamilana*. The species most commonly associated with disease is *A. perfoliata*, which attaches near the ileocaecal valve. The other two species are less common and attach in the small intestine. All three species have an indirect life cycle involving a mite that is accidentally ingested by horses during grazing.

Figure 9.11. Tapeworm (*Anoplocephala perfoliata*) adults at the ileocecal junction.

9.4.2 Distribution and seasonality

Widespread distribution with significantly lower numbers in arid locations. *Anoplocephala perfoliata* and *Paranoplocephala mamilana* occur worldwide and *A. magna* is more common in the Americas and less so in Europe. They are most common in the summer through winter months (Rehbein *et al.*, 2013; Tomczuk *et al.*, 2015).

9.4.3 Description

Tapeworms are difficult to identify ante mortem as egg flotation is ineffective due to the eggs being released in egg packets (Tomczuk *et al.*, 2014). Serological tests are available but preventative treatment may be more cost-effective. Eggs are D-shaped (Figure 9.12), with eggs of *A perfoliata* being larger (80 μm) than those of *A. magna* (64 μm) (Bohórquez *et al.*, 2014). Adult worms are 8 to 25 cm (3.1 to 9.8 in) long, yellow-green, somewhat triangular in shape, and have an unarmed head with four suckers.

9.4.4 Life history

Eggs of these species are passed in the faeces of infected horses and are ingested by oribatid mites (*Galumna* spp.) in the water film on vegetation (Denegri, 1993). Oribatids are tiny (<1 mm or 0.05 in) mites with hard shells that are common in pastures and forest floors (Figure 9.13). The larvae hatch while in the mites and develop to cysticercoids, or the larval stage of the parasite. Mites on the pasture are accidentally ingested by a horse and the cysticercoids are released when the mites are digested. The cysticercoids attach at the ileocecal junction where they mature. Once mature, egg packets are released during defecation. Larvae mature to the adult stage in 2 months. Eggs are shed into faeces in the form of proglottids, which are tapeworm segments that are packed with eggs.

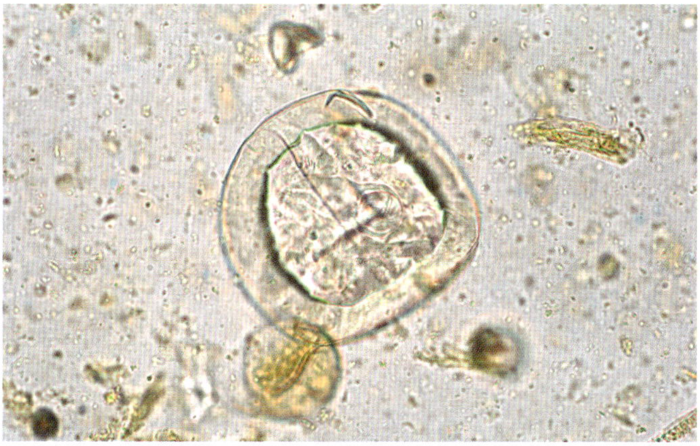

Figure 9.12. Egg of *Anoplocephala perfoliata*, with the characteristic D-shape.

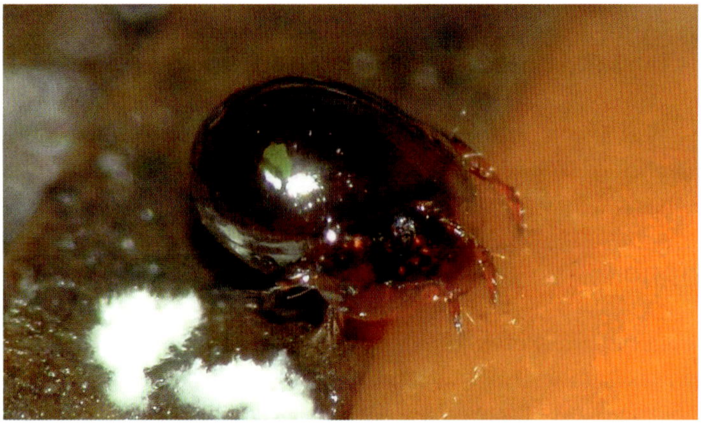

Figure 9.13. Oribatid mites are microscopic mites that horse tapeworms use as intermediate hosts.

9.4.5 Control

Control of this group is difficult due to the wide variety of species involved. The best practice is to manage the parasitic worms in the horse to reduce environmental contamination.

Chemical control

Anoplocephala species are managed through a yearly administration of a praziquantel dewormer. Due to their long life cycle, this is all that is required for effective management.

9.4.6 Veterinary concerns

Anoplocephala species attach to the intestinal wall using the suckers on the head causing ulceration. Symptoms are generally worse with *A. perfoliata* than with *A. magna*. The severity of the disease is directly related to the number of parasitic worms. Signs of infection include recurring diarrhoea and weight loss. *Anoplocephala perfoliata* specifically is associated with mild to moderate spasmodic colics due to inflammation at the ileocaecal junction and ileal impaction. Symptoms of infection are recurring colic episodes with no obvious cause. On veterinary examination, these horses will have normal physical and rectal examinations, yet display intermittent symptoms of colic, including rolling, and pawing. They generally respond well to conservative therapy of oral fluids and anti-inflammatories only to colic again in 10 to 14 days. On rare occasions, pain will be unmanageable to the point of requiring surgical intervention (Figure 9.14). They are further implicated in caecocolic intussusception. This is a rare but life-threatening event that requires surgical correction.

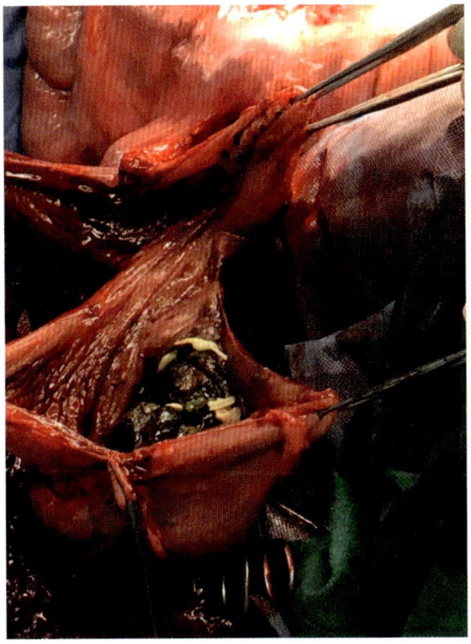

Figure 9.14. In rare cases, surgery may be required because of complications due to tapeworm infections. Tapeworms are visible here as off-white rectangular shapes on digesta in the caecum.

Continuing education

Talks by Dr Martin Nielsen on YouTube

Martin K. Nielsen Equine Parasitology Youtube channel
https://www.youtube.com/channel/UC_pDqB6sRvi1XJ8AHUapPCw

Internal parasite control guidelines

American Association of Equine Practitioners, 2021.
https://aaep.org/guidelines/internal-parasite-control-guidelines

- A detailed guide on treatment of gastrointestinal parasites of horses from a North American perspective.

A guide to the treatment and control of equine gastrointestinal parasite infections

European Scientific Counsel Companion Animal Parasites, 2018.
https://www.esccap.org/link-document/142/

- A detailed guide on the gastrointestinal parasites of horses from a European perspective.

Chapter 17 Parasitology

In: The working equid veterinary manual. The Brooke Hospital for Animals, 2016.
https://www.thebrooke.org/sites/default/files/Professionals/Working%20Equid%20Veterinary%20Manual/WEVM-chapter-17.pdf

- A guide aimed at owners of working horses, donkeys, and mules in developing countries. Contains information on internal and external parasites including how to do a faecal egg count and design an effective deworming program.

How to control worms in horses

World Horse Welfare

https://www.worldhorsewelfare.org/advice/health/how-to-control-worms-in-horses

- A guide aimed at the UK and worldwide to preventing worms in your horse including using cultural and chemical options. This guide also covers diagnosis and treatment of each main parasitic worm pest of horses.

Parasites and deworming

Veterinary Teaching Hospital, Masses University, New Zealand.

http://www.massey.ac.nz/massey/fms/Colleges/College%20of%20Sciences/IVABS/Equine%20Clinic%20and%20Hospital/PDFs/Worming.pdf

- A guide to parasites, cultural management, and deworming in New Zealand.

Systematic review of gastrointestinal nematodes of horses from Australia

Saeed, M.A., Beveridge, I., Abbas, G., Beasley, A., Bauquier, J., Wilkes, E., Jacobson, C., Hughes, K.J., El-Hage, C., O'Handley, R., Hurley, J., Cudmore, L., Carrigan, P., Walter, L., Tennent-Brown, B., Nielsen, M.K. and Jabber, A., 2019. Parasites and Vectors 12: 188.

https://doi.org/10.1186/s13071-019-3445-4

- Review of current literature on pathogeneses, epidemiology, diagnoses, treatment and control of parasitic worms in Australia.

Equine parasite control

Horse Council, British Columbia, 2020.

https://hcbc.ca/wp-content/uploads/2020/04/Equine-Parasite-Control-version-2.pdf

- Slides on equine parasite control from a Canadian perspective with useful terminology, important parasites, prevention and control.

Resources

BioWorma®. https://www.bioworma.com/

References

Behera, H.K., Jena, G.R., Kumar, D., Mohapatra, P., Mishra, S.K., Sahoo, R. and Dalei, N., 2017. Infestation of *Oxyuris equi* in horse and its successful therapeutic management. The Pharma Innovation Journal 6: 387-389. https://www.thepharmajournal.com/archives/2017/vol6issue11/PartF/6-11-21-277.pdf

Bohórquez, A., Meana, A., Pato, N.F. and Luzón, M., 2014. Coprologically diagnosing *Anoplocephala perfoliata* in the presence of *A. magna*. Veterinary Parasitology 204: 396-401.

Braga, F.R., Ferraz, C.M., da Silva, E.N. and de Araújo, J.V., 2020. Efficiency of the Bioverm® (*Duddingtonia flagrans*) fungal formulation to control *in vivo* and *in vitro* of *Haemonchus contortus* and *Strongyloides papillosus* in sheep. 3 Biotech 10(2): 1-5.

Bredtmann, C.M., Krücken, J., Murugaiyan, J., Kuzmina, T. and von Samson-Himmelstjerna, G., 2017. Nematode species identification – current status, challenges and future perspectives for cyathostomins. Frontiers in Cellular and Infection Microbiology 7: 283. https://doi.org/10.3389/fcimb.2017.00283

Clayton, H.M., 1986. Ascarids: recent advances. Veterinary Clinics of North America: Equine Practice 2: 313-328.

Clayton, H.M. and Duncan, J.L., 1979. The development of immunity to *Parascaris equorum* infection in the foal. Research in Veterinary Science 26: 383-384.

Corning, S., 2009. Equine cyathostomins: a review of biology, clinical significance and therapy. Parasites and Vectors 2: S1. https://doi.org/10.1186/1756-3305-2-S2-S1

Denegri, G.M., 1993. Review of oribatid mites as intermediate hosts of tapeworms of the Anoplocephalidae. Experimental and Applied Acarology 17: 567-580.

Healey, K., Lawlor, C., Knox, M.R., Chambers, M. and Lamb, J., 2018a. Field evaluation of *Duddingtonia flagrans* IAH 1297 for the reduction of worm burden in grazing animals: tracer studies in sheep. Veterinary Parasitology 253: 48-54.

Healey, K., Lawlor, C., Knox, M.R., Chambers, M., Lamb, J. and Groves, P., 2018b. Field evaluation of *Duddingtonia flagrans* IAH 1297 for the reduction of worm burden in grazing animals: pasture larval studies in horses, cattle and goats. Veterinary Parasitology 258: 124-132.

Hernández, J.Á., Arroyo, F.L., Suárez, J., Cazapal-Monteiro, C.F., Romasanta, Á., López-Arellano, M.E., Pedreira, J., de Carvalho, L.M.M., Sánchez-Andrade, R., Arias, M.S. and de Gives, P.M., 2016. Feeding horses with industrially manufactured pellets with fungal spores to promote nematode integrated control. Veterinary Parasitology 229: 37-44.

Hutchinson, G.W., Abba, S.A. and Mfitilodze, M.W.,1989. Seasonal translation of equine strongyle infective larvae to herbage in tropical Australia.Veterinary Parasitology 33: 251-263.

Leathwick, D.M., Sauermann, C.W., Geurden, T. and Nielsen, M.K., 2017. Managing anthelmintic resistance in *Parascaris* spp.: a modelling exercise. Veterinary Parasitology 240: 75-81.

Nielsen, M.K., 2012. Sustainable equine parasite control: perspectives and research needs. Veterinary Parasitology 185: 32-44.

Nielsen, M.K., 2016. Evidence-based considerations for control of *Parascaris* spp. infections in horses. Equine Veterinary Education 28: 224-231.

Nielsen, M.K., Banahan, M. and Kaplan, R.M., 2020. Importation of macrocyclic lactone resistant cyathostomins on a US thoroughbred farm. International Journal for Parasitology: Drugs and Drug Resistance 14: 99-104.

Nielsen, M.K., Kaplan, R., Thamsborg, S., Monrad, J. and Nautrup Olsen, S., 2007. Climatic influences on development and survival of free-living stages of equine strongyles: Implications for worm control strategies and managing anthelmintic resistance. The Veterinary Journal 174: 23-32.

Papini, R.A., Orsetti, C. and Sgorbini, M., 2020. A controlled study on efficacy and egg reappearance period of ivermectin in donkeys naturally infected with small strongyles. Helminthologia 57: 163-170. https://doi.org/10.2478/helm-2020-0017

Poynter, D., 1954. Seasonal fluctuation in the number of strongyle eggs passed by horses. Veterinary Record 66: 74-78.

Rehbein, S., Visser, M. and Winter, R., 2013. Prevalence, intensity and seasonality of gastrointestinal parasites in abattoir horses in Germany. Parasitology Research 112: 407-413.

Reinemeyer, C.R. and Nielsen, M.K., 2014. Review of the biology and control of *Oxyuris equi*. Equine Veterinary Education 26: 584-591.

Tatz, A.J., Segev, G., Steinman, A., Berlin, D., Milgram, J. and Kelmer, G., 2012. Surgical treatment for acute small intestinal obstruction caused by *Parascaris equorum* infection in 15 horses (2002-2011). Equine Veterinary Journal 44: 111-114.

Tomczuk, K., Kostro, K., Szczepaniak, K.O., Grzybek, M., Studzińska, M., Demkowska-Kutrzepa, M. and Roczeń-Karczmarz, M., 2014. Comparison of the sensitivity of coprological methods in detecting *Anoplocephala perfoliata* invasions. Parasitology Research 113: 2401-2406.

Tomczuk, K., Kostro, K., Grzybek, M., Szczepaniak, K., Studzińska, M., Demkowska-Kutrzepa, M. and Roczeń-Karczmarz, M., 2015. Seasonal changes of diagnostic potential in the detection of *Anoplocephala perfoliata* equine infections in the climate of Central Europe. Parasitology Research 114: 767-772.

Wolf, D., Hermosilla, C. and Taubert, A., 2014. *Oxyuris equi*: lack of efficacy in treatment with macrocyclic lactones. Veterinary Parasitology 201: 163-168.

Photo credits

Title page	Rattiya Thongdumhyu, Shutterstock no. 730748662.
Figure 9.1.	Erica Lacher.
Figure 9.2.	Erica Lacher.
Figure 9.3.	Erica Lacher.
Figure 9.4.	Erica Lacher.
Figure 9.5.	Aric Adams, Equine Medical Center of Ocala.
Figure 9.6.	Olgaru79, Shutterstock no. 385656697.
Figure 9.7.	Erica Lacher.
Figure 9.8.	Image provided courtesy of the National Center for Veterinary Parasitology, ncvetp.org.
Figure 9.9.	Sue Hamilton.
Figure 9.10.	Erica Lacher.
Figure 9.11.	Image provided courtesy of the National Center for Veterinary Parasitology, ncvetp.org.
Figure 9.12.	Image provided courtesy of the National Center for Veterinary Parasitology, ncvetp.org.
Figure 9.13.	Steve Kerr, license NC BY https://inaturalist.nz/photos/1261578?size=original.
Figure 9.14.	Meghan Abbott.

Chapter 10. Guidelines for equine pest control

Chapter overview

In the previous chapters, different methods of control, from cultural practices to chemical control options, were provided for each group of pests. However, there is additional information that could improve pest control success that the authors felt should be expanded upon. For example, monitoring (Section 10.1) is an important part of an integrated pest management (IPM) plan and can be incorporated into pest control to better understand how control methods are working over time. The use of pest protection on horses, like fly sheets, boots, and masks (Section 10.2) could be improved by understanding what to consider when selecting a product. Chemical use for pest and parasite control is often used and there are many guidelines for safe and effective use (Sections 10.3-10.5), from selecting pesticide active ingredients to understanding chemical resistance, to how to use repellents and topical pesticides. Finally, pest insect traps are developed based on the behaviour of the target insect. There are some tips provided for trapping different groups of pests (Section 10.6) that could improve trap success. Pest control is challenging and the hope is that these guidelines will improve the success of an IPM plan targeting pests of horses.

10.1 Guidelines for pest monitoring

Monitoring pests is an important component of IPM. Monitoring is the regular observation of pests, measuring pest abundance, and sometimes keeping records of pest levels. Monitoring can help you identify how many pests are present, where they are located, what type of control is needed based on the species of pest, and if the control methods you are using are working. Keeping monitoring records can also help with identifying pest problems early before they become unmanageable and necessitate immediate reduction methods like chemical control options. Records provide historical data that can help improve control techniques over time. For example, what month flies are likely to become a problem so control is implemented before that time, or when mites become a problem in the cooler months so feathers can be clipped or treatments started before horses are aggravated.

Monitoring can be challenging, and as with control, is often species-specific. Before you can start monitoring, you must be able to recognise the type of pest that may be present. Use the quick reference guide in Chapter 2 to help determine which pest is a problem, keeping in mind there may be more than one. Once you have identified the pest or pests, establish a monitoring and record-keeping program that is appropriate for the pest and works for you. Some pests, especially the filth flies discussed in Chapter 3, have well-established monitoring methods that can be used to keep track of relative numbers of pests over time. Others, such as mange mites and keds, are mainly monitored by presence/absence. Table 10.1 is a list of monitoring resources for the major groups of pests that tend to be repetitive pests on equine farms. Whichever method of monitoring is chosen, the key is to be consistent. Using the same monitoring methods over time at consistent intervals can give you the best estimate of pest numbers.

10.2 Guidelines for fly sheets, boots, and masks

The variety of fly sheets, boots, and masks that can be purchased continues to increase. With the many options, it may be challenging to determine which product will be most suitable for your horse and purpose. However, there are some key considerations to help with decision-making.

10.2.1 Target pest

Some protective equipment designs are for specific pests. If horse and deer flies (Chapter 4) are the primary pests of concern, fly sheets with zebra stripes may reduce bites (Figure 10.1). For biting midges, maximum coverage should be considered and sweet itch-specific sheets can be purchased (Chapter 4) (Figure 10.1). Most fly pests are interested in any exposed area of the horse's body, so fly sheets with extended necks or detachable hoods and belly bands should be considered.

The durability and wearability of fly masks and boots are generally more important than the target pest as these types of fly protection will work similarly for most fly pests. The exception is fly masks with extended nose guards. These may be more effective against face flies (Chapter 3), eye gnats (Chapter 6), some bot flies (Chapter 7), or other pests associated with the nose.

Table 10.1. Resources for monitoring some pests associated with horses.

Equine Pest[1]	Monitoring resource (see references)
Chapter 3	
House fly, face fly, *Fannia* fly	Geden *et al.*, 2021; Gerry, 2012, 2020; Murillo *et al.*, 2021; Trout-Fryxell *et al.*, 2021
Horn fly, buffalo fly, stable fly	Brewer *et al.*, 2021; Hogsette and Kline, 2017; Kaufman and Weeks, 2018; Rochon *et al.*, 2021; Smythe *et al.*, 2017; Taylor *et al.*, 2020
Chapter 4	
Mosquitoes	Drago *et al.*, 2012; Hoel *et al.*, 2009; Lühken *et al.*, 2014
Biting midges	Goffredo and Meiswinkle, 2004; McDermott and Lysyk, 2020
Horse and deer flies	Bučanović *et al.*, 2020; Kline *et al.*, 2018; Van Hennekeler *et al.*, 2008
Black flies	Otabil *et al.*, 2020; Shipp, 1985
Chapter 5	
Ticks	Salomon *et al.*, 2020
Lice	Lowden *et al.*, 2007
Chapter 6	
Fire ants	Drees *et al.*, 2013
Eye gnats	Bethke *et al.*, 2013
Chapter 7	
Sand flies	Hoel *et al.*, 2014; Rodríguez-Rojas *et al.*, 2016; Somwang *et al.*, 2021
Chapter 9	
Parasitic worms	Nápravníková *et al.*, 2019

[1] Many pests discussed in this book do not have a standardised method of monitoring established, and so they have been omitted from this table.

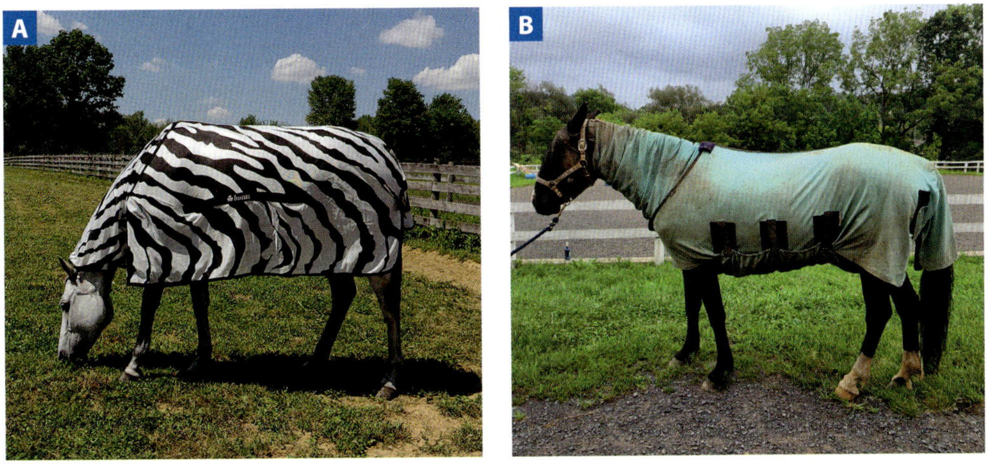

Figure 10.1. Fly sheets come in many different styles. The selection of a fly sheet should consider the pest. Zebra fly sheets (A) reduce bites from several types of large biting flies, like horse and deer flies, while fine-mesh full coverage sheets (B) are more effective for biting midge protection.

10.2.2 Fit and durability

Fly sheets, masks, and boots fit should fit comfortably and be chafe-free. Before purchasing a fly sheet, measure your horse to get the correct size. To measure a horse, place the end of a measuring tape at the centre of the chest at the base of the neck. Run the tape over the point of the shoulder and closely along the body to the centre of the tail (Figure 10.2). That number can be used to check against the manufacturer's sizing chart. However, just like human clothing, sheet sizing will vary slightly by manufacturer. The specific conformation of an individual horse may also influence how any specific product will fit but fly sheets can include darts, pleats, and gussets to allow for extra movement, especially in the chest and shoulder area. Nylon lining in the same areas may help to reduce rubbing. Fly masks and fly boots are generally less specific in size and are grouped based on the height or breed of the horse.

The fabric that the sheet is made of may be more or less durable, depending on the activity of the horse. If the horse is turned out with destructive pasture mates, playful bites at the sheets can cause tears. The use of a hot pepper anti-bite product or similar to discourage biting of the sheets may be necessary to eliminate this behaviour.

Fly masks and fly boots come in several sizes and many different fabrics. Fly masks can be purchased without ears, with extended nose guards, made out of canvas or stretch fabrics (Figure 10.3). The key is to determine which masks and boots are appropriate for the pest and which masks or boots will remain on the horse, especially if the animal frequently rolls or rubs, or is turned out with other horses that may bite and pull on the fabric. Sometimes this can be a bit of trial and error. With the correct fit, fly sheets, masks, and boots can be worn the entire day and will dry quickly after rainfall. Horses should be checked regularly to ensure the continued fit of fly sheets, masks, and boots.

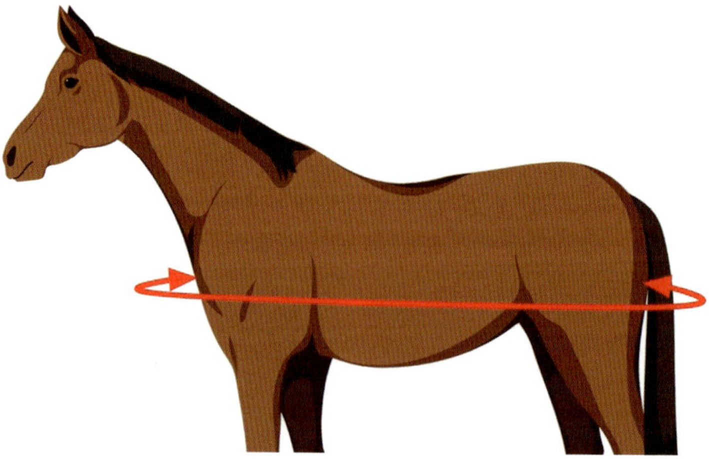

Figure 10.2. To measure a horse for a fly sheet, place the end of a soft measuring tape in the centre of the chest at the base of the neck and measure to the centre of the tail. A long piece of string can also be used to wrap around the horse and the string length measured.

Figure 10.3. A variety of fly masks are available and their use depends on the pest and the tolerance of the horse. In some cases, fly masks without ear covers may be necessary (A). In other cases, fly masks without (B) or with (C) nose guards may be necessary.

10.2.3 Breathability

Breathability is an important consideration since most fly problems occur in the warmer summer months. The smaller the mesh size of the fly sheet, the less airflow the sheet will have, but small mesh size is necessary for some biting pests like mosquitoes, sand flies, black flies, and biting midges. Newer polyester fabrics and mesh of fly sheets will allow more breathability than older fabrics or some lower-cost fabrics. Be sure to check what material the fly sheet is made of before purchase. Materials vary in their breathability but also in their ultraviolet light protection, mesh size, durability, and cost.

For fly masks and boots, fabrics range from stretchy spandex to thick canvas. Breathability is not as much of a concern on the face, but the ability of the horse to see through the mesh fabric should be considered. For fly boots, breathability is important to reduce heat stress on the tendons and ligaments.

It is also important to consider the temperature and humidity where the horse will be when wearing these protective products, and access to shade. Horses should be observed on hot days to avoid overheating. This is especially true for larger horses, horses with extra fat, non-sweating conditions, thick coats, or when heat-index temperatures approach or exceed 30 °C (86 °F).

10.2.4 Washability

Fly sheets, masks, and boots are designed to be washed according to the manufacturer's recommendations. Using a hose to quickly spray the items with water can be useful for cleaning soiled material. Permethrin-treated products should be washed separately from other fabrics.

10.3 Guidelines for using chemical control

Chemical control is one of the types of control used for pests in an integrated pest management plan. As mentioned in Chapter 1 and all preceding chapters, cultural control practices should be the foundation of all pest control when suitable for the pest. Physical and mechanical control and biological control options also should be considered. Chemical control should be used judiciously as part of a pest management plan. Registration of pesticides is managed by governmental agencies and can have further restrictions imposed by individual states or territories. The specific pesticides that are registered are always changing as new products are introduced and older ones are phased out. The discussion below is intended as general information. The toxicant in pesticides is referred to as the 'active ingredient' in North America and as the 'active substance' in Europe, Australia, and New Zealand. In the sections that follow we use the former term.

Before using any pesticide, check to be sure that it is legal to use in the area, labelled for the intended use (e.g. applied directly to horse, sprayed in the barn or stable, etc.) and the intended pests, and all safety precautions are followed.

10.3.1 How to select a pesticide

The prevalence of parasites of horses continues to increase globally and, unfortunately, a limited number of products are approved for use in horses (see Karasek *et al.* 2020 in 'Continuing education'). When selecting a pesticide, read the label and make sure that the label lists the target pest (e.g, flies, ticks, etc.) and the animal (e.g. horses) or site (e.g. stable or pasture) to be treated. Also check that the formulation (e.g. dust, gel, cream, spray, wipe, etc.) is suitable for your intended use and that you have the right equipment to apply it. Residual sprays will require an application method like a backpack sprayer, whereas baits will need some method to contain the bait. Finally, make sure that you have the correct personal protective equipment to safely complete the application. In some cases this may mean disposable gloves, in others, it may require eye or respiratory protection. The label directions will provide this information (see Section 10.3.2 for more information on how to read a label). The label will also include important toxicology information. If you have any questions about worrisome symptoms after exposure of horses to a pesticide call a veterinarian. If human exposure is the concern, call a poison control hotline or local emergency phone number.

10.3.2 How to read a pesticide label

A pesticide label is your guide to using pesticides safely and effectively. The information provided is given so that you can achieve maximum pest reduction with minimal risk to non-target animals and the environment. The label should be read before purchasing the pesticide and before each use. The label instructions also will provide information on storing and disposing of the pesticide. It is imperative to understand that the use of any pesticide in any way that does not comply with the label directions and precautions is illegal unless under the direction of a veterinarian.

The content of a pesticide label may differ based on the country of pesticide registration. Usually, the label will include the brand name of the product, which is specific to the manufacturer. The label also will include the registration number of the product with the country-specific regulatory agency and the active ingredient. Generally, precautionary statements and first aid information also will be included. Other information on the label may include the other ingredients, application instructions, pests that the pesticide will target, required personal protective equipment, environmental hazards, as well as proper storage and disposal instructions.

10.3.3 How to slow pest chemical resistance

While IPM allows for judicious use of chemicals, other control methods should be used as much as possible. Besides protecting the environment, another reason is to slow or prevent resistance in the target pests. The more you use pesticides, the more likely pesticide resistance will develop. Chemical resistance is becoming increasingly common in many pests. This is because pesticides have been the go-to method for control for many years, likely because of the ease of application and perceived effectiveness for reducing adult pest numbers. It is important to note that while rotating chemical classes is an effective strategy for arthropods, it is not an effective strategy for parasitic worms.

10.3.3.1 Arthropod pests

Pesticide resistance is a frequent and legitimate concern when using synthetic pesticides, especially with filth flies. Resistance is an inherited physiological or behavioural change that allows some individuals in a population of pests to continue to survive and reproduce in the presence of a toxin. Cross-resistance is when developed resistance to one toxin leads to resistance to a similar toxin because of how these toxins act on the pest or the mode of action.

To understand how to avoid or manage pesticide resistance, you first have to understand how it happens (Figure 10.4). Subtle differences in the genetic code, or DNA, of each living creature can determine how they look on the outside and how they operate on the inside. DNA is copied from adult to offspring, but small mistakes called mutations can make the offspring have different characteristics or features. Sometimes these mutations can make an individual more likely to survive a challenging environment. Because they are more likely to survive, they are also more likely to pass on their DNA to offspring. Those with the most successful traits will become more common over time as they are selected over and over in a process called natural selection. The fly life cycle is short in optimal conditions and many offspring are produced so populations can quickly develop resistance.

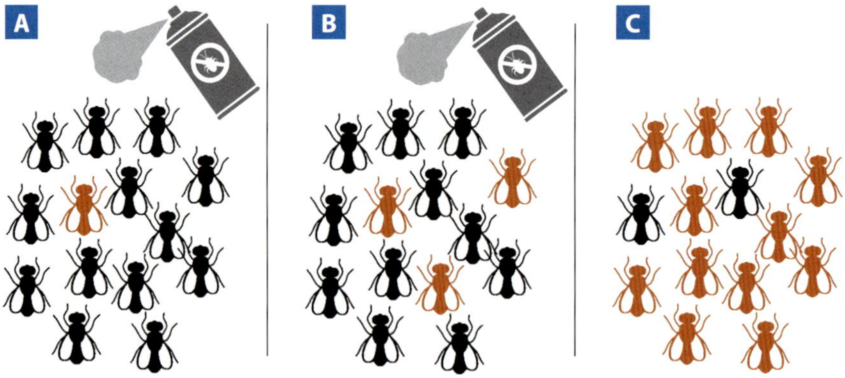

Figure 10.4. Using the same chemical active ingredient too often or using too much can mean that some individual pests that are resistant to the population escape the chemical and live to reproduce (represented here in orange) while others that are not resistant die (represented here in black) (A). In the next generation, there are more resistant individuals than the previous generation of pests (B), until the majority of the pest population is resistant to the chemical used (C).

Chemicals used in pesticides act on an insect with a specific mode of action. This is the way the pesticide works on the insect to produce the desired effect, which is death in most cases. There are several modes of action for pesticides; for example, the chemical may change the way a particular protein is produced, or it could cause neurological dysfunction in the pest.

The best way to combat pesticide resistance is to prevent it from occurring. There are two important strategies in pesticide resistance management: using chemicals only when necessary, and rotating active ingredient classes that have different modes of action. The timing of pesticide application is very important. The margin for error in dose, timing, coverage, and other factors is reduced as pests become more tolerant, and the risk of resistance increases. It is important to follow all manufacturer and label recommendations as well as any relevant regulations in your area or country. The mode of action in a pesticide is different depending on the chemical. Some chemicals inhibit pests from developing; others dose them with toxins or prevent some critical process in their body from occurring. To reduce the risk of resistance it is important to not expose multiple generations of a pest population to pesticides with the same mode of action. This is important whether you are using residual pesticides, feed-throughs, baits, or other methods of application.

With a better understanding of how pesticide resistance happens, it is easy to see how automatic spray systems can be detrimental to pest control. Not only do these systems spray droplets of pesticides into the environment, which are inhaled by humans, horses, and other animals, but a proportion of the flies present at the time of spraying will receive a sub-lethal dose because they happen to be in protected locations that pesticide does not fully reach when the system is activated. Over time, this allows the survivors to reproduce, with each generation having a higher number of spray-tolerant flies. The automatic aspect of the application also allows for more frequent applications than might be possible with a manual approach, which will increase the pressure on the population to develop resistance. This can eventually lead to product failure.

Fannia canicularis and *Musca domestica* collected from poultry facilities utilising pyrethrins and piperonyl butoxide automatic space spray systems have been found to be 109 times (twice daily sprays) and 12 times (twice weekly sprays) less sensitive to those active ingredients than flies that had not been previously exposed, respectively (Meyer *et al.*, 1990).

To reduce the risk of resistance the best option is to create an IPM plan as proposed by this book (Chapter 1, Appendix B) and, if using chemical control methods, by rotating chemical classes. When you decide on a pesticide that you want to use, look at the active ingredients and compare them to the Insecticide Resistance Action Committee (IRAC) chart found in 'Resources.' Rotate that choice with another pesticide that contains an active ingredient that is in a different IRAC group, which may have a different mode of action. Keep in mind that rotating repellents and topical pesticides is not recommended. There are limited options for active ingredients for topical repellents and pesticides. These products are not meant to control pests but to protect animals from contact with pests.

Unfortunately, it is difficult or impossible to determine how quickly resistance will develop in a pest population. In some areas, resistance may take a single year, and in other areas, it might be many years. When you rotate chemical classes, the effectiveness of the active ingredients is increased and the likelihood of resistance is decreased.

10.3.3.2 Parasitic worms

The principles described above in Section 10.3.3.1 also apply to resistance in internal parasites. Widespread anthelmintic (dewormer) resistance is common in most parasitic worms affecting horses. Historically, veterinary recommendations to avoid resistance included rotation of deworming products over the year. This has not been effective and current guidelines are to conduct assessments of horses and other equids yearly to direct management recommendations. The first step toward managing internal parasites is to know which species are being targeted with faecal egg counts. Faecal egg counts involve collecting a sample of manure and using a microscope to count parasitic worm eggs (Figure 10.5). Regional weather patterns will help determine the

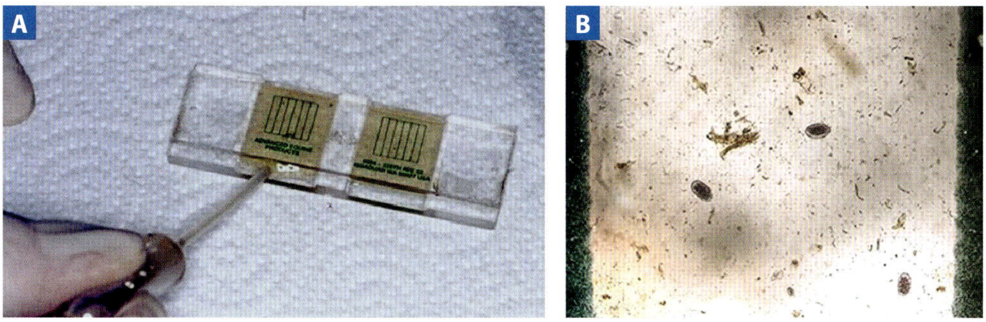

Figure 10.5. Eggs are separated from manure, placed in a known volume of salt solution, and counted under a microscope on special slides that have grid marks (A); eggs (B) can be categorised into different species groups by their shape as discussed in Chapter 9.

best time for faecal egg counts, but spring is generally the best time to make egg counts to assess parasitism.

For adult horses, faecal egg counts primarily involve monitoring the *Strongyloides* species. If a horse is classified as a 'high shedder' of parasite eggs, meaning this animal has a high number of parasitic worm eggs (>350 eggs/gram of manure), deworming is recommended using an ivermectin product. A faecal egg count should be performed again in 14 days to ensure the efficacy of ivermectin. As of publication current best practices also deworm every adult (more than 2 years old) horse with an ivermectin/praziquantel product once yearly. Timing for this is dependent on location and weather patterns.

Horses less than 2 years old should be monitored by a veterinarian, with deworming protocols created based on farm history, faecal egg counts, and the individual horse. Young horses are more likely to carry significant loads of *Parascaris* species. These parasites are notorious for developing resistance to parasiticides, and protocols are continually adjusting based on current knowledge of best practices. Pyrantel salts, macrocyclic lactones, and benzimidazoles are all used when treating *Parascaris* species and their use should be directed by faecal egg count monitoring under the guidance of a parasitologist or veterinarian.

Resistance is evolving so quickly in *Parascaris* that what works one year, may not work the next. Careful monitoring and use in the horse population will help minimise the resistance of deworming product active ingredients in the equine population.

10.3.4 How to store pesticides

The pesticide label directions will specify special considerations for the individual products. In general, always store pesticides in a well-ventilated room, preferably locked. Ensure that horses and other livestock, pets, children, and wild animals cannot reach the pesticides. Post a poison sign near the entrance to the room or on the cabinet where the pesticides are stored. The building and cabinet should be fire-resistant and insulated to prevent large changes in temperature that may affect the stability of the pesticides, which may affect their action and safety when used.

Pesticides should be stored in the containers in which they were purchased. Any leaking containers should be placed into a second larger container. Any spills should be cleaned up quickly, carefully, and completely. If you store pesticides on your property you should have a spill kit in case of any accidents. Spill kits should include an absorbent material (e.g. sawdust, cat litter) to put over a spill, a shovel, a watertight container for the waste product and absorbent material, and personal protective equipment (e.g. coveralls, boots, gloves, goggles, respirator). Understand how to properly contain a spill when you purchase the product by checking the label. Safe disposal of pesticides can be challenging. Before you make the purchase calculate how much you will need and do not purchase more than this amount. Pesticides can degrade over time, even in their original containers. Empty pesticide containers should be disposed of as directed on the label.

10.4 Common active ingredients used in chemical control

The type and concentration of active ingredients is a critical consideration, particularly for those that are applied directly on the animal. The active ingredient is the chemical that produces the desired effect, either repellency or mortality. This information can be found on the label below the brand or trade name as the active ingredient and the percent concentration. Regardless of the product, repellents and topical pesticides (pesticides applied directly on the animal) can only be expected to be effective for a fairly short time (See Mottet *et al.* 2018 in 'Continuing education'). Most are only effective for a few hours, and even the most potent repellents and topical pesticides will be washed off the animals by rainfall, bathing, or sweating, or inactivated by rolling in dirt or mud. Common pesticide active ingredients are presented below in alphabetical order, which is not indicative of the relative importance of each active ingredient. In addition, this list is not comprehensive and there may be additional active ingredients that are developed or tested that are not included in this list.

10.4.1 Common active ingredients in topical pesticides and repellents.

Topical pesticides and repellents include those products that are applied directly to the horse. Methods of application include spot-ons, pour-ons, dips, or sprays. Some products have pesticidal activity and are meant to kill ectoparasites. Other products have a primarily repellent effect and are formulated to prevent the pest from reaching the animal but may not kill the pest.

10.4.1.1 Benzyl benzoate

Benzyl benzoate is often used to treat horses for mites and lice and to repel the biting midges that cause sweet itch. The repellency effect is short-lived and only lasts a day or so. The mode of action is not fully understood. Benzyl benzoate is considered to be toxic to cats.

10.4.1.2 Natural repellents and topicals

In addition to products containing chemicals with well-defined toxic or repellent effects, there are many products marketed as natural repellents. Many of these are botanical extracts or some part of a botanical extract that show varying degrees of effectiveness. Examples of these are essential oils of neem, citronella, geraniol, aloe, lemon eucalyptus, peppermint, and others and fatty acid mixtures (e.g. octanoic, nonanoic, and decanoic acids). Some of the products are as effective as those containing permethrin in laboratory trials, but reliable information on their efficacy in the field and duration is limited (Tuorinsky and Machtinger, 2020). See Mottet *et al.* (2018) in 'Continuing education' for an evaluation of several repellents on horses.

10.4.1.3 N-N-Diethyl-3-methylbenzamide (DEET)

N-N-Diethyl-3-methylbenzamide or DEET is a widely used insect repellent that is highly effective against mosquitoes, biting midges, ticks, and other biting arthropods. DEET in a 15 to 17% oil and water emulsion was found to be effective against horse and deer flies for up to 4 hours (Herholz *et al.*, 2016). While DEET repellents labelled for use on horses are not common in the USA, they are fairly common in other countries and can be purchased as a spray, roll-on, and cream. DEET should not be applied with other active ingredients because of the risk of interaction and toxicity.

10.4.1.4 Organophosphates

Organophosphates are an old class of pesticides that were the most commonly used pesticides before the development of synthetic pyrethroids. Their mode of action is to act as acetylcholinesterase inhibitors that result in the uncontrolled firing of nerves and eventual death. The only active ingredient pesticide in this class that is used to treat horses directly in the USA is coumaphos, which is used to control ticks and lice. Coumaphos is not effective against flies.

10.4.1.5 Pyrethrins and synthetic pyrethroids

Pyrethrins and synthetic pyrethroids are the most common pesticides used for pest control on horses. Active ingredients in this class of pesticide can be used as repellents, topical pesticides, and residual sprays. The most commonly used pesticides for direct application to horses are permethrin and pyrethrins. Synthetic pyrethroids and pyrethrins are toxicants that act on the channels in the nervous system that are important in cell activation. This results in uncontrolled excitation of nerve cells causing paralysis and death. Some insects and ticks are repelled by the odour of these pesticides and will not land or crawl on a treated animal. Pyrethrins, sometimes called natural pyrethrins, are compounds extracted from the flowers of some *Chrysanthemum* species that have insecticidal and repellent properties. These typically provide knockdown benefits (i.e. they knock the pest off the animal or to the ground). However, they may not kill the insect pests, just stun or repel them. These compounds tend to be quite sensitive to ultraviolet light and will break down quickly in the sun. Pyrethroids such as permethrin are synthetic chemicals based on pyrethrins that are more stable than pyrethrins and have longer-lasting effectiveness. Piperonyl butoxide is a chemical that is commonly combined with pyrethrins and synthetic pyrethroids that acts as a synergist to make the products more effective.

The risk of poisoning with synthetic pyrethroids and pyrethrins to horses or people is low as long as the label instructions include horses and are followed. Pyrethroids and pyrethrins can cause seizures or even death in cats, so barn cats should be kept away from areas where these chemicals are being sprayed including areas with automatic spray systems using a pyrethrin-based product. (Figure 10.6). Overapplication of synthetic pyrethroids or applying synthetic pyrethroids not labelled for horses can result in severe injury.

10.4.2 Common active ingredients used as space sprays, residual sprays, or in baits

Space sprays refer to applications through a misting or fogging device that dispenses fine particles of spray in a closed space such as a stable. This method of application is helpful when flies become so bothersome indoors that there is an immediate need to reduce their numbers. The effect is immediate but temporary. Premise or residual sprays are applied to building walls, rafters and other structures where flies rest.

Figure 10.6. Barn or stable cats should be kept away from areas where pyrethrins and synthetic pyrethroids are applied. These chemicals are toxic to cats.

10.4.2.1 Carbamates

The carbamates are toxins that inhibit enzymes (cholinesterases) that are involved in the transmission of signals. This is an old group of chemicals that have mostly been retired or banned. Their use around horses is limited to methomyl (used in some baits) and carbaryl to treat vegetation for mosquitoes and ticks in countries where its use is allowed.

10.4.2.2 Diamides

Cyantraniliprole is a relatively new active ingredient in the diamides class of pesticides. The mode of action in insects is the unregulated activation of ryanodine receptor channels, leading to internal calcium store depletion that interferes with control of muscle contraction, resulting in lethargy and death. Cyantraniliprole is a slow-acting pesticide. It is only used around animals in a sugar bait product. Although it has low mammalian toxicity, cyantraniliprole is quite toxic to bees and its status and availability vary among countries.

10.4.2.3 Neonicotinoids

Neonicotinoid pesticides are based on nicotine, a natural pesticide. Examples of neonicotinoids include imidacloprid, nithiazine, thiamethoxam and dinotefuran. Neonicotinoids bind to nicotinic acetylcholine receptors in the nervous system. Overstimulation of these receptors causes paralysis and death. Its toxicity to mammals falls in the same range as synthetic pyrethroids.

In the USA, imidacloprid is available as a spray or paint-on formulation that contains sugar to encourage eating by house flies. Imidacloprid and several other neonicotinoids are commonly used in fly baits with sugar and a pheromone attractant. In areas where flies are still susceptible to these pesticides, flies are killed immediately during or after feeding and will form piles of carcasses on and around the bait. However, fly resistance to these materials is a growing problem. Several neonicotinoids have been banned for most applications in European Union countries because of concerns about their effects on bees.

10.4.2.4 Organophosphates

Although pyrethrins and synthetic pyrethroids are the most commonly used materials for premise treatments, some organophosphates still have registrations in some locations. DDVP, also known as dichlorvos or Vapona®, can be used as a space spray for rapid knockdown of flies under certain conditions. Other organophosphates used as premise sprays may include tetrachlorvinphos (Rabon®, stirofos), dibrom (naled), or combinations of DDVP and tetrachlorvinphos. House flies in many places are resistant to organophosphates because of their long history of use. Dibrom (naled) has been banned in the European Union.

10.4.2.5 Pyrethrins and synthetic pyrethroids

Synthetic pyrethroids (discussed above) are by far the most common active ingredient in materials that are applied as space spays and premise treatments. See the above Section 10.4.1.5 for general information about pyrethrins and synthetic pyrethroids. Pyrethrins, often synergised by combining with piperonyl butoxide, are commonly used as space sprays. In addition to permethrin, synthetic pyrethroids that are used as premise treatments may include cyfluthrin, bifenthrin, lambda-cyhalothrin, deltamethrin, esfenvalerate, and pralethrin.

10.4.2.6 Spinosyns

Spinosad is a mixture of chemicals (spinosyns) derived from a bacteria species, *Saccharopolyspora spinosa*, that was originally isolated from sugarcane. Spinosad's mode of action is on the insect nervous system by disrupting acetylcholine neurotransmission and by blocking the neurotransmitter γ-amino-butyric acid (GABA). Unlike synthetic pyrethroids, spinosad is slow-killing and may take a day or more to kill the pest after contact. It is formulated for use as sprays or in combination with sugar as a bait.

10.4.3 Common active ingredients in insect growth regulators (IGRs)

Although they are called 'insect' growth regulators, many active ingredients in this group also work on other arthropods like ticks. IGRs are considered biochemical insecticides because they do not control the pest through direct toxicity, but instead interfere with a critical developmental process that prevents the arthropod from reaching maturity and reproducing. In some cases, IGRs may be mixed with other insecticides to improve control. IGRs are comparatively safer to beneficial insects and the environment than many other forms of chemical control, although they do take longer to have noticeable effects as they target the immature stages of arthropods.

10.4.3.1 Cyromazine

Cyromazine is an IGR developed for the control of fly larvae in manure and other breeding sites in animal housing. This active ingredient interferes with the moulting process of larvae and pupation, leading to deformed and/or dead larvae, pupae or adults. Cyromazine appears to be fairly specific to flies (order Diptera). Granular and spray products have been developed that include cyromazine, but the most commonly used in equine facilities are feed-through products (e.g. Solitude IGR®).

10.4.3.2 Benzoylphenyl ureas (diflubenzuron and novaluron)

Diflubenzuron is used to control the larval stage of many insect pests, including flies. When exposed to diflubenzuron, the process of forming the exoskeleton when a fly pupates is interrupted. The larvae simply dry up and do not survive to adulthood. Formulations include soluble concentrate, flowable concentrate, wettable powder, and pelleted or tableted products. Feed-through products are available to prevent house and stable fly development (e.g. Equitrol II®, SimpliFly®). Novaluron is a related IGR that is available as a feed-through for horses in the USA (e.g. MultiFly®).

10.4.3.3 Fenoxycarb

Fenoxycarb is a non-neurotoxic carbamate. It does not have the same mode of action as other carbamate insecticides. Instead, it prevents immature insects from maturing by mimicking an important hormone required for moulting. This insect growth regulator is used for long-term fire ant control.

10.4.3.4 Methoprene

Methoprene (e.g. Altosid®) is an insect growth regulator that comes in two different forms called (s)-methoprene and (r)-methoprene, and s-methoprene is the one that behaves like juvenile hormone, an important hormone in insects. By acting like this insect hormone, it interferes with insect growth and development. It can prevent normal moulting, egg-laying, egg-hatching, and development from the immature phase to the adult phase. Methoprene is effective on a variety of insect species including horn flies, mosquitoes, beetles, tobacco moths, sciarid flies, fleas (eggs and larvae), fire ants, pharaoh ants, midge flies and Indian meal moths. Methoprene is not found in feed-through products for horses, although there are products for cattle for the control of horn flies. It can also be found in mineral blocks for use with horses, although uptake is always variable with these free-choice products.

10.4.3.5 Pyriproxyfen

Pyriproxyfen is another juvenile hormone mimic that disrupts the growth of arthropods. This chemical affects immature arthropods and eggs. Pyriproxyfen affects many types of arthropods, including fleas, cockroaches, ticks, ants, carpet beetles, and mosquitoes. Pyriproxyfen is a very common active ingredient and is found in more than 300 registered pesticide products, although not in feed-through products for horses.

10.4.4 Common active ingredients in perimeter and landscape treatments

Perimeter and landscape treatments are chemical treatments of environmental spaces and can include sprays, dusts, or granular applications. These are used for pests of horses that are developing outside like mosquitoes and ticks. Synthetic pyrethroids (Section 10.4.1.5) are by far the most common active ingredient in materials that are applied as perimeter or landscape treatments.

10.4.4.1 Carbaryl

Carbaryl is a carbamate pesticide that is commonly used on outdoor pests. This active ingredient can be found in many formulations including sprays, dusts, and granules. When an arthropod contacts carbaryl it over-stimulates the nervous system. This results in the inability of the nerves to rest and thus contract the breathing muscles resulting in arthropod death. Many of these products can be used on crops, home gardens, lawns, and other ornamental plants. Others are used around the outside of homes and on anthills. Carbaryl has been banned in the European Union and its use is restricted in the USA, Australia, and New Zealand.

10.4.4.2 Hydramethylnon

Hydramethylnon kills insects by disrupting energy production in their cells. This active ingredient must be consumed to be effective. Hydramethylnon is an active ingredient used to control fire ants.

10.4.4.3 Indoxacarb

Indoxacarb is active against a wide spectrum of insect pests from at least ten orders and well over 30 families. Indoxacarb is designated by the US Environmental Protection Agency to be a reduced-risk pesticide and is considered an organophosphate replacement. This active ingredient over-stimulates the nervous system resulting in arthropod death. Indoxacarb is a fast-acting chemical for fire ant control and some products are labelled for use in horse pastures.

10.4.4.4 Pyrethrins and synthetic pyrethroids

See previous sections for general information about pyrethrins and synthetic pyrethroids. In addition to permethrin, synthetic pyrethroids that are used as perimeter and landscape treatments may include cyfluthrin, bifenthrin, lambda-cyhalothrin, deltamethrin, and lambda-cyhalothrin, which are commonly used against many pests. This group of chemicals is effective against ticks.

10.4.4.5 Natural extracts and oils

Many products containing plant extracts or oils are produced as alternatives to synthetic chemical applications. Often these target mosquitoes and ticks. There is little evidence that these products are effective and in many cases they are harmful to beneficial arthropods. Many of these mixtures break down quickly in ultraviolet light or wash away with rain or dew limiting their effectiveness. Better formulations of products may increase the ability of these more natural ingredients to work for pest control.

10.4.5 Common active ingredients used to control internal parasites

10.4.5.1 Macrocyclic lactones

Macrocyclic lactones (avermectins and milbemycins) are used as drugs to treat parasitic worms and as pesticides to control insect pests. They are naturally occurring compounds from the fermentation of bacteria (*Streptomyces* spp.). Examples include ivermectin, moxidectin, abamectin, doramectin, eprinomectin, and selamectin. Oral formulations should be used on horses to avoid the potential for severe, potentially fatal, abscess formation at the site of injection. Additionally, the use of injectable anthelmintics is strongly associated with rapidly increased resistance. They work by stimulating nerve receptors, which inhibits nerve conduction, resulting in paralysis and death in insects and worms. While considered to have a wide margin of safety, care must be taken to prevent overdoses due to miscalculations or accidental ingestion. Some animals are also naturally more sensitive to avermectin treatment. For example, macrocyclic lactones should not be used in any equid with low body fat percentages (foals and starvation cases typically) as well as juvenile horses due to an underdeveloped blood-brain barrier.

10.4.5.2 Tetrahydropyrimidines

The first tetrahydropyrimidines were developed in the 1960s. Their mode of action is to block acetylcholinesterase at the neuromuscular junction of helminths, resulting in uncontrolled muscle contraction and death. Pyrantel is the primary drug in this class used in horses as an oral anthelmintic. These drugs have greater activity for nematodes than tapeworms, against the latter higher doses are required and control is limited. Resistance to these drugs is widespread in equine parasites, especially with small strongyles and tapeworms. See Matthew (2014) and Gokbulut and McKellar (2018) in 'Continuing education' for excellent reviews on the status of anthelmintics in horses.

10.4.5.3 Benzimidazoles

Benzimidazoles were discovered in 1961 and have the mode of action of binding to nematode microtubules in a manner that deprives them of energy. They also have antifungal properties, but relatively low activity against tapeworms. This is a large group of drugs but fenbendazole and oxibendazole are the ones that are most commonly used in horses. Resistance to these materials is widespread, especially among the small strongyles.

10.4.5.4 Isoquinoline-pyrozines

Drugs in this class act by disrupting the flow of calcium within the cells of tapeworms and other flatworms, resulting in uncontrolled muscle contraction, paralysis and death. They are specific to flatworms and are not effective against nematodes. Praziquantel, discovered in the 1970s, is widely available for tapeworm control in horses.

10.4.5.5 Imidazothiazoles

This class of anthelmintics act by binding competitively to nicotinic acetylcholine receptors on the body wall of worms, causing uncontrolled spasms and expulsion from the host. First discovered in the mid-1960s, levamisole is the only drug in this class that is used at present. Its use in equines as an anthelmintic is uncommon due to extreme resistance. It has been used in other species in combination with other anthelmintics to manage very resistant nematode populations.

10.4.6 Active ingredients/active substances of concern

There are some active ingredients that should not be used on horses unless directed by a veterinarian because of their toxic effects. In some cases, products are labelled for horses, but additional evaluations have documented negative impacts. This list is not exhaustive and may change as more research is conducted.

10.4.6.1 Active ingredients labelled for other animals/humans

As previously mentioned, products labelled for other animals including humans should not be used on horses unless the label specifies horses as well. In some cases, these products may be toxic to horses or produce other reactions that are veterinary concerns. Even products that contain the same active ingredients but are labelled for other animals like cattle, goats, and sheep and not for horses should not be used. Some products are labelled for other animals as well as horses. In these cases, it is important to read the label to determine the appropriate concentration and amount to apply as it may differ by animal.

10.4.6.2 Concentrated active ingredients

Many topical repellents or topical pesticides come in ready-to-use and concentrated products. Concentrated products have higher amounts of active ingredients and should be diluted according to label directions before applying. Some arthropods have a threshold for pesticides and higher concentrations of the product may not be effective. In addition, using concentrations higher than recommended can increase the risk for skin reactions or toxic effects that may require veterinary care.

Amitraz

Amitraz is an acaricide used in the control of ticks, mites, and lice. Amitraz has rapid action and it persists on hair and wool long enough to control all stages of the parasites. However, amitraz is toxic to horses causing potentially life-threatening complications and should not be applied unless directed by a veterinarian (Auer *et al.*, 1984).

10.4.6.3 Tetrachlorvinphos

The organophosphorus pesticide tetrachlorvinphos (also called stirofos and Rabon®) is used as a feed-through larvicide to prevent the completion of the fly larval life cycle in horse manure. It is often incorporated into feed or mineral blocks. While tetrachlorvinphos is labelled for use in horses, research has demonstrated marked behavioural changes in horses after the use of this active ingredient. Horses become increasingly more reactive and less controllable after ingesting tetrachlorvinphos (Berger *et al.*, 2008).

10.5 Best practices for using chemical control

The proper use and application of a product are as important as selecting the right active ingredient or pesticide. Along with reading and understanding the product label, there are some best practices for using chemical control methods that can maximise their effectiveness and minimise risk.

10.5.1 How to apply topical repellents

Repellents are chemicals applied to exposed skin or clothing that can provide some relief and protection from biting pests. These products are not intended to kill the pest, but rather interfere with the ability to locate the animal including humans to bite or discourage the arthropod from biting.

Some horses are more or less sensitive to repellent or topical product application. It is important to be prepared for and use safe horse handling procedures before applying any product. Read the label before application as well and understand the precautions for both the horse and yourself. Horses should be groomed with excess dirt removed before applying repellents.

Do not use products that are not labelled for horses or that are concentrated. Other livestock, like cattle, can tolerate much higher percentages of active ingredients, which can be damaging to equine skin. Synthetic pyrethroids, in particular, used at too high of a dose can cause skin reactions that can be severe (Figure 10.7). In some cases, other active ingredients labelled for other livestock can be toxic to horses. Similarly, products labelled for horses should never be applied to humans or other animals unless listed on the label.

Repellents should not be mixed for application or combined on a horse (i.e. applying a cream repellent and a spray on top of the cream). This can increase the toxicity of the active ingredients. Refer to the label instructions for each product for the appropriate amounts to apply and the frequency of application.

Recent research in the USA has demonstrated that many spray repellents with synthetic pyrethroid active ingredients marketed for horses have a limited duration of effectiveness for house flies

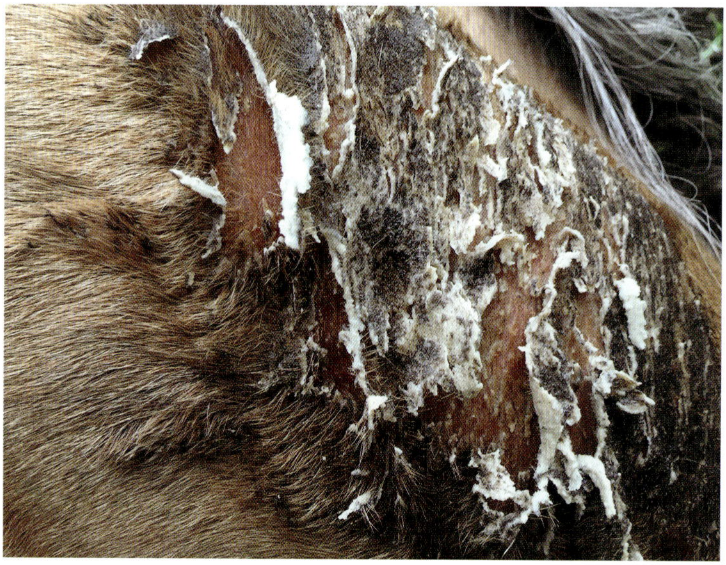

Figure 10.7. Using undiluted synthetic pyrethroids can cause extreme skin reactions, as seen here. Be sure that concentrated products are diluted according to label directions before use.

(Tuorinsky and Machtinger, 2020). In some cases, 4 hours or less. This limited efficacy is likely due to environmental factors like sun exposure, rain, dirt, or sweat, or chemical resistance. While flies in different regions may experience variable levels of chemical resistance, it is important to consider that repellents may only work for a short duration and repellents may be most effective against flies when used only for limited activities (e.g. farrier or vet visits or exercise), and not for long-term pasture protection. Other pests like ticks and keds/louse flies may be less tolerant of repellents but the duration of effectiveness has not been tested on other pests groups. Recent research has suggested that more natural products containing the fatty acids octanoic, nonanoic, and decanoic acid, or those with mixtures of neem or geraniol are as effective as synthetic compounds in repelling house flies and have a longer duration of effectiveness in the laboratory (Tuorinsky and Machtinger, 2020).

It is important to understand that repellent products that are registered under a regulatory agency have been tested for effectiveness against the pests listed on the label and safety to the applicators and the animals to which the repellent will be applied are listed on the label. Homemade mixtures of essential oils or other products have not gone through this testing. These products may not only be ineffective, but they may also be dangerous to horses, people, or other animals. See Strid *et al.* 2018 in 'Continuing education' for more information on how to use repellents safely.

10.5.1.1 Applying spray repellents

The most common method of repellent application is with ready-to-use products or mixtures made from concentrate in spray bottles. Often this involves a few spritzes on the animal before putting it in the paddock or getting on to ride. However, this is not the ideal method of repellent application. Spraying often releases the product into the air with only a small amount getting on the top of the hairs on the horse and increasing the risk of inhalation. In addition, many horses are sensitive to spray noises and the feel of the spray so it can be challenging to get full coverage.

Repellent sprays are most effectively applied to the body by first applying them to a towel, a microfiber application mitt (e.g. self-tanner application mitt or similar), or a brush, and then wiped on the horse (Figure 10.8). This sets the product into the hair and greater coverage is achieved. Full coverage is ideal, and for most products, the hair should be damp after application. Greater amounts of product do not mean that greater protection will be achieved. If using a towel or mitt where there is potential for human skin contact, disposable gloves should be considered when applying repellents.

Remember to apply products to the head and the tail. The tail can distribute the product as the horse swishes. Use caution when applying the product to the head, especially around the eyes. Apply repellent about a finger-width away from the eyes, and cover the jaw, front, and sides of the head as well as the ears.

10.5.1.2 Applying topical pesticides

For other application methods like shampoos, spot-ons, gels, ointments, pre-moistened wipes, liquids, roll-ons, and foams, the label directions will provide suitable instructions for application. These products are particularly useful for protecting the face, around wounds, or other areas where more control over the application method is desired.

10.5.2 How to use space sprays or residual pesticides

Choose the least toxic pesticide that is suitable for the pest. After you have carefully read the product label, be sure to use the recommended personal protective equipment for applying it safely.

Figure 10.8. Using an applicator mitt or towel to apply pesticides instead of a spray bottle can improve coverage of the product, reduce waste, and avoid inhalation of the product.

Usually, this includes gloves, eye protection, a long-sleeved shirt, long pants, and closed shoes. The application equipment may vary according to the application site and the pesticide. After applying pesticides it is a good idea to shower as soon as possible. See Kaufman and Weeks (2013) in 'Continuing education' for more information on how to use pesticides safely around animals.

10.5.2.1 Space sprays

Space sprays, including automatic spray systems and mist foggers, provide a quick knockdown of adult flies in an enclosed air space. The chemical compounds typically used in space sprays have very little residual activity and are designed to provide a rapid relief from active adult flies. These are most effective against house flies, which are the most common indoor pest. Space sprays will not be effective on arthropod pests living on horses (e.g. mites, lice, ticks, etc.) as they are protected by the hair of the horse.

The most common space spray in equine facilities are automatic fly spray systems (Figure 10.9), but other space sprays can be applied using various spray equipment that delivers fine mists, aerosols, or fogs. Motorised backpack sprayers can deliver a fine mist. To avoid respiratory aggravation to humans and other animals, space sprays (including automatic spray systems) should not be used when people or horses are in the stables unless otherwise directed by the product label. Fans should be turned off and space sprays should not be applied when it is windy to minimise droplets moving outside the planned treatment areas. As with other pesticides, all label directions for timing, application rate, and safety precautions should be read and understood prior to application.

Figure 10.9. Automatic spray systems can deliver space sprays with precise timing and amounts, but they are designed to provide a rapid relief from active adult flies and have little residual activity.

10.5.2.2 Residual pesticide sprays

Residual pesticide sprays are applied to fly resting surfaces like walls, fence lines, support posts, and other areas where flies congregate. These sprays can be applied with hand sprayers (Figure 10.10) or power equipment. Generally, sprays should be applied to a clean surface according to label directions. Remove or cover water and feed buckets before application. Animals should be removed from the building according to the label directions.

Some pesticides require diluting a concentrate with water before application. In these cases, measuring tools specific to those pesticides should be used. These should be clearly labelled as pesticide only and stored well away from food preparation areas. Pesticide labels will detail dilution and application rates. Application rate can be modified by the application equipment and simply walking faster or slower.

Spray application should target areas where flies have been seen to rest, or where fly spots have been observed. To reduce the amount of pesticide applied, these areas can receive spot treatments instead of treating the entire area (e.g. an entire wall, fence line, or stable). These sprays may last up to a few weeks, but their effectiveness is based on resistance in the flies, the extent to which they become covered with dust or dandruff, and exposure to ultraviolet light, rain, and wind.

Historically, residual pesticide spray application to stable or barn surfaces has been one of the more popular fly control strategies. As a result, high levels of pesticide resistance to active ingredients in these products are now common. Residual pesticide application should be used sparingly and only as a last resort to control fly outbreaks that cannot be managed with other control tools. If residual pesticide spray use does not reduce the fly population after application, either the pest species has been misidentified or the fly population is resistant to the active ingredient being used and they should not be reapplied.

Figure 10.10. Sprayers that are kept pressurised by manual pumping are useful for delivering a relatively coarse spray for residual premise treatments.

10.5.3 How to use perimeter and landscape treatments

Landscape pesticides are similar to space sprays or premise sprays except they are applied in the environment. For equine pests, these are primarily targeting mosquitoes and other biting flies and ticks. Like residual treatments and space sprays, it is important to use the least toxic pesticide suitable for the pest, read the label carefully, and use personal protection. Like residual pesticides, having good mixing and storage practices is necessary.

Environmental contamination should be minimised as much as possible. Spot treatments, or treatment of areas that tend to have more pests, will limit exposure risk for the surrounding area. The applied pesticide should not be able to run off or be blown into drains or water bodies. Do not apply pesticides on rainy or windy days. Pesticides should also not be applied to impervious surfaces like driveways.

In some regions, it is possible to hire a pest control professional to apply perimeter or landscape treatments for pests like mosquitoes and ticks. This may help save time and make applications as controlled as possible. Licensed pest control professionals often have access to products that are not available to the homeowner and the equipment and training necessary to apply them.

10.6 Guidelines for using baits and traps for pest flies

Traps and baits have been developed for several pests that attract and/or capture insects by understanding their behaviour and using it against them. Often, traps and baits are used with little consideration of where to place them in the environment or how to use them for maximum effectiveness. The guidelines and tips below will help increase the effectiveness of traps and will focus on traps and baits for house flies, stable flies, horse and deer flies, and mosquitoes, biting midges, and black flies.

10.6.1 House fly traps and baits

House flies are one of the most common pests around horse farms. Sticky tapes and ribbons can be helpful, as can traps that use ultraviolet light to lure flies to glue boards or electrocuting grids. Perhaps the most widely used fly control method on equine facilities is the many varieties of traps and baits. These come in two broad categories: (1) traps that attract and collect flies into a liquid reservoir without using toxicants (jug or jar traps); and (2) baits that contain sugar, a toxicant, and attractant(s).

10.6.1.1 Jug and jar traps for house flies

Jug and jar traps work on the principle of using a liquid reservoir containing attractants to lure flies into a trap from which they cannot escape. The flies eventually die or fall into the liquid. There are many products in this category. After adding water to the traps, some baits are fully operational right away while others require a period of fermentation. One of the limitations of these traps is the number of flies that they catch before becoming saturated. Small bag-style traps can collect only a few thousand flies, whereas larger-capacity models can collect 15,000 to 20,000. A few tips for using these kinds of traps for best effect are:

- Replace the trap or refresh the liquid at least once per week. The liquid baits are only attractive for a limited time.
- It is better to place a larger number of small traps than a smaller number of large ones.
- Trap performance is highly site-specific, so placement is critical. Move the traps around to find out which locations are the most productive at catching flies.
- Place the traps low to the ground. House flies usually fly less than 0.5 m (1.5 ft) off the ground and will find a low trap more easily than one near the ceiling.

You can also buy traps that do not include an attractant (Figure 10.11). If making bait, do not use meat as the bait even if it is recommended by the manufacturer. Meat will only attract and collect blow flies and flesh flies, which are not pests and can be beneficial (Chapter 8). For these traps, an effective bait can be made by preparing a 25% solution of blackstrap molasses in water in a cup or dish.

10.6.1.2 Sugar baits for house flies

These baits are usually formulated containing sugar, a pesticide, and an attractant. All three components may be applied as a layer to a plastic strip that is hung upright (a 'bait strip'). Most baits, however, are formulated as granules. The pesticide may be methomyl, any of several neonicotinoids, or cyantraniliprole. Below are some specific tips on using baits for house flies:

- The temptation is strong to stick with a product that is working. The best way to keep it working is to rotate with other products with different modes of action to slow the development of resistance.

Figure 10.11. Example of a trap that is sold without an attractant. Do not use meat or other foods that will decompose, as these odours will attract blow flies instead of house flies.

- Contain the granules. Place the granules in a bait station (see Chapter 3) in jugs, pans, plates, or boxes and place where flies can be seen aggregating. Ensure that children, pets, and other animals cannot access the bait.
- Check the bait frequently and replace it when needed. Observe the bait stations. Under hot conditions, the bait may liquefy and need replacing. Even if the granules remain solid, old bait in bait stations should be replaced with fresh material because products can lose attractiveness and potency over time.
- Retire bait strips when their colour changes. Bait strips usually include a coloured pigment, typically yellow, in the material placed on the strip. As the flies consume the bait on the strip the colour changes to that of the underlying colour (often white). Replace the strips when they no longer have their original colour.
- Keep baits and bait strips out of the sun. Sunlight, especially ultraviolet light, accelerates the breakdown of the pesticides used in baits, especially neonicotinoids.

10.6.2 Stable fly and horse and deer fly traps

Stable flies and horse and deer flies are not attracted to odours like house flies. Instead, these flies use vision to track down horses and other animal hosts. Several traps have been developed that target stable flies (provided as resources in Chapter 3) and horse and deer flies (provided as resources in Chapter 4) specifically. Here are a few tips for placement and use of these traps:

- Placement of stable, horse and deer fly traps is critical. Traps are most effective for stable flies where stable fly activity is observed. For horse and deer flies, this may take trial and error but traps are also likely most effective around areas horses are being attacked.
- Traps should be placed in sunny areas.
- In many cases, traps can be set up outside of fence lines. However, if traps are placed in horse pastures, they will need to be protected by solid or electric fencing to prevent access to them (see Chapter 3, Figure 3.26 for an example).
- Stable fly traps should be placed no higher than 1 m (3.3 ft) from the ground.
- Several traps are usually needed, depending on the size of the area where the flies are a problem. Trapping may be most effective when protecting a few horses in a small pasture.
- Stable fly traps typically have a sticky substance that also will capture other non-target arthropods (e.g. butterflies, grasshoppers, and others), or even lizards, birds, and bats attempting to eat the trapped insects. If repeated captures of vertebrates occur, traps should be immediately moved or removed from the site.
- Stable fly traps are less effective as more arthropods become trapped on them. They should be checked frequently and replaced when ¾ of the available area is filled. Traps also lose effectiveness when covered with dust, pollen, or grass clippings from mowers. Traps should be replaced when they are no longer sticky, even if few insects have been trapped.
- Epps Biting Fly Trap® for horse and deer flies uses soapy water to drown the flies. These may dry out quickly in warmer weather so should be routinely checked and liquid replaced.
- The Nzi, Horse Pal® Trap, and BiteLite® H-Trap Professional Horse Fly Control System traps for horse and deer flies collect flies in jars. Routinely inspect these jars for captures because they will stop collecting flies once they are full (Figure 10.12). Jars should be emptied once they are about ¾ full.

Figure 10.12. Some horse and deer fly traps collect the flies in jars and bottles, which stop working once they are full (left trap).

10.6.3 Mosquito, biting midge, and black fly traps

Mosquitoes, biting midges, and black flies use vision, heat, carbon dioxide, and other odours when locating a host. However, the odour is detected from further away than heat and so is a better attractant to use in traps for these flying, biting insects. Control of these three pest groups by trapping is not well established. Results can depend on the pest species, other control methods implemented, and the ability of insects to move into the area from their larval development sites. However, it has been demonstrated to reduce pest abundance and relieve animals of biting pressure. It is important to know the pest that is being targeted to know which type of trap to use and when the traps need to be used effectively. Black flies are active in the day, biting midges at dawn and dusk, and there are various species of mosquitoes that might be active at different times of the day. While CDC miniature light traps (with carbon dioxide as a lure) and Mosquito Magnets® are fairly general in which species they capture, the BG-sentinel trap is particularly effective at trapping day-biting *Aedes* species. Autocidal gravid oviposition traps or lethal oviposition traps catch mosquitoes that lay eggs in containers, such as some *Aedes* species. The female mosquitoes are trapped as are any mosquitoes that hatch from the eggs that are laid in the container. Resources for these traps are provided in Chapter 4. Here are some suggestions for placing traps for control purposes:

- Place traps away from animals so as not to attract more insects into the area surrounding the animals.
- Placing traps between fly development sites (if known) and the animals can be a good way to intercept host-seeking insects and prevent biting.
- If the height can be altered, place the entrance of the trap approximately 1.5 m (5 ft) from the ground.
- If targeting biting midges or black flies, use a trap catch bag with a fine mesh. Traps are often supplied with catch bags for mosquitoes that smaller flies like biting midges and black flies will pass through.

- If the trap has a light, remove it or disable it. Lights are attractive to a wide range of nocturnal insects so if you use a trap with a light you will catch many moths and other night-active insects. Removing the light and just using odours and/or carbon dioxide will make the trap selective to blood-feeding insects.
- Using carbon dioxide will increase the trap catch. Carbon dioxide is in animal breath and is used by most blood-feeding insects to locate their hosts. Carbon dioxide can be provided as dry ice, carbon dioxide gas canisters, burning butane, or fermentation.
- Using odours instead of or in addition to carbon dioxide will increase the trap catch. Several are available that are based on animal odour including lactic acid, octanol, and other proprietary blends of animal odours.
- Some traps need batteries and replacement of lures regularly such as the CDC miniature light trap and the BG-sentinel. Other traps, such as the Mosquito Magnet®, can run for up to 3 weeks.
- Autocidal gravid oviposition traps or other lethal oviposition traps should be placed in areas where mosquitoes like to lay eggs. Usually, these are cool, shaded spots that are out of the wind. This may be different for some species.
- Combine trapping with other control methods to have the best chance of success. Although mass-trapping might reduce pest abundance in an area where traps are used, it will not affect the source populations of the insects. Pest numbers will rebuild as soon as trapping stops.

Continuing education

A review on the treatment and control of ectoparasite infestations in equids

Karasek, I., Butler, C., Baynes, R. and Werners, A., 2020. Journal of Veterinary Pharmacology and Therapeutics 43: 421-428. https://doi.org/10.1111/jvp.12874

- Ectoparasites infestations are a growing concern to horse owners and equine veterinarians alike. In this review, the currently available evidence for the prevention and treatment of ectoparasite infestations in horses is described.

How effective are fly protectants for horses?

Mottet, R., Martinson, K., and Moon, R., 2018.
https://extension.umn.edu/horse-care-and-management/protecting-your-horse-flies#results%3A-which-fly-protectants-work-best%3F-1542561

- In this summary, the results of a study evaluating repellents and protection for horses is described. Of the tested products, a combination of fly boots, leg bands, and citronella spray was the most effective at reducing fly avoidance behaviours, although no one treatment could reduce all behaviours in horses.

Anthelmintic drugs used in equine species

Gokbulut, C. and McKellar, Q.A., 2018. Veterinary Parasitology 261: 27-52. http://doi.org/10.1016/j.vetpar.2018.08.002

- A summary of the use of historic and current anthelmintic drug use in equids and resulting resistance of parasites to many modern active ingredients.

Anthelmintic resistance in equine nematodes

Matthews, J.B., 2014. International Journal for Parasitology: Drugs and Drug Resistance 4: 310-315. https://doi.org/10.1016/j.ijpddr.2014.10.003

- A review of the current active ingredients with documented parasite resistance, grazing management, and progress towards developing novel diagnostics for parasites in horses.

Insect repellents fact sheet

Strid, A., Hanson, W., Cross, A., Bond, C. and Jenkins, J., 2018.
National Pesticide Information Center, Oregon State University Extension Services. http://npic.orst.edu/factsheets/repellents.html.

- Insect repellent fact sheets produced by the National Pesticide Information Center to educate about what repellents are and how to use them safely and effectively.

Pesticide safety around animals

Kaufman, P.E. and Weeks, E.N.I., 2013. Publication #ENY-272/IG128, 10/2012. EDIS 2013(1). https://doi.org/10.32473/edis-ig128-2012

- Pesticides are commonly used to protect animals from pests. However, it is important to remember that any pesticide should be considered an active poison. The successful control of pests requires careful mixing and application of recommended pesticides according to label directions. This guide provides tips and resources for safer and effective use of pesticides around animals.

Resources

Resources are listed in alphabetical order and may not be inclusive of all companies, materials, or products available.

Equitrol II®. https://www.farnam.com/all-products/fly-insect-control/equitrol-ii
Insecticide Resistance Action Committee mode of action table. https://irac-online.org/modes-of-action/
MultiFly®. https://horselogicpro.com/shop/multifly/
SimpliFly®. https://www.farnam.com/all-products/fly-insect-control/simplifly
Solitude IGR®. https://www.zoetisus.com/products/horses/solitude-igr.aspx

References

Auer, D.E., Seawright, A.A., Pollitt, C.C. and Williams, G., 1984. Illness in horses following spraying with amitraz. Australian Veterinary Journal 61: 257-259.

Berger, J., Valdez, S., Puschner, B., Leutenegger, C.M., Gardner, I.A. and Madigan, J.E., 2008. Effects of oral tetrachlorvinphos fly control (Equitrol®) administration in horses: physiological and behavioural findings. Veterinary Research Communications 32: 75-92.

Bethke, J.A., Vander May, B. and Bates, L.M., 2013. Eye gnats. UC IPM Pest Notes no. 74164. Available at: http://ipm.ucanr.edu/PDF/PESTNOTES/pneyegnat.pdf.

Brewer, G.J., Boxler, D.J., Loftin, K.M., Domingues, L.D., Talley, J.L., Trout Fryxell, B.T., Holderman, C., Machtinger, E.T., Smythe, B. and Watson, W., 2021. Horn fly, *Haematobia irritans irritans* (Diptera:

Muscidae) – biology, management, and future research directions. Journal of Integrated Pest Management 12: 42. https://doi.org/10.1093/jipm/pmab019

Bučanović, T., Kovačević, J. and Krčmar, S., 2020. Efficiency of six different octenol-baited traps for collecting horseflies (Diptera: Tabanidae). Medical and Veterinary Entomology 34: 493-497.

Drago, A., Marini, F., Caputo, B., Coluzzi, M., della Torre, A. and Pombi, M., 2012. Looking for the gold standard: assessment of the effectiveness of four traps for monitoring mosquitoes in Italy. Journal of Vector Ecology 37: 117-123.

Drees, B.M., Calixto, A.A. and Nester, P.R., 2013. Integrated pest management concepts for red imported fire ants *Solenopsis invicta* (Hymenoptera: Formicidae). Insect Science 20: 429-438.

Geden, C.J., Nayduch, D., Scott, J.G., Burgess IV, E.R., Gerry, A.C., Kaufman, P.E., Thomson, J., Pickens, V. and Machtinger, E.T., 2021. House fly (*Musca domestica* [Diptera: Muscidae]) – biology, pest status, current management prospects, and research needs. Journal of Integrated Pest Management 12: 39. https://doi.org/10.1093/jipm/pmaa021

Gerry, A.C., 2012. Stable fly. Available at: https://www.veterinaryentomology.org/stable-fly.

Gerry, A.C., 2020. Monitoring house fly (Diptera: Muscidae) activity on animal facilities. Journal of Insect Science 20: 15. https://doi.org/10.1093/jisesa/ieaa109

Goffredo, M. and Meiswinkel, R., 2004. Entomological surveillance of bluetongue in Italy: methods of capture, catch analysis and identification of *Culicoides* biting midges. Veteranaria Italiana 40: 260-265.

Herholz, C., Kopp, C., Wenger, M., Mathis, A., Wägeli, S. and Roth, N., 2016. Efficacy of the repellent N, N-diethyl-3-methyl-benzamide (DEET) against tabanid flies on horses evaluated in a field test in Switzerland. Veterinary Parasitology 221: 64-67.

Hoel, D.F., Kline, D.L. and Allan, S.A., 2009. Evaluation of six mosquito traps for collection of *Aedes albopictus* and associated mosquito species in a suburban setting in north central Florida. Journal of the American Mosquito Control Association 25: 47-57.

Hoel, D.F., Kline, D.L., Hogsette, J.A., Bernier, U.R., El-Hossary, S.S., Hanafi, H.A., Watany, N., Fawaz, E.Y., Furman, B.D., Obenauer, P.J. and Szumlas, D.E., 2014. Efficacy of commercial mosquito traps in capturing phlebotomine sand flies (Diptera: Psychodidae) in Egypt. Journal of Medical Entomology 47: 1179-1184.

Hogsette, J.A. and Kline, D.L., 2017. The Knight Stick trap and Knight Stick sticky wraps: new tools for stable fly (Diptera: Muscidae) management. Journal of Economic Entomology 110: 1384-1389.

Kaufman, P.E. and Weeks, E.N.I., 2018. Horn fly management. University of Florida IFAS Publication #ENY288. Available at: https://edis.ifas.ufl.edu/publication/in952.

Kline, D.L., Hogsette, J.A. and Rutz, D.A., 2018. A comparison of the Nzi, Horse Pal® and Bite-Lite® H-traps and selected baits for the collection of adult Tabanidae in Florida and North Carolina. Journal of Vector Ecology 43: 63-70.

Lowden, S., Gray, S. and Dawson, K., 2007. Treatment of natural infestations of the biting louse (*Werneckiella equi*) on horses using triflumuron, a benzoylurea derivative insect growth regulator. Veterinary Parasitology 148: 295-300.

Lühken, R., Pfitzner, W.P., Börstler, J., Garms, R., Huber, K., Schork, N., Steinke, S., Kiel, E., Becker, N., Tannich, E. and Krüger, A., 2014. Field evaluation of four widely used mosquito traps in Central Europe. Parasites & Vectors 7: 268.

McDermott, E.G. and Lysyk, T.J., 2020. Sampling considerations for adult and immature *Culicoides* (Diptera: Ceratopogonidae). Journal of Insect Science 20: 17. https://doi.org/10.1093/jisesa/ieaa094

Meyer, J.A., Georghiou, G.P., Bradley, F.A. and Tran, H., 1990. Filth fly resistance to pyrethrins associated with automated spray equipment in poultry houses. Poultry Science 69: 736-740.

Murillo, A.C., Hubbard, C.B., Gerry, A.C. and Hinkle, N.C., 2021. Big problems with the little house fly, *Fannia canicularis* (Diptera: Fanniidae). Journal of Integrated Pest Management 12: 40. https://doi.org/10.1093/jipm/pmaa023

Nápravníková, J., Petrtýl, M., Stupka, R. and Vadlejch, J., 2019. Reliability of three common fecal egg counting techniques for detecting strongylid and ascarid infections in horses. Veterinary Parasitology 272: 53-57.

Otabil, K.B., Gyasi, S.F., Awuah, E., Obeng-Ofori, D., Tenkorang, S.B., Kessie, J.A. and Schallig, H.D., 2020. Biting rates and relative abundance of *Simulium* flies under different climatic conditions in an onchocerciasis endemic community in Ghana. Parasites & Vectors 13: 229.

Rochon, K., Hogsette, J.A., Kaufman, P.E., Olafson, P.U., Swiger, S.L. and Taylor, D.B., 2021. Stable fly (*Stomoxys calcitrans* [Diptera: Muscidae]) – biology, management, and research needs. Journal of Integrated Pest Management 12: 38. https://doi.org/10.1093/jipm/pmab029

Rodríguez-Rojas, J.J., Arque-Chunga, W., Fernández-Salas, I. and Rebollar-Téllez, E.A., 2016. Comparative field evaluation of different traps for collecting adult phlebotomine sand flies (Diptera: Psychodidae) in an endemic area of cutaneous leishmaniasis in Quintana Roo, Mexico. Journal of the American Mosquito Control Association 32: 103-116.

Salomon, J., Hamer, S.A. and Swei, A., 2020. A beginner's guide to collecting questing hard ticks (Acari: Ixodidae): a standardized tick dragging protocol. Journal of Insect Science 20: 11. https://doi.org/10.1093/jisesa/ieaa073

Shipp, J.L., 1985. Comparison of silhouette, sticky, and suction traps with and without dry-ice bait for sampling black flies (Diptera: Simuliidae) in central Alberta. The Canadian Entomologist 117: 113-117.

Smythe, B.G., Urias, S., Wise, M.E., Scholljegerdes, E.J., Summers, A.F. and Bailey, D.W., 2017. Comparing visual and digital counting methods to estimate horn fly (Diptera: Muscidae) populations on cattle. Journal of Medical Entomology 54: 980-984.

Somwang, P., Khositharattanakool, P., Pathawong, N., Pongsiri, A., Davidson, S.A. and Ponlawat, A., 2021. Field evaluation of four commercial light traps, trap placement, and effect of carbon dioxide on phlebotomine sand fly collection in northern Thailand. Acta Tropica 220: 105953. https://doi.org/10.1016/j.actatropica.2021.105953

Taylor, D.B., Harrison, K. and Zhu, J.J., 2020. Methods for surveying stable fly populations. Journal of Insect Science 20: 17. https://doi.org/10.1093/jisesa/ieaa094

Tuorinsky, E.V. and Machtinger, E.T., 2020. Behavioral inhibition of the house fly (Diptera: Muscidae) when exposed to commercial equine fly repellents. Journal of Economic Entomology 113: 518-526.

Trout Fryxell, R.T., Moon, R., Boxler, D. and Watson, W., 2021. Face fly (*Musca autumnalis* [Diptera:Muscidae]) – biology, pest status, current management prospects, and research needs. Journal of Integrated Pest Management 12: 5. https://doi.org/10.1093/jipm/pmaa020

Van Hennekeler, K., Jones, R.E., Skerratt, L.F., Fitzpatrick, L.A., Reid, S.A. and Bellis, G.A., 2008. A comparison of trapping methods for Tabanidae (Diptera) in North Queensland, Australia. Medical and Veterinary Entomology 22: 26-31.

Photo credits

Title page Rob Kemp, Shutterstock.
Figure 10.1. Erika Machtinger, Pennsylvania State University.
Figure 10.2. Erika Machtinger, Pennsylvania State University.
Figure 10.3. (A) Pete Markham, Flickr; (B,C) Erika Machtinger, Pennsylvania State University.

Figure 10.4. Erika Machtinger, Pennsylvania State University.
Figure 10.5. Bob Storey, University of Georgia College of Veterinary Medicine.
Figure 10.6. Fotomuse, Shutterstock.
Figure 10.7. Karen Poh, Pennsylvania State University.
Figure 10.8. Erika Machtinger, Pennsylvania State University.
Figure 10.9. Heather Venter.
Figure 10.10. (A) Foto Cat, Shutterstock no. 619408388; (B) Eric Buermeyer, Shutterstock no. 245479492.
Figure 10.11. Erika Machtinger, Pennsylvania State University.
Figure 10.12. Dan Kline, USDA, ARS, Gainesville, FL.

Chapter 11. Summary

There are numerous pests affecting horses, donkeys, mules, zebras, and other equids all over the world. Some of these pests are common while others are less so, and pest prevalence may depend on region and equine husbandry practices. Nevertheless, many of these pests are associated with a decline in animal welfare. In many countries, horses are a major part of the economy so pest and parasite control should be a priority.

Pest management can be a challenge on and around horses. Unlike livestock production, equine husbandry practices can vary widely by country, region, or even discipline. Horse facilities may range from a single horse to hundreds. The result is that there is not a single formula for pest control on equine facilities. Each horse owner, rider, or property owner needs to evaluate the pest control tools that will work within their constraints. For most pests, elimination is not a realistic goal. Therefore, time, money, labour, and other factors may be important in making pest control decisions to keep pests at tolerable levels.

Every year horse owners spend millions of dollars, euros, or pounds sterling on pest control. In some areas, applications of repellents, fly sheets, boots, or masks, or other pest control options are routine throughout the year. Unfortunately, in many cases, these pest management tools are used incorrectly and indiscriminately not only not impacting the pest of interest, but potentially contributing to insecticide resistance and negative impacts to beneficial arthropods as well. Understanding and identifying the target pest is the first step in an integrated pest management

plan to control pests around horses. In many cases, control tools are species-specific, or at least arthropod group-specific. For example, although house flies and stable flies look almost identical, house fly traps will not help control stable flies. The tools provided in this book have given the reader some practical ways to use physical and behavioural characteristics to identify the pest or pests found on horses. Help with identifying pests can also come from university systems and government specialists, depending on the area.

Climate change and global travel will have an increasing influence on pest population distributions. Arthropods rely on suitable environmental conditions like temperature and humidity for development. As climates shift, it is likely that regional shifts of equine pests will occur as well. In many cases, like with ticks in North America, this is already occurring. The increase in equine travel can also impact pest distributions when pests hitchhike on horses travelling overseas, or even regionally. Exposure to pests is directly tied to the risk of diseases caused by pathogens they may transmit, or conditions they may cause. Therefore, it will be increasingly important to have up-to-date working knowledge of the local pest risk.

Successful management of pests and parasites on and around horses requires some planning and effort. Collaboration with equine veterinarians, especially when there are concerns regarding equine welfare, is encouraged for the best results. However, it is the authors' hope that this book has provided useful and practical guidance on pest management that will help keep horses healthier, people happier, and the environment safer.

Photo credits

Title page Bettina Calder, Shutterstock.

Appendix A. Glossary and word usage

Words with different American and British English spelling

British	American
Anaemia	Anemia
Autumn	Fall
Behaviour	Behavior
Centre	Center
Channelled	Channeled
Colour	Color
Diarrhoea	Diarrhea
Faeces	Feces
Favour	Favor
Fibre	Fiber
Grey	Gray
Harbour	Harbor
Haemolymph	Hemolymph
Kilometre	Kilometer
Labelling	Labeling
Labour	Labor
Levelling	Leveling
Litre	Liter
Metre	Meter
Minimising	Minimizing
Moult	Molt
Neighbouring	Neighboring
Odour	Odor
Oesophagous	Esophagus
Ploughing	Plowing
Toxaemia	Toxemia
Travelling	Traveling
Tumour	Tumor
Tyre	Tire
Vapour	Vapor

The terms below are defined based on their usage in this book

Abdomen (adjective abdominal)	The abdomen is the last of the three-body segments of insects, which contains the heart, reproductive and digestive organs.
Acaricide	An acaricide is a pesticide that targets acarids, such as mites and ticks.
Accidental pest	An accidental pest is a pest that accidentally causes injury to a host not in its normal life cycle.
Active ingredient	The active ingredient in a pesticide is the ingredient that is biologically active (USA, Canada).
Active substance	The active substance in a pesticide is the ingredient in a pesticide that is biologically active (Europe, Australia, New Zealand).
Adulticide	An adulticide is a pesticide that is used to target the adult life stage of a pest.
Aggregation	A number of arthropods forming a group is an aggregation, often for mating or resting purposes.
Anaphylactoid	An anaphylactoid reaction resembles anaphylaxis, it is a severe, potentially life-threatening allergic reaction.
Antennae	The sense organs of insects are called antennae, they usually originate in the head and are thin, moveable, and made up of segments.
Aposematic	An aposematic organism uses bright colours to advertise that it is toxic or bad tasting.
Aquatic	An arthropod that spends its life submerged in water is known as aquatic.
Arachnid	An arthropod in the class Arachnida is known as an arachnid, it includes spiders, mites, and ticks. Arachnids have eight legs as adults, a two-segmented body, and no antennae or wings.
Arbovirus	Arbovirus is the informal name for any virus that is transmitted by arthropod vectors.
Arid	An area (or climate) with little to no rain is known as arid, it is usually too dry to support most vegetation.
Arthropod	An invertebrate animal in the phylum Arthropoda is known as an arthropod, includes insects and arachnids. Arthropods have an exoskeleton, a segmented body, and paired jointed legs or appendages.
Autocidal	When used in reference to traps, autocidal denotes that the insect actively orients to and enters a device where it or its offspring will be killed.
Biodegradable	A biodegradable substance can be broken down by bacteria or other living organisms.

Biological control	The use of biological organisms, such as predators, parasites, and pathogens, to reduce the presence of a pest is known as biological control.
Caecocolic intussusception	Caecocolic intussusception is a rare cause of colic that results from uncoordinated peristalsis that allows the apex of the caecum (pouch at the beginning of the large intestine) to invaginate into the base. The condition is often associated with parasitic worm infestations.
Carrion	The decaying flesh of a dead animal is known as carrion.
Caste	The word caste is used to describe the individuals in a colony of social insects with different roles, such as queen, worker, drone, etc.
Cellular respiration	The conversion of oxygen and nutrients into energy followed by the excretion of waste products is cellular respiration.
Cement cone	The cement cone is the anchor produced when an adhesive substance is produced by the tick and injected into the bite site that holds the tick's mouthparts in the wound. It is this cement that makes it difficult to remove an attached tick.
Cerci	The cerci are a pair of appendages on the last abdominal segment, they may be sensory or used in copulation or defence, but most are functionless.
Chemical control	The use of chemicals, synthetic or natural, to reduce the presence of a pest is known as chemical control. For the purpose of this book, these chemicals may have a temporary repellent action or a permanent toxic effect on the pest.
Cocoon	A cocoon is a silky case that is spun by the larvae to protect the pupae.
Colony (plural colonies)	A group of individuals of the same species that live together in a group are a colony.
Compound eye	An eye consisting of many small visual units is a compound eye, in insects these units are called ommatidia.
Cultural control	The use of sanitation, hygiene, and other preventative measures to reduce the presence of a pest is known as cultural control.
Cuticle	The cuticle is the exoskeleton of insects, which is made up of many layers.
Cysticercoids	Cysticercoids are the larval stage of some tapeworms.
Desiccation	Desiccation is the loss of water. Arthropods are susceptible to desiccation so control techniques often attempt to promote water loss for control.

Detritus	Organic matter produced by the decomposition of organisms is known as detritus.
Detritivore	An organism that feeds on detritus is known as a detritivore.
Diurnal	A diurnal organism is active during daylight hours.
Dorsoventrally flattened	A dorsoventrally flattened organism is flattened from top to bottom such as a louse, not side to side such as a flea.
Drench	A liquid pesticide used to inundate an ant colony is known as a drench.
Drone	A male honeybee whose role is to mate with the queen is a drone.
Dung (synonym manure)	Dung is the excrement of animals.
Ectoparasite	A parasite that lives on the outside of its host, such as a tick, mite, louse, or flea, is an ectoparasite.
Ectotherm	An ectotherm is an animal whose regulation of body temperature depends on external sources, such as sunlight or air temperature increases.
Elytra	Modified hardened forewings of beetles and a few true bugs are known as elytra.
Endemic	An endemic organism is native (of indigenous origin) to a restricted area.
Endoparasite	An endoparasite lives on the inside of its host, such as parasitic worms.
Engorgement	Some arthropods engorge or swell in size when taking large liquid meals. Most female blood-feeding arthropods, including ticks and mosquitoes, engorge to obtain enough blood to produce eggs.
Entomopathogenic	Organisms that act as a pathogen and kill or severely injure insects (and sometimes other arthropods) are known as entomopathogenic. The term usually refers to microorganisms such as fungi, bacteria, and protozoa.
Facultative	When something occurs optionally in response to circumstances rather than by nature it is known as facultative.
Filth flies	Filth flies feed on and develop as immatures in manure and other decaying organic material.
Flatworm	Flatworms are a type of worm with a simple flattened body in the phylum Platyhelminthes.
Flight corridor	In entomological terms, an opening or break in a forested area (e.g. power lines, logging trails, etc.) that is used by flying insects is known as a flight corridor.
Fluke	Flukes are a group of parasitic flatworms in the class Trematoda.
Fogger	A fogger is a device that creates a fog of small pesticide particles that float in the air.

Genus (plural genera)	Genus is the taxonomic rank that comes above species. The scientific name of an organism is made up of the genus and species. For example, for the house fly, *Musca domestica*, the genus is *Musca*.
Gestation (adjective gestational)	The period of time between conception and birth is known as gestation.
Gill	A gill is a respiratory organ, found in immature aquatic insects. Gills extract dissolved oxygen from water and excrete carbon dioxide.
Gravid	When an arthropod is carrying a full complement of mature eggs that are ready to be laid it is referred to as gravid.
Grub	A grub is an insect larva. The name is typically reserved for those with a short fat body form, that is usually somewhat C-shaped.
Haemolymph	Haemolymph is the blood equivalent found in most invertebrates, it fills the haemocoel or the interconnected spaces within the body of an arthropod.
Herbivore	An organism that eats plants is known as a herbivore.
Hibernate	To hibernate is to spend the winter in a dormant state.
Hybrid	When individuals from two different species mate and produce viable offspring that offspring is known as a hybrid.
Hyperkeratosis	Hyperkeratosis is the thickening of the outer layer of skin.
Hypovolaemia	Hypovolaemia is the term used when there is insufficient amounts of fluid between the cells of an animal.
Ileocaecal junction	The ileocaecal junction is a sphincter muscle situated at the junction of small intestine and ceacum.
Immigration	When individuals move into an area it is known as immigration.
Indicator species	Any living organism that can be used to measure certain environmental conditions is known as an indicator species.
Insect	Invertebrates in the class Insecta are insects. Insects have a chitinous exoskeleton, a three-section body (head, thorax, and abdomen), three pairs of legs, compound eyes, and one pair of antennae.
Pesticide classes	The active ingredients within a pesticide class that have the same mode of action.
Insect growth regulators (IGRs)	Pesticides that mimic the hormones of immature insects are known as insect growth regulators or IGRs, they typically interfere with the moulting process. There are IGRs that work on other arthropods too, such as ticks.
Insectivorous	An organism that eats insects is known as insectivorous.

Integrated pest management (IPM) The integration of different pest management techniques for economic, sustainable, and effective control of a pest species is known as integrated pest management or IPM.

Larva (plural Larvae, adjective larval) A larva is an immature stage of an arthropod, different groups have other names such as maggots, grubs, bots, and caterpillars, etc.

Larvicides A pesticide used to target the immature larval stage is known as a larvicide.

Maggot A maggot is a type of insect larva, the name is typically reserved for dipteran larvae. Maggots are soft-bodied and legless.

Mandibles The mandibles are a pair of appendages near an arthropod mouth that form part of the mouthparts, usually used for chewing, grasping, crushing, or cutting, and sometimes used in defence.

Mange Mange is a skin disease caused by parasitic mites.

Manure (synonym dung) Manure is the excrement of animals, it is typically applied to the land as fertiliser.

Mating swarms When male insects form an aggregation in flight for mating purposes it is referred to as a mating swarm.

Mechanical control The use of mechanical tools, such as traps and barriers, to reduce pest presence and pest exposure is known as mechanical control.

Mechanical vector An animal that carries a pathogen from one host to another without becoming infected itself is a mechanical vector. In arthropods, the pathogen is usually on the mouthparts.

Metabolic inhibitor A chemical that blocks an enzyme to slow or prevent a chemical process is called a metabolic inhibitor.

Metamorphosis Metamorphosis is the development of an organism after birth or hatching from an immature to an adult form, which may be complete or incomplete in arthropods. In incomplete metamorphosis, there is a pupal stage and an abrupt change whereas in incomplete metamorphosis there is no pupal stage and the immatures look like smaller less developed versions of the adults.

Microbes (synonym microorganism) A microbe or microorganism is a microscopic organism that can be viewed only under a microscope such as bacteria, protozoa, viruses, and fungi.

Monitoring Scouting or using traps to detect the presence and abundance of pest populations is referred to as monitoring.

Moulting (noun moult)	The process of shedding the cuticle (called a moult) to make way for new growth is known as moulting.
Mouth hooks	The hook-like mouthparts of some immature insects are known as mouth hooks.
Mucosal areas	Mucosal areas are areas with thin skin covering the inside areas of some parts of the body such as the nose and mouth. Mucus is produced to protect these areas from damage.
Myiasis	The parasitic infestation of the body of a live animal by fly larvae or maggots is known as myiasis.
Naiad	The aquatic larva or nymph of a dragonfly, mayfly, or stonefly, is called a naiad.
Necrosis	The death of most or all of the cells in an organ or tissue is a process known as necrosis.
Necrotic	Dead or dying animal or plant tissue is referred to as necrotic.
Nematode	A worm of the phylum Nematoda is a nematode, including the roundworms and threadworms. Nematodes are small, slender worms with digestive systems that are open at both ends.
Nest parasite	A nest ectoparasite lives in the nest or resting area of its host.
Nocturnal	A nocturnal organism is active during the night or when it is dark.
Non-target organism	A non-target organism is not the target of the control measure. The term is usually used when discussing the negative effects of pesticides on fauna other than the pest species.
Nymph (adjective nymphal)	A nymph is an immature stage of an arthropod with a life cycle with incomplete metamorphosis.
Obligate parasites	Obligate parasites rely on their hosts for the completion of their life cycle.
Overwinter	To overwinter is a strategy that an arthropod species uses to survive the winter. The organism may delay development and stay as eggs, larvae, pupae, or adults, and it might involve hibernation.
Oviposition	The laying of eggs is known as oviposition.
Ovipositor	An ovipositor is a modified appendage, a tube-like organ used for laying eggs.
Parasite	A parasite lives on or in another in a relationship that results in some degree of harm to the other organism.
Parasitic nematodes	A parasitic nematode has a stage in its life cycle that is parasitic (see parasite).

Parasitoid	A parasitoid is an insect whose larvae live as parasites (see parasite) and will eventually kill their host.
Pathogen	A pathogen is a microbe or microorganism that can cause diseases in its host, such as a bacteria, virus, protozoa, or fungi.
Pericarditis	Inflammation of the tissues surrounding the heart is known as pericarditis.
Pheromone	A chemical that is produced and released by one individual that affects the behaviour or physiology of another individual of the same species is called a pheromone. Types of pheromone include aggregation, alarm, and sex pheromones.
Phoresy (adjective phoretic)	When one organism travels on the body of another without being a parasite, this behaviour is known as phoresy. The individual that practices the behaviour is called phoretic.
Physical control	Modification of physical aspects of the environment to reduce pest presence as well as physical removal of organisms from the environment is known as physical control.
Physiology	The way that an organism or specific body part functions is its physiology.
Pour-on	Pour-on is a way of applying pesticides directly to animals. Pour-on products usually come with an applicator so it is easy to apply a set dose to an animal to the target location, usually along the backline.
Predator (adjective predaceous)	A predator kills and eats other animals and is by nature predaceous.
Proboscis	A proboscis is a tube-like feeding apparatus in some insects such as mosquitoes, used for sucking up fluids for feeding.
Prolegs	The fleshy structures found on the abdomen of some larvae are prolegs, particularly prevalent on lepidopteran (butterflies and moths) caterpillars. They help support the body of the caterpillar and enable it to move.
Pupa (plural pupae, adjective pupal)	A pupa is an immature stage of an arthropod that has complete metamorphosis. It is the stage where the larvae develop into adults. The stage is usually non-feeding and fairly inactive, but many are able to move to escape predation such as mosquito pupae.
Puparium	The hardened larval skin that encloses the pupae, particularly in some flies, is known as the puparium.
Pupate	When a larva pupates it becomes a pupa.

Purulent	When a wound or a sore contains or discharges pus it is known as purulent. The term is also used for body secretions containing pus.
Questing	Questing is a type of host location behaviour unique to ticks. When questing, ticks climb a surface, usually vegetation, then stretch out the first pair of legs that contain their sense organ (the Haller's organ) and grasp a passing host.
Regurgitate	When an animal regurgitates it brings swallowed food back up into the mouth.
Renal	The term renal is used when referring to any conditions affecting the kidneys.
Residual pesticide	A pesticide that is applied to surfaces in order to kill arthropods that land or crawl on the treated surface is a residual pesticide.
Resistance	Resistance is said to occur when there is a heritable change in the sensitivity of a pest population to an active ingredient that is reflected in repeated failure to achieve the expected level of control when used according to the label.
Rhabdomyolysis	The destruction of striated muscle cells is known as rhabdomyolysis; in horses, this is most often seen as stiffness and pain in the muscles of the hindquarters and back.
Sanitation	The prevention of disease by proper management of faeces and other organic waste is known as sanitation.
Sclerotised	Insect cuticle that is hardened and usually dark in colour is said to be sclerotised.
Scutum	The hard chitinous shield behind the head that is present in ticks is the scutum. In male ticks, the scutum covers the entire body (dorsally) but in females, it is reduced to allow for expansion during engorgement.
Secondary infection	A secondary infection occurs during or after treatment for an earlier condition that has made the patient more susceptible to infection, often bacteria or fungi.
Self-limiting	A disease that typically resolves without treatment is referred to as self-limiting; usually due to the action of the animal's immune system.
Setae	Setae are stiff hair-like structures or bristles, they often have sensory roles such as detection of pheromones, other odours, temperature, or touch.
Space spray	A pesticide sprayed into the air, usually indoors, to control flying insects.

Species	Species is the principal taxonomic rank. Organisms within a species are capable of mating and producing viable offspring. The scientific name of an organism is made up of the genus and species. For example, for the house fly, the scientific name is *Musca domestica*, this is composed of the genus *Musca* and species *domestica*.
Spiracles	The external respiratory openings in arthropods are known as spiracles.
Spot-on	Spot-on is a way of applying pesticides directly to animals. Spot-on products usually come in a measured single dose in an applicator so it is easy to apply a set dose to an animal to the target location(s).
Sterile insect technique	Sterile insect technique is a type of control where the environment is inundated with sterile insects, typically males, which outcompete the fertile wild-type males and prevent the production of offspring.
Terrestrial	An arthropod that spends its life on land is known as terrestrial, could also be a flying insect.
Thorax (adjective thoracic)	The thorax is the middle segment of the three-body segments of insects, which contains the muscles to operate the legs and wings (if present). Legs and wings originate in the thorax.
Thrush	Thrush is a common bacterial infection that occurs on the hoof of a horse, specifically in the region of the frog. Thrush can cause distinctive unpleasant odours that are attractive to blow flies.
Topical	A topical application is one that is applied directly to the body, with the aim of treating the skin.
Toxicity	The quality of being toxic or poisonous is known as toxicity.
Trematode	A parasitic flatworm of the class Trematoda is a trematode.
Understory	An underlying layer of vegetation beneath the tree canopy is the understory.
Ungulates	An ungulate is a large hoofed mammal; includes horses, cattle, pigs, sheep, and deer.
Urticating	When something is referred to as urticating it has hairs that cause a stinging or prickling sensation when the hair penetrates the skin of a mammal. These hairs may also cause welts to form on the skin.
Vestigial	A part of the body that has become functionless over time due to evolution is known as vestigial.
Viable	A viable organism is capable of surviving or living successfully. For example, larvae can only hatch from viable eggs.

Appendix B. Example integrated pest management plan for equine pests

Pests may affect horse health, public health, impede operations, or damage property. Integrated pest management (IPM) is a sustainable approach to managing pests that combines biological, cultural, physical and mechanical, and as a last resort, chemical tools in a way that minimises economic, health, and environmental risks. The basis of IPM is reducing the causes of pest problems by minimising their ability to grow and thrive in the environment, and their access to food and harbourage sites.

IPM plans can range from informal or simple records in a notebook to more complex and detailed records logged in spreadsheets or a dedicated binder. In this Appendix, an example IPM plan has been developed that the reader can use or modify to fit their facility. The purpose of a plan is to document the pest control strategies to improve control over time.

Objectives of IPM include:

- ❑ minimising significant threats caused by pests to the health and safety of animals, staff, visitors, and the community, including pathogen transfer and annoyance;
- ❑ protecting the environmental quality inside and outside buildings by reducing the application of chemical insecticides;
- ❑ recording progress in control through monitoring and review.

General IPM Strategies

IPM strategies for preventing pests include cultural control methods, physical and mechanical control methods, biological control methods, and site-appropriate chemical control if needed.

An IPM plan consists of the following components:
- ❑ regular inspection for pests;
- ❑ accurate identification of the pest;
- ❑ if control is necessary, selecting the appropriate option(s);
- ❑ frequent evaluation of the plan to ensure continuous control in changing circumstances;
- ❑ insecticide storage and application following label directions and any other country, region, state, commonwealth, and/or local ordinances.

For facilities newly adopting IPM practices, successful implementation of this IPM plan will be tracked over time according to the following metrics:
- ❑ reduced amount of insecticide application;
- ❑ increased types of control methods applied;
- ❑ fewer pest numbers as indicated by monitoring.

For facilities with an established IPM plan, successful implementation of this IPM plan can be tracked according to the following metrics:
- ❑ no increase in pest sightings;
- ❑ no increase in the amount of insecticide applied.

Facility information

The inclusion of facility information can improve the appearance and formality of an IPM plan, and also provide important information on contacts that are responsible for executing the IPM plan.

Facility information

Farm name: ______________________________

Owner: ______________________________

Manager: ______________________________

IPM coordinator: ______________________________

Address: ______________________________

____

____

Phone: ______________________________

Email: ______________________________

Pest identification records

In the areas where you are finding pests, identify them using behavioural characteristics, general appearance, or with the help of academic or other institutional experts. Remember, you may have more than one pest at a time.

Write the date you scouted (i.e. looked) for pests and where you looked. Scouting should be done at regular intervals. It is recommended at least once a month and before pest populations typically increase. Pest populations change over time and the best control methods are those that are used before the pest becomes a problem. Scouting should also happen if the control methods used did not produce satisfactory results because that may mean the pest species was misidentified. Pest or horse behaviours can be recorded as general behaviour or more specific behaviours. For example, '10 flies landed on fence line in 1 minute' or 'Luna shook her head 10 times in 30 seconds.' more specific records can also serve as a monitoring tool (see sections below and Chapter 10 for more information on monitoring).

This scouting report example can be adapted to your own needs and circumstances.

Scouting report		Pest behaviours observed	Horse behaviours observed	Other observations	Pest ID
Date	**Location**				
Example April 12	*Example* Mare field	*Example* Flies appear on legs and flying around belly; flies do not seem to be in the barn; flies rest on fence line at an angle	*Example* Horses are stomping and turning their heads to bite their sides; Luna had 15 stomps per minute.	*Example* None	*Example* Stable flies
Example June 11	*Example* Paddock 1	*Example* Flies congregating on the body and head of the horse; some flies in the stable, on fence lines, and water troughs; some small flies on the back or belly area	*Example* Dragon had 10 head shakes in 30 seconds; also saw relentless tail swishing, twitching body frequently	*Example* Blood spots on the neck	*Example* House flies, horn flies, horse or deer flies

Pest control methods

After you have identified your pest, review the options for pest control that are best suited for your facility. Remember, cultural control and sanitation practices are the most effective at stopping pests from becoming a problem.

Remember, before you choose your control methods it is important to know which pest species you have and understand that there are multiple life stages of each of the major pest species and they all have somewhat different development area preferences and length of development times that may affect control. Make sure that when you are choosing control methods, you are using a combination of different methods that will break the life cycle of your pest flies at multiple points in their development. An example would be good moisture management to reduce larval development of filth flies, biological control using parasitoids to kill pupae, and sticky traps to eliminate adult flies.

These records are also helpful so you can keep notes about the frequency that control is applied, active ingredients used so you can rotate, and to visually see that you are using a mix of control options that could include cultural, physical and mechanical, biological, and chemical methods.

This example control table is based on the 'June 11' entry from the pest identification section.

This is an example that can be adapted to your own needs and circumstances.

Scouting report		Pest	Type of control	Control methods	Notes
Date	Location				
June 11	*Example* Paddock 1	*Example* Horse and deer flies	Physical/ mechanical	Installed Epps traps by the paddock	Traps changed out weekly
			Physical/ mechanical	Horses turned out in this paddock given zebra-striped fly sheets and masks	
	Example Throughout[1]	*Example* House flies	Chemical	Stable started on insect growth regulator	
			Cultural	Round bales put in a metal ring feeder	
			Chemical	Residual spray to run in shed walls	Used active ingredient cypermethrin
			Physical/ mechanical	Manure pit covered with a tarp	
			Physical/ mechanical	Sticky traps put up in the barn	Traps changed weekly
		Horn flies	Physical/ mechanical	Horses are turned out with fly sheets	

[1] 'Throughout' in this case means that control was not applied to animals only in one area, but throughout the facility.

Pest control evaluation

Evaluation of monitoring data is a critical component of a successful IPM plan and the only way to assess if your control methods are working. Keep observation records with the date of the evaluation, the areas being assessed, and whether the records demonstrate changes to the pest population. There are no requirements for evaluation intervals, but monthly is a good rule of thumb during the spring and summer months. Evaluation can be less frequent when pest populations are lower.

For the most accurate records, monitoring of the pest should be included in a pest IPM plan. Evaluations of pest numbers can be based on horse behaviour (e.g. number of stomps, twitches, tail swishes in a given amount of time), or on pest numbers from traps or direct counting, or both. More details on how to monitor many equine pests are in Chapter 10.

This corrective action table example can be adapted to your own needs and circumstances.

Date	Location	Pest	Pest population changes			Corrective actions
			Increase	Decrease	Level maintained	
Example July 12	*Example* Paddock 1	House fly		x		*Example* None
		Horn flies		x		None
		Horse and deer flies		x		None
Example July 12	*Example* Mare field		x			*Example* Round bale waste had not been cleaned; this was cleaned to reduce fly breeding areas

Photo credits

Title page Erika Machtinger, Pennsylvania State University.

About the authors

Erika T. Machtinger is a veterinary entomologist at the Pennsylvania State University, Pennsylvania, USA. She studied Wildlife Conservation and Ecology at the University of Delaware before receiving her PhD in Entomology at the University of Florida. She was a Postdoctoral Scholar at the United States Department of Agriculture working on areawide control of ticks. Her current research focus is on the ecological relationships between vertebrates and their ectoparasites and other related arthropod pests to facilitate the development of control methods. Part of her current research is developing and evaluating fly and tick control methods for horses to improve equine welfare. Horses have been a part of Dr Machtinger's life since she was four years old. She has competed in eventing in the USA through the intermediate level and Prix St. George's dressage. She has owned her own farm and is familiar with the many pests of horses. Currently, Erika competes in dressage with her homebred mare Bright Opposition (Luna), who is featured heavily throughout this book.

Emma N.I. Weeks is a medical and veterinary entomologist. She is courtesy faculty at the University of Florida Entomology and Nematology Department, editor in chief of Florida Entomologist, and editor in chief of Medical and Veterinary Entomology. She studied a BSc in Zoology at the University of Wales (Swansea, UK), an MSc in Biology and Control of Disease Vectors and a PhD in Medical Entomology at the London School of Hygiene and Tropical Medicine (London, UK). She was a faculty member at the University of Florida Entomology and Nematology department in Gainesville, Florida (USA) from 2011 to 2019. Emma now lives in Spain working as a self-

Erika Machtinger and Bright Opposition (Luna) (photo: GRC Photography, LLC)

employed entomologist. She continues to collaborate with faculty and students at the University of Florida on ongoing projects in the management of pests of medical and veterinary importance. Emma has loved horses since she was a child and has owned a horse or ridden or admired other horses as much as she could throughout her life.

Christopher J. Geden is a veterinary entomologist for the United States Department of Agriculture (USDA), Center for Medical, Agricultural and Veterinary Entomology in Gainesville, Florida, USA. A native of eastern Massachusetts, Chris received a BSc in Biology from Boston College before attending the University of Massachusetts at Amherst, where he received his MSc and PhD. He did postdoctoral research at North Carolina State University for three years and was a Research Associate at Cornell University for five years before joining

Emma Weeks (photo: Suzy Rodriguez, University of Florida)

Chris Geden (photo: Sandy Allan)

USDA. His research has focused on the biology, behaviour and management of pests associated with livestock, poultry and horses. His particular emphasis has been on flies and their diseases, predators, and parasites.

Erica Lacher is a veterinarian at Springhill Equine Veterinary Clinic in Gainesville, Florida (USA). As a horse owner she has a keen understanding of the pests of horses and how they affect the quality of life for horses. As a veterinarian, she focuses on educating horse owners about best practices for maintaining healthy, happy horses. She is the co-host of the podcast Straight from the Horse Doctor's Mouth, hosts frequent seminars at her veterinary clinic, and posts regular educational content on social media platforms. A lifelong rider, Dr Lacher competes in the Jumper arena in the USA.

Erica Lacher and Vespa (photo: Erica Lacher)

Acknowledgements

The authors would like to first thank the reader for caring enough about their horses, donkeys, mules, and other equids and sustainable pest management practices to try to learn more about pests and pest control. We would like to thank the horse owners, caretakers, and barn owners for the opportunity to photograph specific pests and pest control options to present to the reader. Similarly, we would like to thank all the photographers that allowed us to use their wonderful images in this book, either with permission or through a creative commons license. It cannot be overstated how much we appreciate the hard work and time it took to take these original photographs and the book would not be the same without them. As they say, a picture is worth a thousand words.

We would also like to thank our many reviewers, both scientific and non-scientific, from the UK, Europe, North America, and Australia, that critically evaluated each chapter for content. Namely: Dr Justin Talley (Oklahoma State University) Aleksandra Pimenides, Dr Diana Leemon (AgriScience Queensland, Dept of Agriculture & Fisheries), Dr Daniel Martín-Vega Universidad de Alcalá), Dr Emmanuelle Cabannes, Ms Juliet Motteram, Dr Steffen Rehbein (Kathrinenhof Research Center), Ms Jodie Toy, Ms Chlöe Jolley, Ms Alison Weiner, Ms Lesley Stone, and Ms Becky Johnston. These folks made sure that we covered as many major pests of hoses as we could and were critical to our understanding of pests from regions from which we were not as familiar. In particular, we would like to thank Professor Myron Zaluki, Professor Wayne Bryden, Dr Lynda Perkins, and the rest of the equine amnionitis and foetal loss team at the University of Queensland (Australia) for careful review and provision of images of the bag-shelter moth. We would also like to express our gratitude to Dr Martin J.R. Hall (Natural History Museum, UK) for thoughtful comments on the book contents, sourcing great photos, and reminding us to include other myiasis-causing flies. We would like to thank Mia Escudo for her careful review and cross checking of photograph captions and credits and section consistency. We would also like to thank Wageningen Academic Publishers for all their support and design expertise. Specifically, we would like to thank Renate Smallegange for believing in us and giving us the push to get started and the support to keep going and Mike Jacobs for leading us to the finish line.

We would like to finally thank the global equine community for their love of horses. Having an idea and turning it into a book is just as challenging as it sounds. Without this community and their animals, this book would not have been written. Similarly, the experience of producing this book has been a labour of love from all the authors who have a bond to these wonderful animals in one way or another. As Winston Churchill once said, 'no hour of life is wasted that is spent in the saddle.'